Matlab for Engineering

Matlab for Engineering

Berardino D'Acunto

University of Naples Federico II, Italy

World Scientific

EW JERSEY · LONDON · SINGAPORE · BEIJING · SHANGHAI · HONG KONG · TAIPEI · CHENNAI · TOKYO

Published by

World Scientific Publishing Co. Pte. Ltd.

5 Toh Tuck Link, Singapore 596224

USA office: 27 Warren Street, Suite 401-402, Hackensack, NJ 07601

UK office: 57 Shelton Street, Covent Garden, London WC2H 9HE

Library of Congress Cataloging-in-Publication Data
Names: D'Acunto, Bernardino, author.
Title: MATLAB for engineering / Berardino D'Acunto, University of Naples Federico II, Italy.
Description: New Jersey : World Scientific Publishing Co. Pte. Ltd., [2022] |
　　Includes bibliographical references and index.
Identifiers: LCCN 2021033978 | ISBN 9789811240669 (hardcover) |
　　ISBN 9789811240676 (ebook for institutions) | ISBN 9789811240683 (ebook for individuals)
Subjects: LCSH: Engineering--Data processing. | MATLAB.
Classification: LCC TA345 .D24 2022 | DDC 620.001/51--dc23
LC record available at https://lccn.loc.gov/2021033978

British Library Cataloguing-in-Publication Data
A catalogue record for this book is available from the British Library.

For any available supplementary material, please visit
https://www.worldscientific.com/worldscibooks/10.1142/12380#t=suppl

Desk Editor: Amanda Yun

Typeset by Stallion Press
Email: enquiries@stallionpress.com

Preface

This book provides an introduction to Matlab for engineering students and engineers. However, the book can also be useful for students, technicians and researchers in all scientific areas whenever they need to apply Matlab to solve problems arising from their areas of studies, work and research. It is assumed that they do not have any initial Matlab knowledge and the reader is very interested to learn this powerful tool of scientific computing.

After introducing the essential Matlab commands, mainly devoted to matrix manipulation, most of the other commands are presented in action, within programs aimed at solving specific problems. In the first part of the book, the *function file* is presented. This concept plays a basic role in Matlab programming and, therefore, is illustrated with many examples.

Since physical processes occur very often in space and time, the related governing equations are partial differential equations. Therefore, most Matlab programs in this book are devoted to solve these types of equations. Both the Finite Element Method and Finite Difference Method are introduced and applied. Generally, a problem is discussed globally: the mathematical model from the physical phenomena is derived and the equations are solved with Matlab.

The book presents more than a hundred listings and number of exercises at the end of each chapter.

I wish to thank World Scientific Publishing for inviting me to write this book. Special thanks are due to Shaun Tan Yi Jie, who helped me while I was writing the book.

Berardino D'Acunto   Naples, March 2021

Contents

Preface   v

Chapter 1.   Function Files   1

    1.1   Matrices   1
        1.1.1   Creating Matrices   1
        1.1.2   Matrix Indexing   2
        1.1.3   Matrix Manipulation   4
        1.1.4   Tridiagonal Matrices   6
        1.1.5   Matrix Operations   8
        1.1.6   Right and Left Divisions   9
    1.2   Script Files   10
        1.2.1   For Loop   10
        1.2.2   Examples of Script Files   11
    1.3   Introduction to Function Files   15
        1.3.1   Structure of Function Files   15
        1.3.2   Function with a Multiple Output Variable   17
        1.3.3   Flow Control Structures   19
        1.3.4   Local Functions, Anonymous Functions   25
        1.3.5   Logical Operators and Logical Functions   27
    1.4   Exercises   33

Chapter 2.   The Finite Difference Method   39

    2.1   Finite Difference Approximations of Derivatives   39
        2.1.1   Forward, Backward and Central Approximations .   39

2.1.2   Approximation of Functions Depending on Two Variables 46

2.1.3   Approximation of Higher Order Derivatives 47

2.2   Diffusion 49

     2.2.1   Fourier's Law and Heat Equation 49

     2.2.2   Fick's Law and Diffusion 55

     2.2.3   Free Boundary Value Problems 56

2.3   Finite Difference Method 58

     2.3.1   Explicit Euler Method 58

     2.3.2   Stability, Convergence, Consistence 64

     2.3.3   Boundary Value Problems 68

     2.3.4   Diffusion in a Multi-layer Medium 76

     2.3.5   Implicit Euler Method 80

     2.3.6   Crank–Nicolson Method 84

     2.3.7   Von Neumann Stability Criterium 89

2.4   Exercises 93

Chapter 3.   Diffusion and Convection   99

3.1   Convection-diffusion Equation 99

     3.1.1   Upwind Method 99

     3.1.2   Other Finite Difference Methods for the Convection-Diffusion Equation 107

     3.1.3   Advection Equation 111

3.2   Method of Lines 118

     3.2.1   Heat Equation 118

     3.2.2   Nonlinear Equations 127

     3.2.3   Variable Diffusivity Coefficient 133

     3.2.4   Convection-Diffusion Equation 136

3.3   Saving Data and Figures 140

     3.3.1   Save Function 140

     3.3.2   Load Function 142

     3.3.3   Saving Figures 143

3.4   Exercises 143

Chapter 4.   Introduction to the Finite Element Method   153

4.1   Numerical Integration 153

4.2   Finite Element Method 164

     4.2.1   Axial Motion of a Bar 164

     4.2.2   Weak Solution 167

4.2.3 Shape Functions 169
4.2.4 Boundary Value Problems 171
4.2.5 Axial Displacement and Stress in a Bar 180
4.2.6 Concentrated Force and Dirac Function 184
4.3 Partial Differential Equations 189
4.3.1 Diffusion Equation 189
4.3.2 Wave Equation 197
4.4 Exercises 205

Chapter 5. Introduction to the Finite Element Method
in Two Spatial Dimensions   213

5.1 Elliptic Partial Differential Equations 213
5.1.1 Green's Identities 213
5.1.2 Boundary Value Problems 214
5.2 Finite Element Method in Two Spatial Dimensions 217
5.2.1 Shape Functions 217
5.2.2 Weak Form of the Poisson Equation 225
5.2.3 Dirichlet–Neumann Problem 230
5.2.4 Applications to the Dam and Sheet Pile Wall . . . 234
5.3 Finite Difference Method 240
5.3.1 Five-Point Method 240
5.3.2 Model of a Dam 247
5.4 Exercises 257

Chapter 6. The Euler–Bernoulli Beam   267

6.1 Finite Element Method 267
6.1.1 Euler–Bernoulli Beam Equation 267
6.1.2 Shape Functions 270
6.1.3 Weak Form 274
6.2 Statics 277
6.3 Beam Subjected to Concentrated Forces 296
6.4 Exercises 305

Bibliography   311

Index   313

Chapter 1

Function Files

Matlab[1] derives its power from its extensive capability of manipulating matrices. Therefore, this topic is discussed in the first section of this chapter. The name "Matlab" cames from Mat(rix) lab(oratory). Matlab programs are written in files with an *.m* extension. There are two kinds of m-files: *script files* and *function files*. The latter are much more interesting than the former. Script files are illustrated in Sec. 1.2 while function files are discussed in Sec. 1.3. See Moler (2011), which could be useful for the topics in Sec. 1.1.

1.1 Matrices

1.1.1 *Creating Matrices*

The command

   A = [1 2 3; 4 -5 6; 7 8 -9]

creates the matrix

$$A = \begin{bmatrix} 1 & 2 & 3 \\ 4 & -5 & 6 \\ 7 & 8 & -9 \end{bmatrix}.$$

The command A' generates the transpose matrix of A. Therefore, the command

   B = A'

---

[1]Matlab is a registered trademark of The MathWorks, Inc.

produces the matrix

$$B = \begin{bmatrix} 1 & 4 & 7 \\ 2 & -5 & 8 \\ 3 & 6 & -9 \end{bmatrix}.$$

A matrix with one row is a row vector and a matrix with one column is a column vector. For example, the command

   rv = [10 11 12 13]

produces

$$rv = \begin{bmatrix} 10 & 11 & 12 & 13 \end{bmatrix}$$

and the command

   cv = [pi; cos(pi)]

produces

$$cv = \begin{bmatrix} 3.1416 \\ -1.000 \end{bmatrix}.$$

As noted, π is introduced in Matlab with the notation pi.

1.1.2 *Matrix Indexing*

To access specific elements of a matrix, use indexing. To refer to a single element, use the A(i,j) command. For example, if A is the matrix introduced beforehand, then the command

   A(2,3)

returns 6 and the command

   A(2,3) = 16

replaces 6 with 16 in A

$$A = \begin{bmatrix} 1 & 2 & 3 \\ 4 & -5 & 16 \\ 7 & 8 & -9 \end{bmatrix}.$$

To access a submatrix of a matrix, use the colon operator. The A(i,:) command returns the i-*th* row of A. For example, the command

   A(2,:)

returns

$$\begin{bmatrix} 4 & -5 & 16 \end{bmatrix},$$

and the command

A(2,:) = 2*A(2,:)

replaces the preceding row in A with the row 8 −10 32

$$A = \begin{bmatrix} 1 & 2 & 3 \\ 8 & -10 & 32 \\ 7 & 8 & -9 \end{bmatrix}.$$

The A(i:h,:) command, where $i \leq h$, produces the submatrix formed by the rows: $i, i+1, \ldots, h$. For example, the command

A(2:3,:)

produces

$$\begin{bmatrix} 8 & -10 & 32 \\ 7 & 8 & -9 \end{bmatrix},$$

and the command

A(2:3,2:3)

produces

$$\begin{bmatrix} -10 & 32 \\ 8 & -9 \end{bmatrix}.$$

The previous command can also be used to extract submatrices with non-consecutive rows (or columns). See Exercises 1.4.1 and 1.4.2.

The row vectors can be created with specific commands too. The x = linspace(x1, x2, n) command creates a row vector of n equally spaced elements from x1 to x2 with step $(x2 - x1)/(n - 1)$. For example, the command

x = linspace(0, 10, 6)

generates the vector

$$x = [0 \ 2 \ 4 \ 6 \ 8 \ 10].$$

The same vector is generated with the x = 0:2:10 command, where the initial value, step and final value are specified. Step 1 can be omitted. For example, the command

y = 0:10

produces

$$y = [0 \ 1 \ 2 \ 3 \ 4 \ 5 \ 6 \ 7 \ 8 \ 9 \ 10].$$

The v(end) command returns the last element of the vector v. For example, the command

    y(end)

returns 10. This command can be useful when a vector is crested dynamically during the program execution and the vector length is not known *a priori*.

The A(i,:) = [ ] command deletes the i-*th* row of the matrix A. For example, the A(3,:) = [ ] command deletes the third row and the A([1 3],:) = [ ] command deletes the first and third rows. More generally, the A(i:h,:) = [ ] command, where $i \leq h$, deletes the rows from i to h. This is similar for columns. The previous command can be used in other situations. See Exercise 1.4.3.

1.1.3 *Matrix Manipulation*

A matrix can be appended to another, provided the dimensions match. After creating the matrices C and D, and the vector x

    C = [1 2 3; 4 5 6]; D = [7 8; 9 10]; x = [11; 12; 13];

the commands

    [C D]
    [C; x']

produce the following matrices

$$\begin{bmatrix} 1 & 2 & 3 & 7 & 8 \\ 4 & 5 & 6 & 9 & 10 \end{bmatrix}, \quad \begin{bmatrix} 1 & 2 & 3 \\ 4 & 5 & 6 \\ 11 & 12 & 13 \end{bmatrix}.$$

The [D x] and [C x] commands generate error messages. The reader is asked to do Exercise 1.4.4.

Matlab provides commands to quickly generate specific matrices. The ones(m,n) and zeros(m,n) commands produce m-by-n matrices of ones and zeros, respectively. The eye(m,n) command generates the m-by-n unit matrix. For example, the commands

    eye(2,3)
    eye(2)

return

$$\begin{bmatrix} 1 & 0 & 0 \\ 0 & 1 & 0 \end{bmatrix}, \quad \begin{bmatrix} 1 & 0 \\ 0 & 1 \end{bmatrix}.$$

If x is a vector of n elements, the diag(x) command creates a square matrix and places the elements of x on the main diagonal. For example, if
   x = [1 2 3];
the command
   diag(x)
produces

$$\begin{bmatrix} 1 & 0 & 0 \\ 0 & 2 & 0 \\ 0 & 0 & 3 \end{bmatrix}.$$

The diag(x,h) command creates a square matrix and places the elements of x along the diagonal specified by h, where $h = 0$ indicates the main diagonal, while $h > 0$ specifies a diagonal in the upper triangular part of the matrix and $h < 0$ specifies a diagonal in the lower triangular part. Therefore, the diag(x,0) command generates the same matrix as before and the commands
   diag(x,-2)
   diag(x,1)
produce the following matrices

$$\begin{bmatrix} 0 & 0 & 0 & 0 & 0 \\ 0 & 0 & 0 & 0 & 0 \\ 1 & 0 & 0 & 0 & 0 \\ 0 & 2 & 0 & 0 & 0 \\ 0 & 0 & 3 & 0 & 0 \end{bmatrix}, \quad \begin{bmatrix} 0 & 1 & 0 & 0 \\ 0 & 0 & 2 & 0 \\ 0 & 0 & 0 & 3 \\ 0 & 0 & 0 & 0 \end{bmatrix}.$$

If u is a vector, the A = reshape(u,n,m) command reshapes u into the n-by-m matrix A. The number of elements in u must be $n \times m$, otherwise the command will generate an error message. For example, after creating the vector u with the command
   u = [1; 2; 3; 4; 5; 6];
the command
   A = reshape(u,2,3)
generates the 2-by-3 matrix

$$A = \begin{bmatrix} 1 & 3 & 5 \\ 2 & 4 & 6 \end{bmatrix}.$$

Of course, the command works with matrices too. Indeed, the command

B = reshape(A,3,2)

produces the 3-by-2 matrix

$$B = \begin{bmatrix} 1 & 4 \\ 2 & 5 \\ 3 & 6 \end{bmatrix}.$$

The A(:) command reshapes the matrix A into a column vector. For example, the command

v = A(:)

returns the original vector u. The reader is asked to do Exercise 1.4.5.

1.1.4 *Tridiagonal Matrices*

Tridiagonal matrices can be created with the diag(x,h) command. For example, after creating the vectors

x1 = [1 1 1]; u =[2 2 2 2]; x2 = [3 3 3];

the command

A = diag(x1,-1) + diag(u) + diag(x2,1)

generates the following tridiagonal matrix

$$A = \begin{bmatrix} 2 & 3 & 0 & 0 \\ 1 & 2 & 3 & 0 \\ 0 & 1 & 2 & 3 \\ 0 & 0 & 1 & 2 \end{bmatrix}.$$

Large tridiagonal matrices require a large amount of memory. The A = spdiags(B,d,m,n) command helps to save memory. This command generates an m-by-n matrix and places the columns of B along the diagonals specified by d. For example, if B was created with the B = [-ones(5,1) (1:5)' ones(5,1)] command, then the A = spdiags(B, -1:1, 5, 5) command produces the following tridiagonal matrix

$$A = \begin{bmatrix} 1 & 1 & 0 & 0 & 0 \\ -1 & 2 & 1 & 0 & 0 \\ 0 & -1 & 3 & 1 & 0 \\ 0 & 0 & -1 & 4 & 1 \\ 0 & 0 & 0 & -1 & 5 \end{bmatrix}.$$

The sparse matrix A is displayed by Matlab as follows

$(1, 1)$	1
$(2, 1)$	-1
$(1, 2)$	1
$(2, 2)$	2
$(3, 2)$	-1
$(2, 3)$	1
$(3, 3)$	3 .
$(4, 3)$	-1
$(3, 4)$	1
$(4, 4)$	4
$(5, 4)$	-1
$(4, 5)$	1
$(5, 5)$	5

Only the nonzero elements are shown a column at a time. The C = full(A) command displays A in its usual form.

Try to recreate the matrix A by replacing its size, 5, with 100. Inspect A with the whos A command. Note that 3,980 bytes were used to store A. Next, use the C = full(A) command to save the full form of A in C. While typing the whos C command, note that 80,000 bytes were used to store C. The spy(A) command generates a picture of the matrix A. Using the spdiags command requires attention. See Exercises 1.4.6 and 1.4.7.

The A = repmat(B,m,n) command replicates the p-by-q matrix B as specified by m and n, and generates the mp-by-nq matrix A. For example, the command

    A = repmat(eye(2),2,3)

generates

$$A = \begin{bmatrix} 1 & 0 & 1 & 0 & 1 & 0 \\ 0 & 1 & 0 & 1 & 0 & 1 \\ 1 & 0 & 1 & 0 & 1 & 0 \\ 0 & 1 & 0 & 1 & 0 & 1 \end{bmatrix}.$$

If B is a scalar, say $B = 5$, the commands

    A = repmat(B,2,3)
    u = repmat(B,1,3)
    v = repmat(B,2,1)

produce

$$A = \begin{bmatrix} 5 & 5 & 5 \\ 5 & 5 & 5 \end{bmatrix}, u = \begin{bmatrix} 5 & 5 & 5 \end{bmatrix}, \ v = \begin{bmatrix} 5 \\ 5 \end{bmatrix},$$

respectively. Exercise 1.4.8 is useful.

1.1.5 *Matrix Operations*

Matlab performs all the algebraic operations on matrices. If A and B are m-by-n matrices, the addition and subtraction are produced with the A + B and A − B commands. Moreover, a matrix A can be added to a scalar a with the Matlab command B = A + a. This operation generates the matrix $B_{ij} = A_{ij} + a$. If A and B are m-by-n and n-by-p matrices, respectively, then the matrix product is generated with the A * B command. For example, after creating the matrices

   A = [1 2 3; 4 5 6]; B = [7 8; 9 0; -1 -2];

the commands

   A * B
   B' * A'

generate the following matrices

$$\begin{bmatrix} 22 & 2 \\ 67 & 20 \end{bmatrix}, \quad \begin{bmatrix} 22 & 67 \\ 2 & 20 \end{bmatrix}.$$

Of course, it results in

$$(A * B)' = B' * A',$$

as it is known. If u is a row vector and v is a column vector, the u * v command yields the scalar product. For example, after creating the vectors

   u = [1 2 −3]; v = [4; 5; 6];

the command

u * v

produces -4. A matrix A can be multiplied with a scalar a with the B = A * a command. This operation produces the matrix $B_{ij} = A_{ij}a = aA_{ij}$. Therefore, it is A * a = a * A. Moreover, in Matlab, an original element-by-element product between two same size matrices can be generated. The C = A.*B command creates the matrix C given by $C_{ij} = A_{ij}B_{ij}$. For example, if A and B are the matrices defined beforehand, then the commands

A'.*B
A.*B'

produce the following matrices

$$\begin{bmatrix} 7 & 32 \\ 18 & 0 \\ -3 & -12 \end{bmatrix}, \quad \begin{bmatrix} 7 & 18 & -3 \\ 32 & 0 & -12 \end{bmatrix}.$$

The A.*B command generates an error. The element-by-element product between a scalar a and a matrix A is meaningful. It is indicated with a.*A and the result is the same as a*A. The element-by-element product is very useful in some situations. See Exercise 1.4.9.

1.1.6 *Right and Left Divisions*

The A/B command is named *right division*. The command returns the product of A times the inverse of B. It is the same as A*B^-1, but faster, since a specific program was devoted to this command by Matlab, whereas the A*B^-1 command uses the general program for A*B^-n. For example, after creating the matrices

A = [1 2; 3 4]; B = [5 6; 7 8];

the command

A/B

generates the following matrix

$$\begin{bmatrix} 3.0000 & -2.0000 \\ 2.0000 & -1.0000 \end{bmatrix}.$$

Of course, the command works with scalar variables. For example, if

   a = 4; b = 2;

the command

   a/b

returns 2.

   The A\B command is named *left division*. It returns the product of the inverse of A times B. It is the same as A^-1*B, but faster. For example, if A and B are the matrices defined beforehand, then the command

   A\B

generates the following matrix

$$\begin{bmatrix} -3.0000 & -4.0000 \\ 4.0000 & 5.0000 \end{bmatrix}.$$

The command works with scalar variables too. For example, if a and b are the variables defined beforehand, then the command

a\b

returns 0.5000. The left division is used to solve the algebraic linear systems $Ax = b$. The unknown vector x is found by the simple A\b command. See Exercise 1.4.10. Moreover, Matlab allows element-by-element left and right divisions: A./B and A.\B. See Exercise 1.4.11.

1.2 Script Files

A *script file* is a file containing a set of Matlab commands. To execute this type of file, it is equivalent to writing and executing the commands in sequence at the Command Window. A new script file is created by the New Script button and can be saved into any directory. To execute a script file, press the Run button or type its name at the command line. Examples of script files will be illustrated in Sec. 1.2.2, after introducing the *for loop* in Sec. 1.2.1.

1.2.1 *For Loop*

This section presents a first flow control structure provided by Matla: the *for loop*. Other structures, e.g., the *while loop*, will be introduced in Sec. 1.3.3. The syntax of the *for* loop is outlined below.

For loop

for variable = expression
    code lines
end

For example, in the loop
    for j = 1:9
      j
    end
the variable j goes from 1 to 9 with step 1: $1, 2, 3, \ldots, 9$. In addition, in the loop
    for i=2:-.2:1
      i
    end
the variable i goes from 2 to 1 with step -0.2: $2, 1.8, 1.6, \ldots, 1$.

1.2.2 Examples of Script Files

Example 1.2.1 The following script file plots the function

$$u(x, t) = \sin x \cos t, \ 0 \le x \le \pi, \ 0 \le t \le 10, \tag{1.2.1}$$

at several times. The function u describes the small oscillations of a thin bar with fixed ends (Fig. 1.2.1). For future applications, let us note that

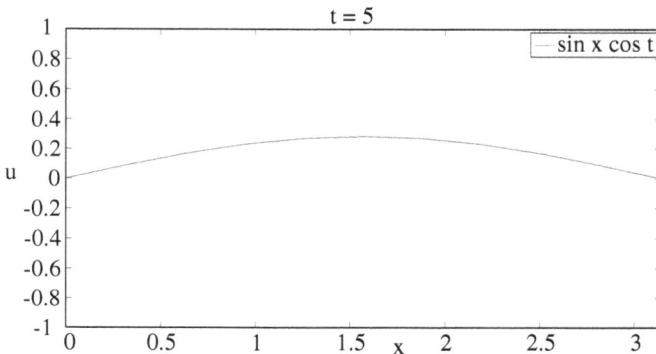

Figure 1.2.1. $u = \sin x \cos t$.

the function u satisfies the partial differential equation

$$\frac{\partial^2 u}{\partial t^2} - \frac{\partial^2 u}{\partial x^2} = 0, \tag{1.2.2}$$

and verifies the initial-boundary conditions

$$u(x,0) = \sin x, \quad u_t(x,0) = 0, \quad u(0,t) = u(\pi,t) = 0. \tag{1.2.3}$$

See Exercise 1.4.12.

```
% This is the script file script_1.m.                        %......[1]
% It plots the function u = sinxcost and prints the matrix u(i,j).

clc;                                                          %......[2]
nx = 10;                                                      %......[3]
x = linspace(0, pi, nx+1);                                    %......[4]
time = 10; nt = 60; t = linspace(0, time, nt+1);
u = zeros(nx+1, nt+1);                                        %......[5]
for j = 1:nt+1
    u(:,j) = sin(x')*cos(t(j));                               %......[6]
    plot(x,u(:,j),'r');                                       %......[7]
    axis([0  pi  -1  1]);                                     %......[8]
    xlabel('x'); ylabel('u');                                 %......[9]
    legend('sin x cos t',1);                                  %....[10]
    title(['t = ', num2str(t(j))]);                          %....[11]
    pause(.1);                                                %....[12]
end
disp(u');                                                     %....[13]
```

———————— Notes ————————

[1] Any word following the % sign is a note, a comment, and is ignored by Matlab.

[2] The clc command cleans the *Command Window*. Information on a command is obtained by writing help name_of_command. The reader is invited to take a look at *See also* where further commands are suggested.

[3] Note the final semi-colon. The value of this variable is saved, but not printed in the Command Window.

[4] The x = linspace(x1, x2, n) command creates a row vector of n elements equally spaced from $x1$ to $x2$ with step $(x2 - x1)/(n - 1)$. Therefore, x = linspace(0, pi, nx+1) generates a vector of $nx + 1$ elements with step $dx = \pi/nx$, i.e., the vector $x = [0 \quad dx \quad 2 * dx...\pi]$.

[5] The matrix u is initialized. Initializing a matrix is not requested by Matlab, but is strongly recommended. Indeed, matrix initialization allows Matlab to allocate the matrix entries in contiguous areas of memory, which results in the script running faster.

[6] u(:,j) indicates the j-th column of the matrix u. Since x is a row vector, the transpose sign ' converts it to a column vector.

[7] The plot command plots (in red) the function (of x) $u(x, t(j))$ at any time $t(j)$.

[8] The axis([x1 x2 y1 y2]) command sets the axis limits. If this command is not introduced, Matlab sets the axis limits automatically. A partial axis control is also possible. For example, the axis([-inf x2 y1 inf]) command sets the maximum limit for the x-axis and the minimum limit for the y-axis.

[9] The xlabel and ylabel commands are optional. The previous commands place the labels beside the corresponding axes. Since the labels are text strings, they must be introduced between two ' signs.

[10] The legend command creates the legend. The user can specify its position. For example, the legend('sin x cos t','Location','northeast') command creates the legend in the upper right corner and the legend('sin x cos t','Location','best') command creates the legend in the best position.

[11] The title command adds text at the top of the graph. In Fig. 1.2.1, it is used to show the time corresponding to the current position of the bar. The text is composed of two strings. The first, $t =$, is statical. The second, related to the current time, is dynamic and changes every time a new plot is made. The num2str(t(j)) command converts the real number $t(j)$ to a text string.

[12] The pause(s) command, where s is a real number, stops the execution for s seconds.

[13] The disp command displays the matrix u'. The transpose matrix is more interesting than u, since its first row is composed by the values corresponding to the initial condition of the bar, the second row contains the values related to the second time, and the last row shows the last values.

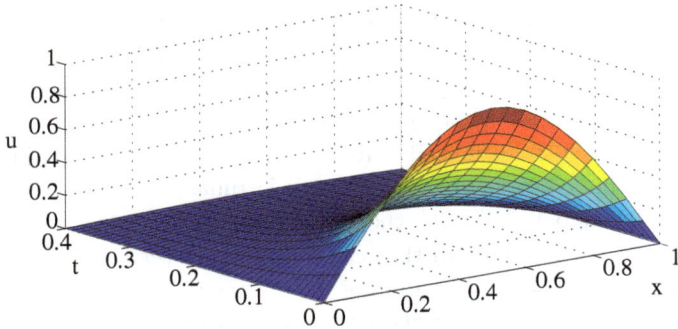

Figure 1.2.2.   3D plot of the function $u = \sin(\pi x)\exp(-\pi^2 t)$.

Example 1.2.2 The following script file produces 2D and 3D plots of the function (Fig. 1.2.2),

$$u = \sin(\pi x)\exp(-\pi^2 t), \ 0 \le x \le 1, \ t \le 0.4. \tag{1.2.4}$$

The function u describes the temperature evolution in a solid where one dimension is prevailing over the others. For future applications, note that u satisfies the partial differential equation

$$\frac{\partial^2 u}{\partial t} - \frac{\partial^2 u}{\partial x^2} = 0, \tag{1.2.5}$$

and verifies the initial-boundary conditions

$$u(x,0) = \sin(\pi x), \quad u(0,t) = u(1,t) = 0. \tag{1.2.6}$$

See Exercise 1.4.13.

```
% This is the script file script_2.m.
% It produces 2D and 3D plots of the function u = sin(πx) exp(−π²t).
clc;

% Initialization
L = 1; nx = 20; x = linspace(0, L, nx+1);
time = .4; nt = 40; t = linspace(0, time, nt+1);
u = zeros(nx+1, nt+1);

% 2D Plot
for j = 1:nt+1
    u(:,j)= sin(pi*x')*exp(-pi^2*t(j));
    plot(x,u(:,j));
```

```
    axis([0  L  0  1]);
    xlabel('x'); ylabel('u');
    legend('sin(pi x) *exp(-pi^2 t)');
    title(['t = ', num2str(t(j))]);
    pause(.1);
end
```

```
% 3D Plot
pause;
    % The pause command stops the execution. Press any key to continue.
figure(2);
surf(x,t,u')
xlabel('x'); ylabel('t'); zlabel('u');
```

```
% Print
disp(u');
```

1.3 Introduction to Function Files

1.3.1 *Structure of Function Files*

The *function file* is an m-file that starts with a line of function definition, where the function name, the input variables passed to the function, and the output variables returned by the function are specified. The first line is followed by comment lines and code lines that form the function body. The last code line is the statement end that ends the function. The syntax is outlined below.

```
                              Function
    function [output] = name_of_function (input)
    % comments
    code lines
    end
```

Example 1.3.1 As the first example, consider the following simple function.

```
function y = sqr(x)                                    %......[1]
% This is the function file sqr.m                      %......[2]
% The sqr function returns x squared. If x is a matrix, %......[3]
% sqr(x) returns the element-by-element product of matrices. %......[4]
```

```
y = x.*x;                                                    %......[5]
end
```

———————————— Notes ————————————

[1] The square brackets are optional when the output consists of one variable, as in this example, or when there is no output variable. The square brackets are necessary for a multiple output variable. The function can be called with a different variable name and the result can be assigned to variables that have different names. In addition, the file name where the function is saved must be the same as the function name.

[2] First line of comment. The comment lines are printed when the user types "help sqr" at the command line. See Exercise 1.4.14.

[3] Second line of comment.

[4] Third line of comment.

[5] Function body. All variables used here, as well as in the function definition, are local and private. See Remark 1.3.1.

The simplest way to call the sqr function is without any output. For example, the sqr(2) and sqr([1 2]) commands return

$$4 \quad \text{and} \quad 1 \quad 4,$$

respectively. The sqr(1,2) command generates an error since the sqr function must be called with one input variable. If the function output has to be used, then the complete syntax must be considered. For example, the command

```
z = sqr(2);
```

assigns the value 4 to the variable z that can be used in other statements.

Remark 1.3.1 Initialize the variable a $= 1$ in the Command Window. Check that a was saved. Consider the sqr function and add two new code lines

```
a = 0;
b = 10;
```

just before end. Matlab outlines the warning that the values of the last variables might be unused. Do not care about that. The new variables will be deleted very soon. Click on the — sign to the left of b $= 10$;. A small gray disk appears that changes to red after saving the sqr.m file. Execute the function by using, for example, the sqr(9) command. Note that the function execution will stop at the code line b = 10;, where there is the disk. The Prompt changes to K>> since we are in the Debug phase. Inspect the variable a by typing a followed by Enter. You may see a = 0. Continue the function execution by pressing the Continue button. You will obtain the result ans = 81. Now, inspect the variable a again and note that a = 1. All this emphasizes the local and private character of the variables defined in the function body. Before executing the function, the value of a was 1. During the execution, it was a = 0. After the execution, it was a = 1. The variables defined in the function body cannot interfere with the variables defined in the external world and vice versa. Moreover, we also learned how to inspect some variables during the Debug phase. Finally, delete the new variables added in the function.

1.3.2 *Function with a Multiple Output Variable*

Example 1.3.2 The following listing presents an example of a function with a multiple output variable. The heat_flux function returns the heat flux vector **q** in a thin solid, according to Fourier's law

$$\mathbf{q}(\mathbf{x}, t) = -k\nabla u(\mathbf{x}, t), \qquad (1.3.1)$$

where k is the thermal conductivity of the material, u is the temperature and ∇u is its space gradient. See Sec. 2.2.1 for details.

```
function [qx, qy] = heat_flux(u,dx,dy,k)
% This is the function file heat_flux.m.
% The heat flux in a thin solid is computed according to Fourier's law. The
% input variable u is the matrix with the temperature values. The input
% variables dx and dy are the spaces among the points along the x- and y-
% direction, respectively. The thermal conductivity k is a positive real
% number, for example, 62.3 (iron), 387.6 (copper), 418.7 (silver), 0.173
% (rubber), 1.177(glass), 2.215 (ice).

[ux, uy] = gradient(u,dx,dy);
    % The gradient(u,dx,dy) function returns the numerical values of the two
    % components of the gradient vector by using dx and dy.
qx = -k*ux; qy = -k*uy;
end
```

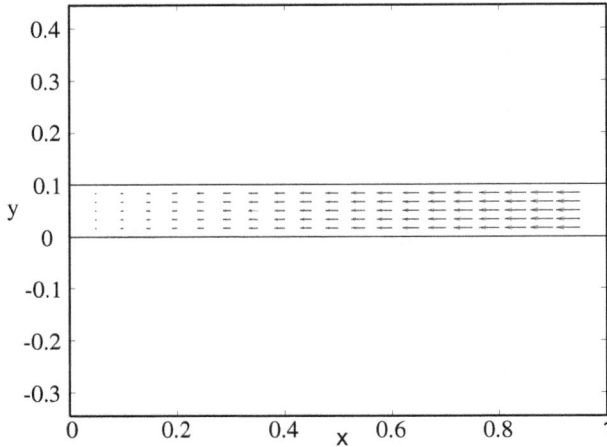

Figure 1.3.1.   Heat flux.

Example 1.3.3 A way to call the heat_flux function is illustrated in the following listing. The heat flux vector is plotted (Fig. 1.3.1) and printed.

```
% This is the script file heat_flux_ex.m.

k = 62.3; Lx = 1; nx = 20; Ly = .1; ny = 6;
x = linspace(0,Lx,nx+1); dx = Lx/nx;
y = linspace(0,Ly,ny+1); dy = Ly/ny;
u = zeros(ny+1,nx+1);
for j=1:ny+1,
    u(j,:)=x.^2;
end
[qx, qy] = heat_flux(u,dx,dy,k);
quiver(x(2:end-1),y(2:end-1),qx(2:end-1,2:end-1),qy(2:end-1,2:end-1));
    % The quiver function plots vectors as arrows with components X and
    % Y at the points with coordinates x and y. See Exercise 1.4.15.
rectangle('position',[0,0,Lx,Ly]);
    % Rectangle with bottom left corner in (0,0), base Lx, and height Ly.
axis('equal'); % Same scale for both axes.
xlabel('x'); ylabel('y');
disp(qx);
```

Remark 1.3.2 The coefficient k is a strictly positive real number. If, by mistake, a negative k is passed to the heat_flux function, then the function output would be wrong. Therefore, data analysis and flow control of the code are indispensable. This topic is discussed in the next section. In the same section, we will provide a modified version of the heat_flux function that is capable of eliminating the problem outlined above.

1.3.3 *Flow Control Structures*

This section presents the flow control structures *if*, *switch* and *while*. The general syntax of the *if-elseif-else* structure is provided below.

If-elseif-else
if logical condition
code lines
elseif logical condition
code lines
else
code lines
end

Special cases include the following: *if*, *if-else*, and *if-elseif*, see below. In addition, all cases can be nested.

If-elseif-else: special cases		
if logical condition	*if* logical condition	*if* logical condition
code lines	code lines	code lines
end	*else*	*elseif* logical condition
	code lines	code lines
	end	*end*

As the first application, the heat_flux function is suitably modified and the problem outlined in Remark 1.3.2 is eliminated.

```
function [qx, qy] = heat_flux(u,dx,dy,k)
% comment lines
if k > 0 % If the coefficient k is positive, the function is executed.
    [ux, uy] = gradient(u,dx,dy); qx = -k*ux; qy = -k*uy;
```

else % Otherwise, a message is sent to the user.
    disp('k must be a positive real number.')
end

Example 1.3.4 When the function has no output, the function definition can be simplified, as in the following listing.

```
function if_1(i)
% This is the function file if_1.m. It is an application on if-elseif-else.
% Note the simplified function definition. It could also be written:
% function [ ] = if_1(i). Call the function by passing 0 or 1 as an argument.

x1 = -pi; x2 = pi; nx = 20; x = linspace(x1,x2,nx+1)
if i == 0 % The  = = sign is a relational operator. It should not be confused
          % with the =  sign that is an assignment operator. The code x = y
          % assigns the value of y to x. Instead, the code x == y compares
          % the values of the variables that retain their values.
          % The Matlab relational operators are provided at the end
          % of the listing.
    plot(x,sin(x));
elseif i == 1
    plot(x,cos(x));
else
    disp('Please call the function by passing 0 or 1 as argument.')
          % This message is sent if the function was called with an
          % argument different from 0 or 1. Beside disp, errordlg('...')
          % can be used too. In this case, the message is shown in a frame.
          % In both cases, the function is executed. A stronger command
          % is error('...') that stops the function execution.
end
end
```

Relational operators	
==	*equal to*
˜=	*not equal to*
<	*less than*
<=	*less than or equal to*
>	*greater than*
>=	*greater than or equal to*

Another example is provided in Exercise 1.4.16.

Example 1.3.5 The following listing considers a function with a multiple output variable.

```
function [max, min] = maxmin_vector(b)
% This is the function file maxmin_vector.m.
% The function returns the maximum and minimum elements of a vector.
% For example, if b = [1 2 43 56 1 3], the command
%           [M  m] = maxmin_vector(b)
% produces
%           M = 56
%           m = 1

min = b(1); max = b(1);
for i=2:length(b)
    % The length(u) command, where u is a vector, returns the number of
    % elements in u.
    if b(i) < min
        min = b(i);
    end
    if b(i) > max
        max = b(i);
    end
end
end
```

Remark 1.3.3 The maxmin_vector function requires a vector as an input. Since the input is not checked, the function is executed even when a matrix is passed. As a result, the function will return a wrong result. For example, the commands

b = [3 2; 3 4]; [M m] = maxmin_vettore(b)

produce the following wrong result

M = 3
m = 3

To avoid such undesirable situations, the function will be modified in Example 1.3.10, Sec. 1.3.5, after introducing a number of logical functions.

If the code flow has a choice of many possibilities, using if may make the program slower. Matlab provides a more efficient command for these situations: the *switch* structure. Its syntax is outlined below.

```
                                Switch
    switch  expression
        case  value_1
            code group 1
        case  value_2
            code group 2
        ...
        case  value_n
            code group n
        otherwise
            last code group
    end
```

The role of expression can be played by a number as well as a text string, see Examples 1.3.6 and 1.3.7. Code group i is executed when value_i matches expression. The last code group is executed when no code group was executed. This code group is optional, but strongly recommended.

Example 1.3.6 The following listing presents a function that returns the graph of $\sin x$ in the color specified by the input variable. For example, calling switch_1('red') returns the graph of $\sin x$ in red. If the passed color is unavailable, a blue color is used and a message is sent about the change.

```
function switch_1(color_name)
% This is the function file switch_1.m.
% It is an application on the switch structure.
x1 = -pi; x2 = pi; nx = 20; x = linspace(x1,x2,nx+1);
switch color_name
    case 'green'
        c = 'g';
    case 'red'
        c = 'r';
    case 'yellow'
        c = 'y';
    case 'black'
```

```
        c = 'k';
    case 'blue'
        c = 'b';
    otherwise
        c = 'b';
    str = upper(color_name);
            % The upper function converts color_name passed by the user
            % to capital letters.
    disp(strcat(str,' color unavailable. Replaced with blue.'));
            % The strcat function concatenates the dynamic string str and the
            % static string 'color ... blue'. The disp function shows the
            % complete message to the user.
end
plot(x,sin(x),c);
end
```

Example 1.3.7 Another example on switch is the following function where the input argument is a number.

```
function switch_2(i)
% This is the function file switch_2.m.
% For example, use switch_2(3) to call the function.

x1 = -pi; x2 = pi; nx = 20; x = linspace(x1,x2,nx+1);
switch i
    case 1
        c = 'g';
    case 2
        c = 'r';
    case 3
        c = 'c';
    case 4
        c = 'y';
    case 5
        c = 'k';
    case 6
        c = 'b';
    otherwise
        c = 'b'; disp(' Unavailable. Replaced with blue.')
end
```

```
plot(x,sin(x),c);
end
```

Matlab provides two commands for loops: *for* and *while*. The *for* loop was introduced in Sec. 1.2.1. Now, the *while* loop is presented. Its syntax is outlined below.

While loop

```
while  condition
      code lines
end
```

First, condition is evaluated. If it is true, the code lines are executed and condition is evaluated again. The process is repeated infinitely until condition becomes false. If condition is initially false, code lines are never executed. Applications on a while loop are provided in Examples 1.3.8 and 1.3.9, and Exercises 1.4.17 and 1.4.18.

Example 1.3.8 Consider the following listing.

```
% This is the script file while_1.m. It is an application on while loop.
```

```
i = 0; a = 10;
while i < a
     i = i + 1;
     disp(i);
end
```

Executing the file yields

```
1
2
...
10.
```

The *break* command forces the program to exit from the *while* loop, even if condition is true. For example, insert the following code just before end and guess what happens. However, using a break is not recommended.

```
if i == 6
     break;
end
```

Example 1.3.9 Another example is illustrated in the following listing.

```
function y = while_2(str)
% This is the function file while_2.m.
% The function returns the number of spaces in the input string.
% For example, the command
%      spaces = while_2('I am from Naples')
% produces
%      spaces = 3.

y = 0; i = 1;
c = isspace(str);                                    %......[1]
while i<= length(c)
    if c(i)>0
        y = y + 1;
    end
    i = i + 1;
end
end
```

——————— Notes ———————

[1] The c = isspace(str) command, where str is a text string, returns a row vector the same size as str containing ones and zeros, where 1 corresponds to a space character and 0 to any other character. For example, if

str = 'I am from Naples',

then

$$c = [0\ 1\ 0\ 0\ 1\ 0\ 0\ 0\ 1\ 0\ 0\ 0\ 0\ 0\ 0].$$

1.3.4 *Local Functions, Anonymous Functions*

Local functions are functions defined within a function file. They are visible only to the *main function* and other local functions. Therefore, they cannot be called by other functions. All variables defined in local functions are private. The local functions are known as *subfunctions* too. A simple example is illustrated in the following listing.

```
function y = local_function(str)
% This is the function file local_function.m.
% It is an application on local functions. The function returns the numbers of
```

```
% characters in the input string different from spaces. For example, the
% command
%      ns = local_function('I am from Naples')
% produces
%      ns = 13.

c = isspace(str);
s = GetSpaces(c);
y = length(c) - s;
end
```

————— Local function —————————

```
% The local function GetSpaces returns the number of spaces.
function s = GetSpaces(c)
s = 0;
for i=1:length(c)
    if c(i) > 0
        s = s + 1;
    end
end
end
```

The *anonymous functions* are a powerful tool provided by Matlab to define simple functions. The syntax is outlined below.

Anonymous function

function_name = @(arg1, arg2,...)   function_expression

As noted, the function name is followed from the = sign, the @ sign that characterizes the anonymous functions, and the input variables in parentheses. Next, after some spaces, the function expression. A simple example of an anonymous function is the following

```
    f = @(x)   x + 2;
```

that defines the function $f(x) = x + 2$, where x can be an array. After defining f, the command

```
    feval(f,3)
```

evaluates f for $x = 3$ and produces

```
    5.
```

Equivalently, the command f(3) can be used. In addition, the command

    fplot(f,[0 1])

returns the graph of $f(x) = x + 2$ on the interval $(0, 1)$. A further example is suggested in Exercise 1.4.20 after introducing the logical functions in Sec. 1.3.5. An anonymous function has an important limitation: it must be defined in one line. However, it can be used in many situations.

1.3.5 *Logical Operators and Logical Functions*

Matlab provides three logical operators that are outlined below.

Logical operators	
&	Logical *AND*
|	Logical *OR*
~	Logical *NOT*

The first two operators work with at least two operands. The third operator needs one operand. In logical expressions related to scalar variables, the following symbols must be used: && and ||, instead of & and |.

    Logical AND evaluates the truth or falseness of the operands and returns true if all operands are true; otherwise it returns false. For example, the expression

    a > 0 && b > 0 && c > 0

returns 1 (true) if a, b and c are strictly positive scalar variables and 0 (false) if at least one is less than or equal to zero. For example, if a, b and c were initialized as

    a = 1; b = -1; c = pi;

then the expressions

    a > 0 && b > 0
    c > 0 && b > 0
    a > 0 && b > 0 && c > 0

return

    0

and the expression

    a > 0 && c > 0

returns

    1.

Note that in Matlab, false is expressed by "0" and true is expressed by "1" or, more generally, any nonzero value. Therefore, the expression

    a && b && c

returns

    1.

The expression

    a - b > 0

returns

    1

since it is true and the expression

    a - c > 0

returns

    0

since it is false. See Exercises 1.4.1 to 1.4.21. If the operands are vectors of the same length, each element of a vector is evaluated with the corresponding elements of the other vectors and a same length vector of zeros and ones is returned. For example, after creating the vectors

    u = [0 1 3]; v = [-1 0 1]; z = [-3 -1 0];

the expression

    u & v & z

returns

    0 0 0.

If a is a scalar, for example,

    a = 1;

then the expression

    a & u

returns

    0 1 1

since Matlab evaluates the scalar with each element of u. If the operands are matrices of the same size, a same size matrix of zeros and ones is returned. For example, after creating the matrices

    A = [0 1 3; 4 5 6]; B = [-1 0 1; -3 -2 0];

the expression

    A & B

returns

```
0 0 1
1 1 0
```

The logical operator & is frequently used in if-else and while structures. For example,

```
if a >= 0 && b < 0
    code lines
end
```

The logical operator | evaluates the operand truth or falseness and returns false if all operands are false; if at least one operand is true, it returns true. For example, the expression

```
a > 0 || b > 0 || c > 0
```

returns "1" if at least one variable assumes a positive value and "0" if all scalars are less than or equal to zero. If the operands are matrices of the same size, then each element of a matrix is evaluated with the corresponding elements of the other matrices and a same size matrix of zeros and ones is produced. In addition, the expression a | A, where a is a scalar and A is a matrix, is a compatible statement. In this case, any element of A is evaluated with a. The OR operator | is often used in flow control code. See Exercise 1.4.22.

Example 1.3.10 As noted in Remark 1.3.3, the maxmin_vector function should be modified. The corrected version is shown below.

```
function [max, min] = maxmin_vector(b)
...
if size(b,1) == 1 || size(b,2) == 1
    % The size(A) command, where A is a matrix, returns a vector of two
    % elements that specify the number of rows and columns in A, respectively.
    Place the old code here.
else
    error('The input variable must be a vector.');
end
end
```

Logical NOT ~ operator works with one operand. It evaluates the operand's truth or falseness and returns "false" if the operand is "true" and "true" if the operand is "false". For example, after creating the vector v

   v = [-1 0 1];

the expression

   ˜ v

returns

   0 1 0.

Matlab provides *logical functions* too. Some of them — *all*, *any*, *find* and *ismember* — will be presented in this section. If u is a vector, the logical function

   all(u)

returns "1" if all the elements of u are different from zero, otherwise it returns "0". For example, after creating the vectors

   u = [0 1 2]; v = [1 2 3];

the command

   all(u)

returns

   0

and the command

   all(v)

returns

   1.

If A is a matrix, the command

   all(A)

evaluates the column vectors of A and returns a row vector of zeros and ones with a length equal to the number of columns of A. For example, after creating the matrix

A = [0 1 2; 1 2 3];

calling

   all(A)

returns

   0 1 1.

The function can also be called with an optional argument: all(A,n). In this case, the function evaluates according to the dimension specified by n. For example, the command

all(A,1)

evaluates according to the first dimension (row) and considers the columns as vectors. The result is the same as all(A). The command

all(A,2)

evaluates according to the second dimension (column) and considers the rows as vectors. It returns

0
1.

The logical function

any(u)

returns "1" if at least one element of vector u is different from zero, otherwise it returns "0". For example, after creating the vector

u = [0 1 2];

the any(u) command returns "1". If a matrix A is passed as an argument, any(A) works exactly the same as all, including the possibility of the optional argument: any(A,n).

If v is a vector and p is a real number, the logical function

find(v > p)

finds the elements of v greater than p and returns their indices. For example, the command

find(-2:3 > 1)

returns

5 6

that are the indices of the two elements, 2 and 3, greater than 1. If A is a matrix, the function

[ri ci] = find(A == p).

finds the elements of A equal to p and returns the two vectors ri and ci containing the row and column indices of the element of A equal to p. For example, after creating the matrix

A = [0 3; -4 0];

the command

$$[ri\ ci] = find(A == 0)$$

produces

ri =

1

2

ci =

1

2

since $A(1,1) = 0$ and $A(2,2) = 0$. The command

$$[ri\ ci] = find(A)$$

returns the indices of the nonzero elements of A. In addition, the command

$$[ri\ ci\ vs] = find(A)$$

also returns the vector vs containing the values of nonzero elements of A. For example, if A is the matrix created beforehand, then the previous command returns

ri =

2

1

ci =

1

2

vs =

-4

3.

If A and B are matrices, the logical function

ismember(A,B)

evaluates if an element of A belongs to B and returns "1" in the positive case, otherwise "0". Therefore, a matrix with the same size as A is generated containing zeros and ones. For example, if A is the matrix created with the command

$$A = [0\ 3;\ -4\ 0];$$

then the command

ismember(A,0)

produces

    1 0
    0 1

and the command

    ismember(A,A)

returns

    1 1
    1 1.

1.4 Exercises

Exercise 1.4.1 After creating the matrix

    A = [1 2 3; 4 5 6; 7 8 9; 1 1 1; 2 2 2];

extract the submatrix formed by the first and third rows.

Answer. A([1 3],:).

Exercise 1.4.2 Extract the submatrix formed by the first, second, third and fifth rows from the matrix A, created in Exercise 1.4.1.

Hint. Clearly, the A([1 2 3 5],:) command works. However, it is not the most efficient when we consider a matrix with 4,000 rows and want to extract the submatrix formed by the first 2,000 rows and the last one. The reader is asked to find a more efficient command.

Exercise 1.4.3 Delete the first, second, third and fifth rows of matrix A.

Hint. Use a command more efficient than A([1 2 3 5],:)=[ ].

Exercise 1.4.4 Let

$$C = \begin{bmatrix} 1 & 2 & 3 \\ 4 & 5 & 6 \end{bmatrix}, \quad D = \begin{bmatrix} 7 & 8 \\ 9 & 10 \end{bmatrix}.$$

The [C'; D] command works to produce the matrix

$$\begin{bmatrix} 1 & 4 \\ 2 & 5 \\ 3 & 6 \\ 7 & 8 \\ 9 & 10 \end{bmatrix}.$$

What is the result of the [C'; D'] command?

Exercise 1.4.5 Convert the matrix

$$B = \begin{bmatrix} 1 & 4 \\ 2 & 5 \\ 3 & 6 \end{bmatrix}$$

to a vector, say z.

Exercise 1.4.6 Use the B = [[1; 2; 3] ones(3,1) -[1; 2; 3]] command to create the matrix

$$B = \begin{bmatrix} 1 & 1 & -1 \\ 2 & 1 & -2 \\ 3 & 1 & -3 \end{bmatrix}.$$

Then, create the matrix A = spdiags(B, -1:1, 3, 3).

Exercise 1.4.7 Create the matrix A = spdiags(B, -1:1, 3, 3), where B is the matrix defined by

$$B = \begin{bmatrix} 1 & 1 & 0 \\ 2 & 1 & -2 \\ 0 & 1 & -3 \end{bmatrix}.$$

Compare the matrix A with the one obtained in the previous exercise.

Exercise 1.4.8 Create the matrix A = spdiags(B, [-4 -1:1 4], 9, 9) by using the matrix B = repmat([ones(3,1) 4*[ones(2,1);0] 2*ones(3,1) 3*[0;ones(2,1)] ones(3,1)], 3, 1). Try to guess the result.

Exercise 1.4.9 Consider the simple function $f(x) = x$, $x \in [1, 20]$. Suppose that $f(x)$ was discretized with the vector $x = [1\ 2\ 3\ \cdots\ 20]$. Write the Matlab command that discretizes the function $f^2(x)$ and produces the vector $[1\ 4\ 9\ \cdots\ 400]$.

Answer. x.*x, or x.^{2}.

Exercise 1.4.10 Consider the three-hinged arch in Fig. 1.4.1 (left). Use the left division to calculate the constraint reactions. Assume: $F = 4N$, $q = 2N/m$ and $L = 4m$. Use the free body diagram (or the Lagrangian model), e.g., D'Acunto and Massarotti (2016), illustrated in Fig. 1.4.1 (right).

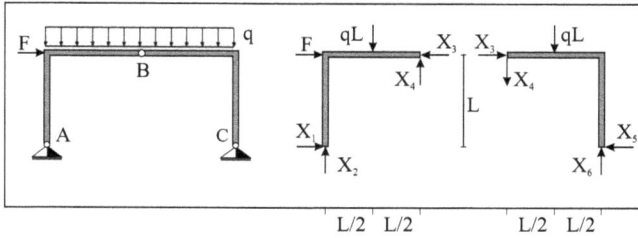

Figure 1.4.1. Three-hinged arch (left) and free body diagram (right).

Hint.

$$
\begin{aligned}
F + X_1 - X_3 &= 0 \\
X_2 + X_4 - qL &= 0 \\
qL^2/2 + LX_1 - LX_2 &= 0 \\
X_3 - X_5 &= 0 \\
X_6 - X_4 - qL &= 0 \\
LX_6 - LX_5 - qL^2/2 &= 0
\end{aligned}
\Leftrightarrow
\begin{bmatrix}
1 & 0 & -1 & 0 & 0 & 0 \\
0 & 1 & 0 & 1 & 0 & 0 \\
4 & -4 & 0 & 0 & 0 & 0 \\
0 & 0 & 1 & 0 & -1 & 0 \\
0 & 0 & 0 & -1 & 0 & 1 \\
0 & 0 & 0 & 0 & -4 & 4
\end{bmatrix}
\begin{bmatrix}
X_1 \\ X_2 \\ X_3 \\ X_4 \\ X_5 \\ X_6
\end{bmatrix}
=
\begin{bmatrix}
-4 \\ 8 \\ -32 \\ 0 \\ 8 \\ 16
\end{bmatrix}.
$$

Exercise 1.4.11 Execute the element-by-element right division A./B, where

$$
A = \begin{bmatrix} 1 & 2 \\ 6 & 8 \end{bmatrix}, \quad B = \begin{bmatrix} 1 & 2 \\ 3 & 4 \end{bmatrix}.
$$

Exercise 1.4.12 Verify that the function u, defined in (1.2.1), satisfies Eq. (1.2.2) and initial-boundary Conditions (1.2.3).

Exercise 1.4.13 Verify that the function u, defined in (1.2.4), satisfies Eq. (1.2.5) and initial-boundary Conditions (1.2.6).

Exercise 1.4.14 Type help sqr at the command line and press Enter.

Exercise 1.4.15 Replace end with the specific values.

Hint. Consider the matrix u.

Exercise 1.4.16 Call the following function. The function output is shown in Fig. 1.4.2.

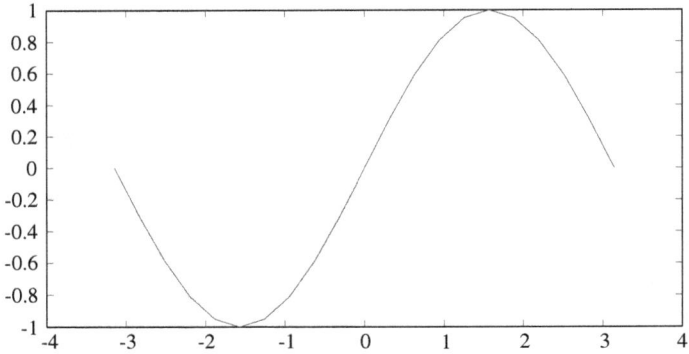

Figure 1.4.2.   Function output.

```
function if_2(i)
% This is the function file if_2.m.
x1 = -pi; x2 = pi; nx = 20; x = linspace(x1,x2,nx+1);
c = 'b';
if i == 0
    c = 'g';
elseif i == 1
    c = 'r';
end
plot(x,sin(x),c);
end
```

Exercise 1.4.17 Try to guess the value of y after executing the following file.

```
% This is the script file while_3.m. It is an exercise on the while loop.
y = 2;
while y > 2
    y = y - 1;
    disp(y);
end
```

Exercise 1.4.18 Replace *while* with *for* in the function while_2, see Example 1.3.9.

Exercise 1.4.19 Consider the function u on the interval $[0, L]$

$$u(x) = \begin{cases} 0 \text{ if } x \in [0, x_1] \cup]x_2, L], \\ 1 \text{ if } x \in]x_1, x_2[. \end{cases} \tag{1.4.1}$$

Write a script file where Function (1.4.1) is defined by using logical operators. Next, plot the function.

Answer.

```
% This is the script file logical_1.m. It is an exercise on logical operators.
L = 2; n =101; i1 = 31; i2 = 61;
x = linspace(0,L,n); x1 = x(i1); x2 = x(i2);
u(1:n) = (x(1:n) - x1 > 0).*(x2 - x(1:n) > 0);
plot(x,u);
axis('equal');
```

Exercise 1.4.20 Write a script file where Function (1.4.1) is defined by using an anonymous function. Next, plot the function.

Exercise 1.4.21 Write a script file where Function (1.4.1) is defined by using a *for* loop. Next, plot the function.

Exercise 1.4.22 Consider the following code.

```
a = 1; L = 2; b = 0; T = 3; n = 10;
if a <= 0 || L <= 0 || T <= 0 || n <= 2
     b = 1;
end
disp(b);
```
What is the value of b after executing the listing?

Chapter 2

The Finite Difference Method

The chapter presents the Finite Difference Method (FDM). This method dates back to Euler[1] who introduced it in *Institutiones calculi Differentialis* (1755). The modern researches on the FDM started after the paper by Courant, *et al.* (1928), where the method was used to obtain approximated solutions to Partial Differential Equations (PDEs). In this field, the method was improved mainly after the Second World War when powerful computers were available. The books by Collatz (1966); Forsythe and Wasov (1960), and Richtmyer and Morton (1967) had a great role in stimulating research on the FDM. Other books by Cooper (1998); Kharab and Guenther (2002) considered Matlab applications too. Today, the FDM is considered a consolidated tool that is able to provide reliable solutions of PDEs and is used by scientists and technicians in many scientific areas, e.g., D'Acunto (2004); de Vahl Davis (1986). In this chapter, FDM will be applied to the heat equation by introducing these noteworthy methods: *Explicit Euler Method Implicit Euler Method* and *Crank–Nicolson Method.*

A section is devoted to the equation governing heat propagation and diffusion.

2.1 Finite Difference Approximations of Derivatives

2.1.1 *Forward, Backward and Central Approximations*

Let $f(x)$ be a function defined on the interval $[0, L]$. A finite set of points x_i, $i = 0, ..., n$, $x_i \in [0, L]$, forms a *grid* or *mesh*. Of special importance are the grids with a constant step indicated with h or Δx (Fig. 2.1.1). If

---

[1]Leonard Euler, a Swiss scientist, 1707–1783. He formulated the laws of solid and fluid dynamics. He published *Institutiones Calculi Differentialis* and introduced the Euler's angles for rotating bodies.

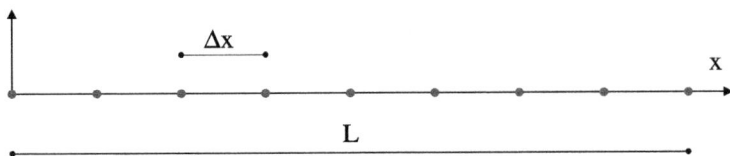

Figure 2.1.1.   Grids with a constant step.

we restrict $f(x)$ to $x = x_i$, we obtain the discrete counterpart, or discrete version, of f. The value of f for $x = x_i$ will be indicated with f_i

$$f_i = f(x_i) = f(ih), \quad i = 0, \ldots, n. \tag{2.1.1}$$

Consider the Taylor[2] series

$$f(x + \Delta x) = f(x) + f'(x)h + f''(x)\frac{(h^2}{2} + f'''(x)\frac{h^3}{6} + \cdots, \tag{2.1.2}$$

evaluated at $x = x_i$

$$f_{i+1} = f_i + f_i'h + f_i''\frac{h^2}{2} + f_i'''\frac{h^3}{6} + \cdots, \tag{2.1.3}$$

where Notation (2.1.1) was used. The previous formula can be written in a more concise way by using the symbol O (capital o)

$$f_{i+1} = f_i + f_i'h + O(h^2). \tag{2.1.4}$$

The symbol $O(h^n)$ indicates a quantity going to zero as h^n, i.e., a quantity bounded by a positive constant times h^n. Solving (2.1.4) with respect to f_i' yields

$$f_i' = \frac{f_{i+1} - f_i}{h} + O(h). \tag{2.1.5}$$

The ratio $(f_{i+1} - f_i)/h$ approximates the derivative f_i' with an error of order h

$$f_i' \approx \frac{f_{i+1} - f_i}{h} \tag{2.1.6}$$

---

[2]Brook Taylor, a British scientist, 1685–1731. He published *Methodus Incrementorum Directa et Inversa* (1715). He stated Taylor's theorem, which was valorized only many years later by Lagrange.

and defines the *forward approximation* of the derivative f_i'. The *backward approximation*

$$f_i' \approx \frac{f_i - f_{i-1}}{h}, \qquad (2.1.7)$$

is inferred similarly. See Exercise 2.4.2. From Formula (2.1.6), we realize that the forward approximation cannot be applied in the last point of the interval where f is defined. Similarly, from Formula (2.1.7), it follows that the backward approximation cannot be applied in the first point of the interval.

Example 2.1.1 The forward function is presented. It returns the forward approximation of the derivative. Two arguments are passed to forward when it is called: the vector u, containing the values of the function to be derived, and the step h.

```
function y = forward(u,h)
% This is the function file forward.m.
% It returns the forward approximation of derivatives. Since the forward
% approximation cannot be applied in the last point, the vector length
% returned by the forward function is equal to that of vector u minus 1.
% Example
% a = 0; b = 1; nx = 20; x = linspace(a,b,nx+1); dx = (b - a)/nx;
% u = x.^2;
% dfu = forward(u,dx)

n = length(u) - 1;
y = (u(2:n+1) - u(1:n))/h;
    % This vector equality is equivalent to
    % y(1) = (u(2)-u(1))/h, ..., y(n) = (u(n+1)-u(n))/h.
    % Note that length(y) = n.
end
```

A way to call forward is suggested in the function comments. Another way is illustrated in the next example.

Example 2.1.2 The forward function is applied to calculate the forward approximation of the derivative of $\sin x$. The exact and approximating derivatives are plotted. See Fig. 2.1.2. The error is evaluated.

```
% This is the script file forward_ex1.m.
% The forward function is called and applied.
a = -pi; b = pi; nx = 32; x = linspace(a,b,nx+1); dx = (b-a)/nx;
Du = cos(x); u = sin(x);
dfu = forward(u,dx);
plot(x,Du,'r',x(1:nx),dfu,'k-*');
legend('Exact','Forward');
xlabel('x'); ylabel('du/dx');
error = max(abs(Du(1:nx)-dfu));
    % If v is a vector, max(v) returns the greatest element of v, and abs(v)
    % returns the vector containing the absolute values of the elements of v.
fprintf('Maximum error = %g\n', error)
    % fprintf formats data and displays the results on the screen.
    % g converts numerical data to a compact format.
    % \n starts a new line.
```

See Exercise 2.4.1, which is related to the error. A function similar to *forward* can be written for the backward approximation too. See Exercises 2.4.2–2.4.4.

Consider the Taylor series for $f(x_i + \Delta x)$ and $f(x_i - \Delta x)$

$$f_{i+1} = f_i + f'_i h + f''_i \frac{h^2}{2} + f'''_i \frac{h^3}{6} + \cdots,$$

$$f_{i-1} = f_i - f'_i h + f''_i \frac{h^2}{2} - f'''_i \frac{h^3}{6} + \cdots.$$

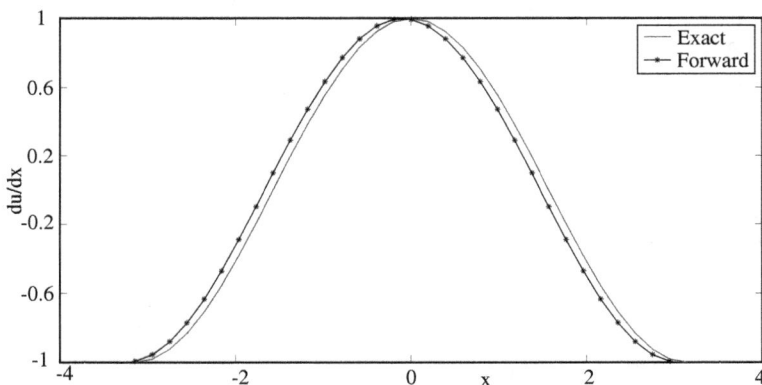

Figure 2.1.2.   Forward and exact derivatives of $\sin x$.

Subtract the second from the first

$$f_{i+1} - f_{i-1} = 2f'_i h + O(h^3),$$

and solve the result with respect to f'_i

$$f'_i = \frac{f_{i+1} - f_{i-1}}{2h} + O(h^2).$$

Hence, the formula for the *central approximation* of f'_i is

$$f'_i \approx \frac{f_{i+1} - f_{i-1}}{2h}, \qquad (2.1.8)$$

with an error of order h^2. The central approximation is more accurate than forward and backward approximations. Clearly, it cannot be applied in the first and last points of the interval where the function is defined. The following example provides the listing of the central function that returns the central approximation of derivatives. Two arguments must be passed to central when it is called: the vector u, containing the values of the function to differentiate, and the step h.

Example 2.1.3

```
function y = central(u,h)
% This is the function file central.m.
% It returns the central approximation of derivatives. Since the central
% approximation cannot be applied in the first and last points, the vector
% length returned by the central function is equal to that of vector u minus 2.
% Example
% a = 0; b = 1; nx = 20; x = linspace(a,b,nx+1); dx = (b - a)/nx;
% u = x.^2;
% dcu = central(u,dx)
n = length(u) - 2;
y = (u(3:n+2) - u(1:n))/h/2;
end
```

A way to call central is suggested in the function comments. Another way is illustrated in the following example.

Example 2.1.4 Forward, backward, central and exact derivatives of $\sin x$ are calculated and compared. The exact and approximating derivatives are plotted. See Fig. 2.1.3. The related errors are evaluated.

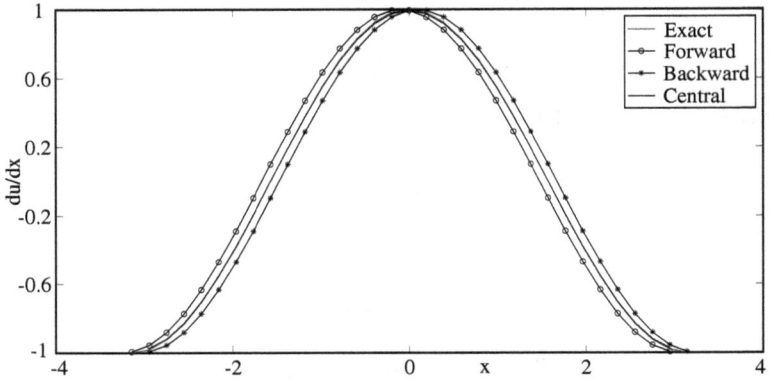

Figure 2.1.3.   Forward, backward, central and exact derivatives of $\sin x$

% This is the script file central_ex1.m.
% The central function is called and applied. The exact and approximating
% forward, backward and central derivatives of u = sin x are plotted. The
% errors are evaluated.

```
a = -pi; b = pi; nx = 32; x = linspace(a,b,nx+1); dx = (b-a)/nx;
u = sin(x); Du = cos(x);
dfu = forward(u,dx); dbu = backward(u,dx); dcu = central(u,dx);
plot(x,Du,'r',x(1:nx),dfu,'k-o',x(2:nx+1),dbu,'k-*',x(2:nx),dcu,'k');
legend('Exact','Forward','Backward','Central');
xlabel('x'); ylabel('du/dx');
errorf = max(abs(Du(1:nx)-dfu));
errorb = max(abs(Du(2:nx+1)-dbu));
errorc = max(abs(Du(2:nx)-dcu));
fprintf('Maximum forward error = %g\n',errorf)
fprintf('Maximum backward error = %g\n',errorb)
fprintf('Maximum central error = %g\n',errorc)
```

See Exercise 2.4.5, which is related to the error. As noted, in the previous example, the points where the approximating derivatives cannot be applied were not plotted. For example, in the case of the central approximation, the first and last points were excluded. When the derivatives in such points are necessary, the forward approximation can be applied in the first point and the backward in the last. The error will increase, since the forward and backward approximations are less accurate than the central one. In these situations the *three-point forward and backward approximations* can be applied. They are accurate to the order 2, like the central approximation, and the error does not grow. The formulas for the three-point forward and backward approximations are the following

$$f_i' \approx \frac{4f_{i+1} - 3f_i - f_{i+2}}{2h}, \tag{2.1.9}$$

$$f_i' \approx \frac{-4f_{i-1} + 3f_i + f_{i-2}}{2h}, \tag{2.1.10}$$

respectively, with an error of order h^2. See Exercise 2.4.6.

Example 2.1.5 The following listing provides the derivative of $u = x^2$ by using the central approximation and Formulas (2.1.9) and (2.1.10) for the first and last points. The result is compared with the exact derivative and the **gradient** function by Matlab. Indeed, the gradient is the same as the derivative for a function of one variable. Central + three-point approximation, gradient and exact derivatives are plotted. See Fig. 2.1.4.

```
% This is the script file central_ex2.m
% The derivative of u = x² is calculated by using the central approximation
% and three-point forward and backward approximations for first and last
% points, respectively. Also, the gradient function by Matlab is applied.

a = 0; b = 1; nx = 20; x = linspace(a,b,nx+1); dx = (b-a)/nx;
u = x.^2; Du = 2*x;
```

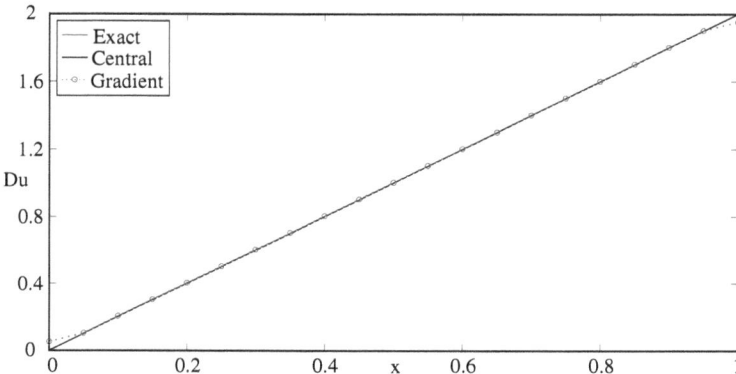

Figure 2.1.4. Central (+ three-point) approximation, Matlab gradient and exact derivatives.

```
g = gradient(u,dx);
dcu = zeros(nx+1,1);
dcu(2:nx) = (u(3:nx+1)-u(1:nx-1))/dx/2;
    % Central approximation
dcu(1) = (4*u(2)-3*u(1)-u(3))/2/dx;
    % Three-point forward approximation
dcu(nx+1) = (-4*u(nx)+3*u(nx+1)+u(nx-1))/2/dx;
    % Three-point backward approximation
plot(x,Du,'r',x,dcu,'k',x,g,'bo:');
xlabel('x'); ylabel('Du'); axis([a b min(Du) max(Du)]);
legend('Exact','Central','Gradient','Location','NorthWest');
```

As it results immediately from Fig. 2.1.4, the derivative provided by gradient presents a greater error in the first and last points. The error tends to vanish for increasing nx. However, it is worth investigating. See Exercise 2.4.8.

2.1.2 *Approximation of Functions Depending on Two Variables*

Consider a function depending on two variables $x \in [0, L]$ and $t \in [0, T]$, and introduce the notation $u_{i,j} = u(x_i, t_j)$. If t denotes time, the notation u_i^j is used too. For a mesh with constant steps Δx and Δt, shown in Fig. 2.1.5, it is

$$u_i^j = u(x_i, t_j) = u(i\Delta x, j\Delta t), \quad i = 0, \ldots, n, \quad j = 0, \ldots, m,$$

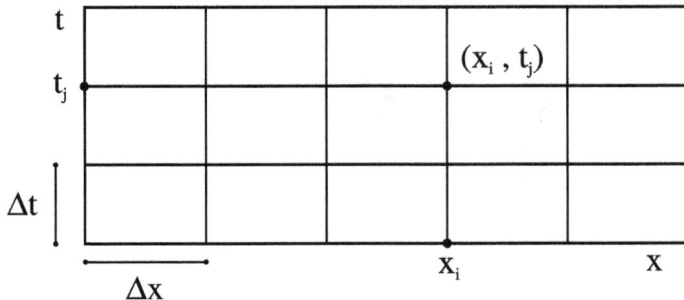

Figure 2.1.5. Space-time mesh.

where $\Delta x = L/n$ and $\Delta t = T/m$. The partial derivative will also be indicated with the following notations

$$u_x = \frac{\partial u}{\partial x}, \quad u_t = \frac{\partial u}{\partial t}.$$

Their approximations follow easily from the approximation formulas of a function in one variable. The forward, backward and central approximations of u_x are expressed as

$$(u_x)_i^j \approx \frac{u_{i+1}^j - u_i^j}{\Delta x}, \quad (u_x)_i^j \approx \frac{u_i^j - u_{i-1}^j}{\Delta x}, \quad (u_x)_i^j \approx \frac{u_{i+1}^j - u_{i-1}^j}{2\Delta x}, \quad (2.1.11)$$

respectively. The forward, backward and central approximations of u_t are expressed as

$$(u_t)_i^j \approx \frac{u_i^{j+1} - u_i^j}{\Delta t}, \quad (u_t)_i^j \approx \frac{u_i^j - u_i^{j-1}}{\Delta x}, \quad (u_t)_i^j \approx \frac{u_i^{j+1} - u_i^{j-1}}{2\Delta x}, \quad (2.1.12)$$

respectively. The reader is asked to derive the previous formulas. See Exercise 2.4.9. Exercise 2.4.10 is devoted to three-point approximations.

2.1.3 *Approximation of Higher Order Derivatives*

This section considers the approximations for the second derivatives. Firstly, the *forward approximation* for the partial derivative u_{tt} is presented. Consider the Taylor series

$$u_i^{j+1} = u_i^j + (u_t)_i^j \Delta t + (u_{tt})_i^j \Delta t^2/2 + O(\Delta t^3), \quad (2.1.13)$$

$$u_i^{j+2} = u_i^j + (u_t)_i^j 2\Delta t + (u_{tt})_i^j 2\Delta t^2 + O(\Delta t^3). \quad (2.1.14)$$

Subtract $(2.1.13) \times 2$ from $(2.1.14)$ and solve the result with respect to $(u_{tt})_i^j$

$$(u_{tt})_i^j = (u_i^{j+2} - 2u_i^{j+1} + u_i^j)/(\Delta t)^2 + O(\Delta t).$$

Hence, the desired approximation with an error of order Δt

$$(u_{tt})_i^j \approx \frac{u_i^{j+2} - 2u_i^{j+1} + u_i^j}{(\Delta t)^2}. \quad (2.1.15)$$

An analogous result holds for the derivative u_{xx}, with an error of order Δx,

$$(u_{xx})_i^j \approx \frac{u_{i+2}^j - 2u_{i+1}^j + u_i^j}{(\Delta x)^2}. \quad (2.1.16)$$

The *backward approximations* are obtained with similar reasoning

$$(u_{tt})_i^j \approx \frac{u_i^{j-2} - 2u_i^{j-1} + u_i^j}{(\Delta t)^2}, \quad (u_{xx})_i^j \approx \frac{u_{i-2}^j - 2u_{i-1}^j + u_i^j}{(\Delta x)^2}, \quad (2.1.17)$$

with an error of order Δt and Δx, respectively. Moreover, consider the Taylor series

$$u_i^{j+1} = u_i^j + (u_t)_i^j \Delta t + (u_{tt})_i^j \frac{(\Delta t)^2}{2!} + (u_{ttt})_i^j \frac{(\Delta t)^3}{3!} + O((\Delta t)^4),$$

$$u_i^{j-1} = u_i^j - (u_t)_i^j \Delta t + (u_{tt})_i^j \frac{(\Delta t)^2}{2!} - (u_{ttt})_i^j \frac{(\Delta t)^3}{3!} + O((\Delta t)^4).$$

Let us sum the two previous formulas and solve the result with respect to $(u_{tt})_i^j$

$$(u_{tt})_i^j = \frac{u_i^{j+1} - 2u_i^j + u_i^{j-1}}{(\Delta t)^2} + O((\Delta t)^2).$$

Hence, the *central approximation* of the partial derivative u_{tt} is

$$(u_{tt})_i^j \approx \frac{u_i^{j+1} - 2u_i^j + u_i^{j-1}}{(\Delta t)^2}, \quad (2.1.18)$$

with an error of order $(\Delta t)^2$. Formula (2.1.18) is more accurate than (2.1.15)-(2.1.17). Of course, an analogous formula holds for u_{xx}

$$(u_{xx})_i^j \approx (u_{i+1}^j - 2u_i^j + u_{i-1}^j)/(\Delta x)^2. \quad (2.1.19)$$

See Exercise 2.4.11. For functions depending on two variables, the approximations for the mixed derivative should be discussed too. Consider the *forward approximation* of u_{xt}. First, calculate the forward approximation for the time derivative $(u_x)_t$ by using the forward approximation $(2.1.12)_1$

$$(u_{xt})_i^j = \frac{(u_x)_i^{j+1} - (u_x)_i^j}{\Delta t} + O(\Delta t).$$

Next, calculate the forward approximations for the space derivatives by using Formula $(2.1.11)_1$

$$(u_{xt})_i^j = \frac{u_{i+1}^{j+1} - u_i^{j+1} - u_{i+1}^j + u_i^j}{\Delta x \Delta t} + O(\Delta x) + O(\Delta t).$$

Hence, the desired formula with an error of order $O(\Delta x) + O(\Delta t)$ is

$$(u_{xt})_i^j \approx \frac{u_{i+1}^{j+1} - u_i^{j+1} - u_{i+1}^j + u_i^j}{\Delta x \Delta t}. \quad (2.1.20)$$

Similar reasonings lead to the *backward approximation*

$$(u_{xt})_i^j \approx \frac{u_{i-1}^{j-1} - u_i^{j-1} - u_{i-1}^j + u_i^j}{\Delta x \Delta t}, \tag{2.1.21}$$

with an error of order $O(\Delta x) + O(\Delta t)$ and to the *central approximation*

$$(u_{xt})_i^j \approx \frac{u_{i+1}^{j+1} - u_{i-1}^{j+1} - u_{i+1}^{j-1} + u_{i-1}^{j-1}}{4\Delta x \Delta t}, \tag{2.1.22}$$

with an error of order $O((\Delta x)^2) + O((\Delta t)^2)$. See Exercise 2.4.12.

2.2 Diffusion

This section presents the equation governing heat propagation and diffusion Cannon (1984); Carslaw and Jager (1959); Crank (1979). The Matlab programs for the mentioned equation will be illustrated in Sec. 2.3. The *heat equation*, introduced by Fourier[3], is the basic tool for solving problems of heat propagation in solids. The heat equation is a parabolic partial differential equation. Its solution depends on initial-boundary conditions, as illustrated in the next section. Fourier's methodology stimulated other scientists to use the mathematical formulation for different physical phenomena. Indeed, some years later, Fick[4] and Darcy[5] introduced similar laws for diffusion and fluid flow in porous media.

2.2.1 *Fourier's Law and Heat Equation*

Fourier's law (1822) follows from observations and experiences that outline that heat flux in homogeneous and isotropous solids is proportional to the thermal gradient and flows from hotter to colder regions

$$\mathbf{q}(\mathbf{x}, t) = -k\nabla u(\mathbf{x}, t). \tag{2.2.1}$$

In Eq. (2.2.1), the vector $\mathbf{q}(\mathbf{x}, t)$ indicates the *heat (or thermal) flux*, the heat flux per unit time per unit isothermal surface, $u(\mathbf{x}, t)$ the *temperature*

---

[3] Jean Baptiste Fourier, a French scientist, 1768–1830. He was taught by Lagrange at the Ecole Normale of Paris. He published *Théorie Analytique de la Chaleur* (1822). He participated in the military expedition to Egypt with Napoleon.
[4] Adolf Eugen Fick, a German scientist, 1829–1901. He was Professor of Physiology at the University of Würburg. He introduced Fick's law of diffusion in 1855.
[5] Henry Philibert Gaspard Darcy, a French scientist, 1803–1858. He was Chief Engineer in Dijon. He published Darcy's law on fluid flow in porous media in *Les Fontaines publiques de la Ville de Dijon* (1856).

and k the *thermal conductivity* of the material. Note that ∇u denotes the gradient of u with respect to the only space variables

$$\nabla u = \left(\frac{\partial u}{\partial x_1}, \frac{\partial u}{\partial x_2}, \frac{\partial u}{\partial x_3} \right). \tag{2.2.2}$$

It is essential to know the gradient properties to better understand Fourier's law (2.2.1). See Exercises 2.4.13 and 2.4.14.

In the thermal process in a solid B, both functions $\mathbf{q}$ and u are unknown. Therefore, Fourier's law is unable to determine both heat flux and temperature. We need a second equation, which can be provided by the principle of conservation of energy: *rate of energy in V = heat flow entering and leaving V through boundary ∂V + energy production in V*, where V is any control volume included in B. See Fig. 2.2.1. The energy balance is formalized as follows

$$\int_V \rho e_t(\mathbf{x}, t) d\mathbf{x} = -\int_{\partial V} \mathbf{q} \cdot \mathbf{n} \, dS + \int_V F(\mathbf{x}, t) d\mathbf{x}, \tag{2.2.3}$$

where $\mathbf{n}$ is the outward unit normal vector to the surface ∂V at the integration point, $\rho(\mathbf{x})$ is the density of the solid at rest, $e(\mathbf{x}, t)$ is the internal energy per unit mass and e_t is the partial derivative of e with respect to time. In addition, $F(\mathbf{x}, t)$ indicates the heat quantity produced per unit time per unit volume by internal heat generators. The internal energy depends on the temperature $e = e(u)$. For most materials and a wide temperature interval, the dependence is linear

$$e = c_p u, \tag{2.2.4}$$

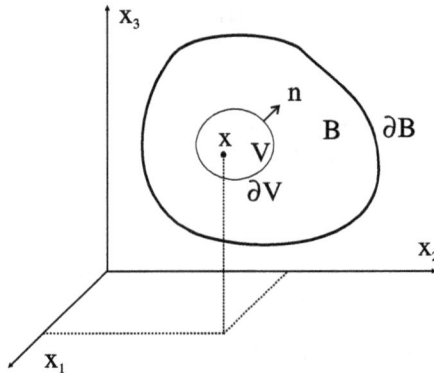

Figure 2.2.1.   Control volume

where c_p indicates the specific heat at constant pressure. Now, consider the Gauss[6] divergence theorem,

$$\int_{\partial V} \mathbf{q} \cdot \mathbf{n} \, dS = \int_V \nabla \cdot \mathbf{q} \, d\mathbf{x}, \qquad (2.2.5)$$

where

$$\nabla \cdot \mathbf{q} = \partial q_1/\partial x_1 + \partial q_2/\partial x_2 + \partial q_3/\partial x_3$$

indicates the divergence of $\mathbf{q}$ with respect to the only space variables. Substituting (2.2.4)–(2.2.5) into (2.2.3) yields

$$\int_V [c_p \rho u_t + \nabla \cdot \mathbf{q} - F] \, d\mathbf{x} = 0, \qquad (2.2.6)$$

that holds for any control volume V. If the integrand function is continuous, then it follows from (2.2.6)

$$c_p \rho u_t + \nabla \cdot \mathbf{q} - F = 0. \qquad (2.2.7)$$

We used the following theorem: *If $f(\mathbf{x})$ is a continuous function on B, then*

$$\int_V f(\mathbf{x}) \, d\mathbf{x} = 0, \ \forall V \subseteq B \Rightarrow f(\mathbf{x}) = 0, \ \forall \mathbf{x} \in B. \qquad (2.2.8)$$

Statement (2.2.8) can be proved as follows. Suppose there exists $\bar{x} \in B$ such that $f(\bar{\mathbf{x}}) > 0$. The continuity hypothesis implies $f(\mathbf{x}) > 0$, $\forall \mathbf{x}$ belonging to a suitable neighbor of $\bar{x}$, say I. Of course, the integral of $f(\mathbf{x})$ on I is positive. This is a contradiction since I is a special V. The contradiction is removed only if $f(\bar{\mathbf{x}}) = 0$. The reasoning is similar if it is supposed there exists $\bar{x} \in B$ such that $f(\bar{\mathbf{x}}) < 0$. See Exercise 2.4.15.

Consider Fourier's law, $\mathbf{q} = -k\nabla u$, in energy balance Eq. (2.2.7) and obtain

$$c_p \rho u_t - \nabla \cdot (k\nabla u) = F, \ \mathbf{x} \in B, \ 0 < t \leq T. \qquad (2.2.9)$$

Partial differential Eq. (2.2.9) is named a *heat equation*. For constant k, (2.2.9) simplifies to

$$c_p \rho u_t - k\Delta u = F, \ \mathbf{x} \in B, \ 0 < t \leq T, \qquad (2.2.10)$$

---

[6] Johann Friedrich Carl Gauss, a German scientist, 1777–1855. He was the greatest mathematician since antiquity. He made significant contributions in Mathematics and Physics.

$$u_t - \alpha \Delta u = f, \ \mathbf{x} \in B, \ 0 < t \le T, \tag{2.2.11}$$

where $f = F/c_p\rho$, $\alpha = k/c_p\rho$ denotes the *thermal diffusivity*, and Δ the Laplace[7] operator

$$\Delta = \nabla^2 = \frac{\partial^2}{\partial x_1^2} + \frac{\partial^2}{\partial x_2^2} + \frac{\partial^2}{\partial x_3^2}.$$

Solving the heat equation yields the unknown function $u(\mathbf{x}, t)$. Next, the heat flux $\mathbf{q}$ is derived from Fourier's Law. However, solving the heat equation requires *initial conditions* and *boundary conditions*. Indeed, the time evolution of the temperature in a solid depends on the initial thermal state and the thermal conditions on the boundary of the solid.

The initial condition is formalized by assigning the function $u(\mathbf{x}, t)$ for $t = 0$

$$u(\mathbf{x}, 0) = \varphi(\mathbf{x}), \quad \mathbf{x} \in B. \tag{2.2.12}$$

Let us illustrate the main types of linear boundary conditions. The *boundary condition of the first type* is related to the situation where the solid surface is kept to a fixed temperature. Therefore, this condition specifies the value of the function $u(\mathbf{x}, t)$ on the boundary

$$u(\mathbf{x}, t) = g(\mathbf{x}, t), \quad \mathbf{x} \in \partial B. \tag{2.2.13}$$

Condition (2.2.13) is also named the *Dirichlet*[8] *boundary condition*.

The *boundary condition of the second type* considers a known heat flux on the solid boundary. Since the heat flux is related to the temperature gradient, this condition is expressed as

$$k\frac{\partial u}{\partial n}(\mathbf{x}, t) = g(\mathbf{x}, t), \quad \mathbf{x} \in \partial B, \tag{2.2.14}$$

where $\partial/\partial n$ indicates the outward normal derivative to the boundary surface. See Exercise 2.4.16. The special case $g = 0$ corresponds to an adiabatic boundary surface. Condition (2.2.14) is also named the *Neumann*[9] *boundary condition*.

---

[7]Pierre Simon Laplace, a French scientist, 1749–1827. He worked on Celestial Mechanics and Probability. He introduced the Laplace equation and Laplace transform.
[8]Pietro Gustavo Dirichlet, a German scientist, 1805–1859. He was Professor at the University of Göttingen. He made deep contributions in Mechanics and Analysis.
[9]Carl Gottfried Neumann, a German scientist, 1832–1925. He was Professor at the University of Leipzig. He worked on Mathematical Physics and Electrodynamics.

The *boundary condition of the third type* is a linear combination of temperature and heat flux

$$k\frac{\partial u}{\partial n}(\mathbf{x}, t) + hu(\mathbf{x}, t) = g(\mathbf{x}, t), \quad \mathbf{x} \in \partial B. \tag{2.2.15}$$

Condition (2.2.15) is also named the *Robin*[10] *boundary condition*. An example of Condition (2.2.15) is provided by *Newton's*[11] *law of cooling*, which states that the rate of heat loss of a body is directly proportional to the temperature difference between a body and the environment

$$-k\frac{\partial u}{\partial n}(\mathbf{x}, t) = h[u(\mathbf{x}, t) - u_{\text{env}}(\mathbf{x}, t)], \quad \mathbf{x} \in \partial B, \tag{2.2.16}$$

where u_{env} indicates the known environment temperature and h is the heat transfer coefficient. Setting $hu_{\text{env}} = g$, Eq. (2.2.16) is reduced to boundary Condition (2.2.15).

When $g = 0$, the corresponding boundary condition of the first, second, or third kind is named *homogeneous*. The heat equation is named *homogeneous* if $f = 0$. An initial-boundary value problem is named *homogeneous* if both the equation and boundary condition are homogeneous.

One-dimensional heat conduction modeling is adopted when the main thermal variables change predominantly in one defined direction, say x. In this situation, all functions in the heat equation depend on x and t only, and Eq. (2.2.10) is reduced to the *one-dimensional heat equation*

$$c_p \rho u_t - (ku_x)_x = F(x, t), \quad 0 < x < L, \quad 0 < t \leq T.$$

For constant k, the previous equation simplifies to

$$u_t(x, t) - \alpha u_{xx}(x, t) = f(x, t), \quad 0 < x < L, \quad 0 < t \leq T,$$

that is used in many applications. The initial condition simplifies to

$$u(x, 0) = \varphi(x), \quad 0 \leq x \leq L.$$

The Dirichlet boundary conditions are reduced to

$$u(0, t) = g_1(t), \quad u(L, t) = g_2(t), \quad t > 0.$$

---

[10]Victor Gustave Robin, a French scientist, 1855–1897. He was Professor of Mathematical Physics at the Sorbonne in Paris. He worked mainly on Thermodynamics.
[11]Sir Isaac Newton, an English scientist, 1642–1727. He was one of the most important scientists of all time. He formulated the laws of Dynamics, published in *Philosophiae Naturalis Principia Mathematica* (1687).

The Neumann boundary conditions assume the following expression

$$-ku_x(0,t) = g_1(t), \quad ku_x(L,t) = g_2(t), \quad t > 0.$$

The - sign for $x = 0$ depends on the fact that the outward normal direction is opposed to that of x. For the same reason, the Robin boundary conditions are written as follows

$$-ku_x(0,t) + h_1 u(0,t) = g_1(t), \quad ku_x(L,t) + h_2 u(L,t) = g_2(t), \quad t > 0.$$

When the thermal process occurs in a moving medium, Fourier's law is modified to take account of the convective term due to the motion

$$\mathbf{q} = -k\nabla u + \rho c_p u \mathbf{v}, \tag{2.2.17}$$

where $\mathbf{v}$ is the velocity of $\mathbf{x}$. Substituting (2.2.17) into the energy equation $c_p \rho u_t + \nabla \cdot \mathbf{q} - F = 0$ yields

$$c_p \rho u_t - \nabla \cdot (k\nabla u) + \nabla \cdot (c_p \rho u \mathbf{v}) = F. \tag{2.2.18}$$

If ρc_p is constant, from (2.2.18), it follows that (see Exercise 2.4.17)

$$c_p \rho u_t - \nabla \cdot (k\nabla u) + c_p \rho (\mathbf{v} \cdot \nabla u + u \nabla \cdot \mathbf{v}) = F, \tag{2.2.19}$$

that is reduced to

$$c_p \rho u_t - \nabla \cdot (k\nabla u) + c_p \rho \mathbf{v} \cdot \nabla u = F, \tag{2.2.20}$$

when $\nabla \cdot \mathbf{v} = 0$ (incompressible flow). Equation (2.2.20) is named a *convection-diffusion equation*, or *advection-diffusion equation*, since it governs thermal processes depending on heat diffusion influenced by convection, or advection. Equation (2.2.20) is very important since it governs a number of physical phenomena. For constant k, the convection-diffusion equation (2.2.20) simplifies to

$$u_t - \alpha \Delta u + \mathbf{v} \cdot \nabla u = f, \tag{2.2.21}$$

where $\alpha = k/c_p \rho$ and $f = F/c_p \rho$. The one-dimensional case equations (2.2.19)–(2.2.21) are expressed as

$$c_p \rho u_t - (ku_x)_x + c_p \rho (vu_x + uv_x) = F, \tag{2.2.22}$$

$$c_p \rho u_t - (ku_x)_x + c_p \rho vu_x = F, \tag{2.2.23}$$

$$u_t - \alpha u_{xx} + vu_x = f, \tag{2.2.24}$$

respectively, where v indicates the only nonzero velocity component.

2.2.2 Fick's Law and Diffusion

Diffusion is the physical process of mass transfer from one region to another in a system, usually fluid or gas. It is modeled by equations very similar to those of heat propagation. The basic law for diffusion is *Fick's Law*, which follows from physical experiences. It states that a diffusing mass moves from regions of higher concentration to regions of lower concentration. The *concentration C* is the dissolving mass per unit volume. Therefore, if the vector **J** indicates diffusive flux per unit area per unit time, the direction of **J** is opposed to the concentration gradient. For isotropous medium at rest, Fick's law is formalized as

$$\mathbf{J} = -D\nabla C, \tag{2.2.25}$$

where D is the *diffusion coefficient*. A second equation involving the concentration can be derived from the principle of mass conservation: *rate of mass in the control volume V = mass entering and leaving V through boundary ∂V*, in absence of internal mass production. This principle leads to the following mass balance equation

$$\int_V C_t \, d\mathbf{x} = -\int_{\partial V} \mathbf{J} \cdot \mathbf{n} \, dS, \quad \forall \, V, \tag{2.2.26}$$

where **n** is the outward unit normal vector to the surface ∂V at the integration point. Using the Gauss divergence Theorem (2.2.5) in Formula (2.2.26) yields

$$\int_V [C_t + \nabla \cdot \mathbf{J}] \, d\mathbf{x} = 0, \quad \forall \, V.$$

This equation holds $\forall \, V$. Therefore, it implies

$$C_t + \nabla \cdot \mathbf{J} = 0, \quad \forall \, \mathbf{x}. \tag{2.2.27}$$

Substituting (2.2.25) into Formula (2.2.27), one arrives at the *diffusion equation*

$$C_t = \nabla \cdot (D\nabla C), \tag{2.2.28}$$

completely similar to the heat equation. Equation (2.2.28) is also named a *diffusive equation*. For constant D, Eq. (2.2.28) simplifies to

$$C_t = D\Delta C.$$

For the diffusion process in a moving medium, Fick's law is replaced by

$$\mathbf{J} = -D\nabla C + C\mathbf{v}, \tag{2.2.29}$$

where $\mathbf{v}$ is the velocity of $\mathbf{x}$. Considering (2.2.29) in (2.2.28) yields

$$C_t = D\Delta C - \nabla \cdot (C\mathbf{v}),$$

$$C_t - D\Delta C + \mathbf{v} \cdot \nabla C + C\nabla \cdot \mathbf{v} = 0, \tag{2.2.30}$$

where D was assumed to be constant. Equation (2.2.30) is similar to Eq. (2.2.19). Therefore, from (2.2.30), equations similar to (2.2.20) to (2.2.24) are immediately derived with the temperature replaced by the concentration.

2.2.3   *Free Boundary Value Problems*

In some problems of noteworthy physical interest, the boundary can move with time during the thermal process. These problems are named *free boundary value problems* Crank (1979, 1984); Rubinstein (1971). Phase transition, e.g., ice turning to water, is a typical example of these problems. Indeed, consider the one-dimensional melting process where the liquid phase is separated from the solid one by a sharp interphase of the equation (see Fig. 2.2.2)

$$x = s(t).$$

As outlined earlier, this function is unknown, since it depends on the thermal process. It cannot be assigned *a priori* and is a further unknown of the problem to be determined together with the temperature field. If the heat transfer occurs by conduction in both phases and there is no internal

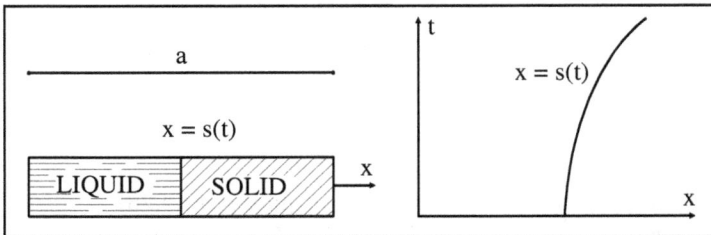

Figure 2.2.2.   Phase change.

heat generation, then the temperature evolution is governed by the following equations

$$u_t = \alpha u_{xx}, \quad 0 < x < s(t), \quad t > 0, \tag{2.2.31}$$

$$u_{st} = \alpha_s u_{sxx}, \quad s(t) < x < a, \quad t > 0, \tag{2.2.32}$$

where the index s is related to the solid phase. At the liquid-solid interphase, the temperature assumes the melting temperature value u_m,

$$u(s(t), t) = u_s(s(t), t) = u_m, \quad t > 0. \tag{2.2.33}$$

Moreover, the initial and boundary conditions must be assigned. For example, reasonable conditions in a melting process are the following

$$u(x, 0) = \varphi(x)(\geq u_m), \quad 0 \leq x \leq s(0), \tag{2.2.34}$$

$$u(0, t) = g(t)(\geq u_m), \quad t > 0, \tag{2.2.35}$$

for the liquid phase and

$$u_s(x, 0) = \varphi_s(x)(\leq u_m), \quad s(0) \leq x \leq a, \tag{2.2.36}$$

$$u_s(a, t) = g_s(t)(\leq u_m), \quad t > 0, \tag{2.2.37}$$

for the solid phase. Since the function $s(t)$ is unknown, a further equation is needed for solving Problems (2.2.31) to (2.2.37). This is provided by the energy balance at the interphase, where the heat flux must equate the absorbed heat given by the product of the latent heat L times the mass of liquid converted from the solid phase

$$Aq(s(t), t) - Aq_s(s(t), t) = A\rho L \dot{s}(t), \quad t > 0, \tag{2.2.38}$$

where ρ is the liquid density and A the sectional area perpendicular to the x-axis. Considering Fourier's law, $q = -ku_x$, in Eq. (2.2.38) yields

$$k_s u_{sx}(s(t), t) - ku_x(s(t), t) = \rho L \dot{s}(t), \quad t > 0, \tag{2.2.39}$$

which is named the *Stefan*[12] *condition*. Equation (2.2.39) allows us to determine the unknown free boundary $x = s(t)$ and the interphase position. The other equations give the temperature time evolution in both phases.

---

[12] Jozef Stefan, an Austrian scientist, 1835–1893. He was Professor at the University of Vienna. He worked on Thermodynamics and the Electromagnetic Theory.

The difficulty is due to the fact that both problems must be solved simultaneously.

A particular case of the previous problem occurs when the solid phase temperature remains constant and equal to u_m during the whole melting process. In mathematical formulation, this happens when conditions (2.2.36) and (2.2.37) are replaced by

$$u_s(x,0) = u_m, \quad s(0) \le x \le a, \tag{2.2.40}$$

$$u_s(a,t) = u_m, \quad t > 0, \tag{2.2.41}$$

respectively. Indeed, the Problems (2.2.32), (2.2.40) and (2.2.41) have the solution

$$u_s(x,t) = u_m, \quad s(t) \le x \le a, \quad t > 0. \tag{2.2.42}$$

In this situation, the phase change process is completely determined by the only liquid phase (*one-phase Stefan problem*) and only the equations related to this phase must be solved

$$u_t = \alpha u_{xx}, \quad 0 < x < s(t), \quad t > 0,$$

$$u(x,0) = \varphi(x) \ge u_m, \quad 0 \le x \le s(0),$$

$$u(0,t) = g(t) \ge u_m, \quad t > 0,$$

$$u(s(t),t) = u_m, \quad t \ge 0,$$

$$-ku_x(s(t),t) = \rho L \dot{s}(t), \quad t > 0.$$

2.3 Finite Difference Method

2.3.1 *Explicit Euler Method*

Consider the one-dimensional heat equation

$$U_t - \alpha U_{xx} = F, \quad 0 < x < L, \quad 0 < t \le T. \tag{2.3.1}$$

As already outlined, Eq. (2.3.1) can be uniquely solved only when the initial-boundary conditions are assigned

$$U(x,0) = \varphi(x), \quad 0 \le x \le L, \tag{2.3.2}$$

$$U(0,t) = g_1(t), \quad U(L,t) = g_2(t), \quad 0 < t \le T. \tag{2.3.3}$$

Conditions (2.3.3) are Dirichlet boundary conditions. Other boundary conditions will be discussed in Sec. 2.3.3. Using the forward approximation for the derivative U_t in (2.3.1) and the central approximation for U_{xx} yields the following finite difference equation

$$\frac{u_i^{j+1} - u_i^j}{\Delta t} - \alpha \frac{u_{i+1}^j - 2u_i^j + u_{i-1}^j}{(\Delta x)^2} = f_i^j, \quad (f_i^j = F(x_i, t_j)). \tag{2.3.4}$$

Note that a solution for Eq (2.3.4) was indicated with u, whereas U is a solution of partial differential Equation (2.3.1). These different notations will always be used in the following. Solving Eq. (2.3.4) with respect to u_i^{j+1} yields (Fig. 2.3.1)

$$u_i^{j+1} = r(u_{i+1}^j + u_{i-1}^j) + (1 - 2r)u_i^j + \Delta t f_i^j, \tag{2.3.5}$$

where

$$r = \alpha \Delta t / \Delta x^2. \tag{2.3.6}$$

Equation (2.3.5) is named the *Explicit Euler Method*. The adjective explicit emphasizes that when the values u_i^j are known for some j, then Eq. (2.3.5) explicitly provides the unknown values u_i^{j+1}. The method is characterized by the forward approximation of the time derivative. Let us show that Method (2.3.5) can be successfully applied to get the solution at any time when initial-boundary Conditions (2.3.2) and (2.3.3) are given. From initial Condition (2.3.2), it follows that the values

$$u_i^0 = \varphi(x_i) = \varphi_i, \quad i = 0, \ldots, n, \quad n\Delta x = L, \tag{2.3.7}$$

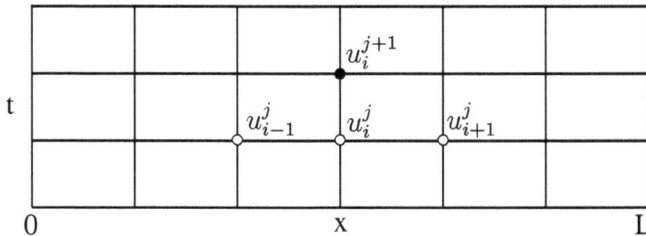

Figure 2.3.1.   Explicit Euler Method.

are known. Using them in Eq. (2.3.5) yields the values u_i^1, $i = 1, \ldots, n-1$. In addition, the first and last values are provided by boundary Conditions (2.3.3): $u_0^1 = g_1^1 = g_1(t_1)$ and $u_n^1 = g_2^1 = g_2(t_1)$. Next, the process is repeated. Lastly, note that Formula (2.3.5) can be arranged in the following matrix form

$$
\begin{bmatrix} u_1^{j+1} \\ u_2^{j+1} \\ \cdot \\ u_{n-1}^{j+1} \end{bmatrix} = \begin{bmatrix} 1-2r & r & & \\ r & 1-2r & r & \\ & \cdot & \cdot & \cdot \\ & & r & 1-2r \end{bmatrix} \begin{bmatrix} u_1^j \\ u_2^j \\ \cdot \\ u_{n-1}^j \end{bmatrix} + \begin{bmatrix} rg_1^j + \Delta t f_1^j \\ \Delta t f_{2,j} \\ \cdots \\ rg_2^j + \Delta t f_{n-1}^j \end{bmatrix},
$$

that incorporates the boundary conditions. Hence, with the apparent meaning of the symbols,

$$\mathbf{u}_{j+1} = A\mathbf{u}_j + \mathbf{a}_j, \tag{2.3.8}$$

where the vector $\mathbf{a}_j$ is known as given by boundary conditions and the source term.

Example 2.3.1   The following function applies the Explicit Euler Method (2.3.5) to solve the Dirichlet problem (2.3.1)–(2.3.3).

```
% function u = euler_e(alpha, L, T, nx, phi, g1, g2, f)
% This is the function file euler_e.m.
% Explicit Euler Method is applied to solve the Dirichlet problem:
% Ut - alpha Uxx = F, U(x,0) = phi(x), U(0,t) = G1(t), U(L,t) = G2(t).
% The input arguments are: thermal diffusivity, length of the solid, final time,
% number of points on the space grid, initial-boundary conditions, source term.
% The function returns a vector with the solution at the final time.

% Check data
if alpha <= 0 || L <= 0 || T <= 0 || nx <= 2
    error('Check alpha, L, T, nx')
end

% Stability
dx = L/nx; nt = 10; st =.5;
while alpha*T/nt/dx^2 > st
    % The stability condition is checked. If it is not satisfied, nt is increased
    % until alpha*T/nt/dx^2 (= r) is less than or equal to 0.5.
    nt = nt+1;
end
```

```
% Initialization
dt = T/nt; r = alpha*dt/dx²;
x = linspace(0,L,nx+1); t = linspace(0,T,nt+1);
u = phi(x);
     % The vector u is initialized with the initial data. The command is
     % equivalent to: u = feval(phi,x);
```

```
% Explicit Euler Method
for j=2:nt+1
     u(2:nx)=(1-2r)*u(2:nx)+r*(u(1:nx-1)+u(3:nx+1))+dt*f(x(2:nx),t(j-1));
     % The computed solution at the next time is saved in the same vector u.
     % At the end of the process, u contains the solution at the final time.
     u(1) = g1(t(j)); u(nx+1) = g2(t(j));
     % Boundary conditions
end
end
```

A way to call euler_e is illustrated in the following example.

Example 2.3.2 Euler_e function is called to solve the Dirichlet problem

$$U_t = U_{xx}, \quad 0 < x < 1, \quad 0 < t \leq T, \tag{2.3.9}$$

$$U(x,0) = \sin(\pi x), \quad 0 \leq x \leq 1, \tag{2.3.10}$$

$$U(0,t) = U(1,t) = 0, \quad 0 < t \leq T. \tag{2.3.11}$$

The solution is plotted in Fig. 2.3.2.

```
function u = euler_e_ex1
% This is the function file euler_e_ex1.m.
% Euler_e function is called to solve the special Dirichlet problem:
% Ut - Uxx = 0, U(x,0) = sin(pi*x), U(0,t) = U(L,t) = 0.
% The approximating solution is plotted together with the exact solution:
% U = sin(pi*x)*exp(-pi²*T). The error is evaluated.

alpha = 1; L = 1; T = 0.4; nx = 40;
phi = @(x)        sin(pi*x);
g1 = @(t)        0;
g2 = @(t)        0;
f = @(x,t)        0;
u = euler_e(alpha, L, T, nx, phi, g1, g2, f);
```

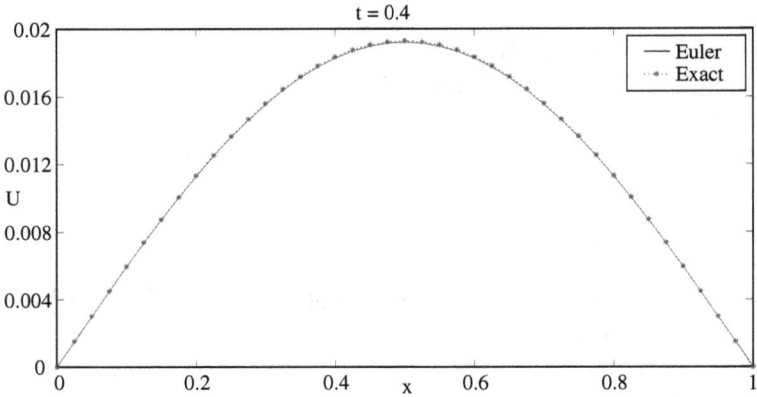

Figure 2.3.2.   Graph of the solution to Problem (2.3.9)–(2.3.11).

```
x = linspace(0,L,nx+1); U = sin(pi*x)*exp(-pi²*T);
plot(x,u,'k',x,U,'r*:');
xlabel('x'); ylabel('U');
title(['t = ',num2str(T)]);
legend('Euler','Exact');
fprintf( 'Maximum error = %g\n',max(abs(U-u)))
end
```

Another application is suggested in Exercise 2.4.19. The euler_e function returns a vector with the solution at the final time. If we are interested to know the approximating solution during the whole computational process for $0 < t \leq T$, then the approximating solution must be saved in a matrix, as illustrated in the following example where the matrix form (2.3.8) is applied.

Example 2.3.3

```
function [u, nt] = euler_em(alpha, L, T, nx, phi, g1, g2)
% This is the function file euler_em.m.
% Explicit Euler Method in matrix form is applied to solve the Dirichlet
% problem: Ut - alpha Uxx = 0, U(x,0)=phi(x), U(0,t)=G1(t), U(L,t)=G2(t).
% The input arguments are: thermal diffusivity, length of the solid, final time,
% number of points on the space grid, initial-boundary conditions. The function
% returns a matrix with the approximating solutions at $t_j$, $j = 1, ..., nt + 1$
% and the number of points on the time grid.
```

```
% Check data
if any([alpha L T nx-2]<= 0)
    error('Check alpha, L, T, nx')
end
```

```
% Stability
dx = L/nx; nt = 10; st=.5;
while alpha*T/nt/dx² > st
    % The stability condition is checked. If it is not satisfied, nt is increased
    % until alpha*T/nt/dx² (= r) is less than or equal to 0.5.
    nt = nt+1;
end
```

```
% Initialization
dt = T/nt; r = alpha*dt/dx²;
x = linspace(0,L,nx+1); t = linspace(0,T,nt+1);
u = zeros(nx+1,nt+1);
u(:,1) = feval(phi, x);
B = [r*ones(nx-1,1) (1-2*r)*ones(nx-1,1) r*ones(nx-1,1)];
A = spdiags(B, -1:1, nx-1, nx-1);
b = zeros(nx-1,1);
```

```
% Explicit Euler Method in matrix form
for j=2:nt+1
    b(1) = r*u(1,j-1); b(nx-1) = r*u(nx+1,j-1);
    u(2:nx,j) = A*u(2:nx,j-1)+b;
    u(1,j) = feval(g1,t(j));
    u(nx+1,j) = feval(g2,t(j));
end
end
```

A way to call and apply euler_em is shown in the following example.

Example 2.3.4

```
function u = euler_em_ex1
% This is the function file euler_em_ex1.m
% Euler_em function is called to solve the special Dirichlet problem:
% Ut - Uxx = 0, U(x,0) = x², U(0,t) = 2*t, U(L,t) = L² + 2*t.
% The approximating solution is plotted together with the exact solution:
% U = x² + 2*t. The error is evaluated.
alpha = 1; L = 1.5; T = 1; nx = 10;
```

```
phi = @(x)          x.^2;
g1 = @(t)           2*t;
g2 = @(t)           L^2+2*t;
[u, nt] = euler_em(alpha, L, T, nx, phi, g1, g2);
x = linspace(0,L,nx+1);
U = x'.^2 + 2*T;
for j = 1:nt+1
    plot(x,u(:,j),'k',x,U,'r*:');
    xlabel('x'); ylabel('u');
    time=(j-1)*T/nt; title(['t = ',num2str( time )]);
    legend('Euler','Exact','Location','NorthWest');
    pause(.01);
end
u = u(:,nt+1); fprintf( 'Maximum error = %g\n',max(abs(U - u)))
end
```

Some code lines of the euler_e and euler_em functions were never discussed: the lines related to the while loop that imposes the constraint $r \leq 1/2$. Using the right value of the parameter r is crucial, as explained in the next section.

2.3.2 *Stability, Convergence, Consistence*

The solutions of difference equations, such as (2.3.5), are the result of a computational process with thousands of operations and inevitable round-off errors. A *stable algorithm* is able to control this kind of error. To better understand the question, let us refer to the Explicit Euler Method (2.3.5) and suppose that a round-off error, say e, arises when the solution related to the point (h, k) of the space-time mesh is computed. Evidently, the error propagates to the next rows during the computational process. Investigate its influence on the solution. Denote with u_i^j the solution without error, and with $\bar{u}_i^j$, the solution perturbed by the error e. This second solution is the same as u_i^j when $j < k$ and it is affected by the error when $j = k$:

$$\bar{u}_i^j = u_i^j \quad \text{for} \quad j < k; \quad \bar{u}_i^k = u_i^k \quad \text{for} \quad i \neq h; \quad \bar{u}_h^k = u_h^k + e.$$

Let us evaluate the error, defined by

$$e_i^j = u_i^j - \bar{u}_i^j$$

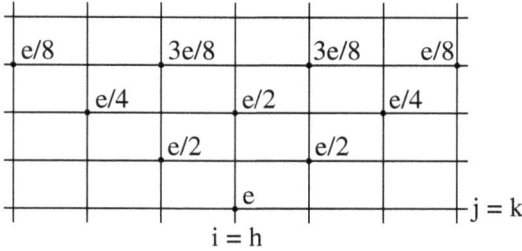

Figure 2.3.3. Error propagation for $r = 1/2$.

for $j \geq k$. Since Eq. (2.3.5) is linear, the error e_i^j satisfies the homogeneous equation

$$e_i^{j+1} = (1 - 2r)e_i^j + r(e_{i+1}^j + e_{i-1}^j). \qquad (2.3.12)$$

Use this equation to investigate the error propagation for some values of the parameter $r = \alpha \Delta t / \Delta x^2$. For $r = 1/2$, Formula (2.3.12) simplifies to

$$e_i^{j+1} = (e_{i+1}^j + e_{i-1}^j)/2,$$

and the error, computed for some values of j greater than k, is shown in Fig. 2.3.3. Note the decreasing error behavior. This fact suggests that Method (2.3.5) is stable for $r = 1/2$. Next, consider $r = 2$. Formula (2.3.12) is reduced to

$$e_i^{j+1} = 2(e_{i+1}^j + e_{i-1}^j) - 3e_i^j,$$

and the error propagation is shown in Fig. 2.3.4. Note the increasing error and oscillating behavior. Solutions obtained with this last value of r are unreliable, since they are affected by out-of-control errors. Method (2.3.5) is unstable when the second value of parameter r is considered. To prevent the error from growing without limits, restrictive actions must be adopted on the mesh size. The *stability analysis* investigates this question.

Consider a finite difference equation that can be expressed in matrix form as

$$\mathbf{u}_j = A\mathbf{u}_{j-1} + \mathbf{a}_{j-1}, \qquad (2.3.13)$$

where the known term $\mathbf{a}_{j-1}$ depends on the boundary conditions and source term. Indicate with $(u_1)_i^j$ and $(u_2)_i^j$, two solutions of (2.3.13) that have same boundary conditions but different initial conditions. If $(u_1)_i^j$ represents the solution without round-off errors and $(u_2)_i^j$ represents the solution

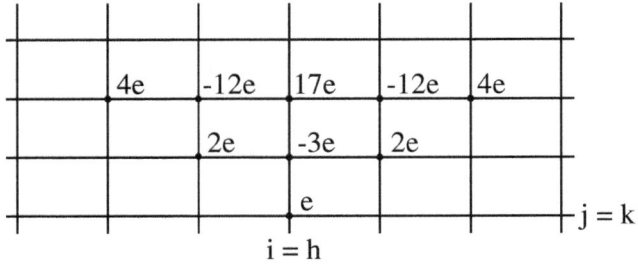

Figure 2.3.4.   Error propagation for $r = 2$.

perturbed by round-off errors at the initial time, the stability analysis leads to a discussion of the behavior of

$$u_i^j = (u_1)_i^j - (u_2)_i^j.$$

This difference satisfies the homogeneous equation

$$\mathbf{u}_j = A\mathbf{u}_{j-1}, \qquad (2.3.14)$$

with homogeneous boundary conditions. Method (2.3.14) is said to be *unconditionally stable* if there exists a positive constant K, independent of Δx, and Δt, such that for Δx, and Δt sufficiently small and $j\Delta t \leq T$, it is

$$||\mathbf{u}_j|| \leq K||\mathbf{u}_0||. \qquad (2.3.15)$$

Formula (2.3.15) means that the error at time t_j is controlled by the initial error. If (2.3.15) holds, but Δt is functionally related to Δx, the method is named *conditionally stable*, or *stable*. This situation occurs in the case discussed in Example 2.3.5, where the maximum norm of a vector is used. The maximum norm of the vector $\mathbf{v} = (v_i)$ is the maximum of the absolute values of its elements:

$$||\mathbf{v}||_\infty = \max_i |v_i|. \qquad (2.3.16)$$

See Exercise 2.4.20 for Matlab commands on norms.

Example 2.3.5   Consider the homogeneous Explicit Euler Method

$$u_i^{j+1} = r(u_{i+1}^j + u_{i-1}^j) + (1 - 2r)u_i^j, \quad i = 1,\ldots,n-1, \qquad (2.3.17)$$

with Dirichlet homogeneous boundary conditions

$$u_0^j = u_n^j = 0. \tag{2.3.18}$$

Prove that the method is stable under the hypothesis

$$0 < r = \alpha \Delta t / (\Delta x)^2 \leq 1/2. \tag{2.3.19}$$

From (2.3.17)–(2.3.19), it follows that

$$|u_i^{j+1}| \leq r(|u_{i+1}^j| + |u_{i-1}^j|) + |1 - 2r| \, |u_i^j|$$

$$\leq r(\max_i |u_i^j| + \max_i |u_i^j|) + (1 - 2r) \max_i |u_i^j|.$$

Using Definition (2.3.16) yields

$$|u_i^{j+1}| \leq ||\mathbf{u}_j||_\infty, \quad i = 1, \ldots, n-1,$$

$$\max_i |u_i^{j+1}| \leq ||\mathbf{u}_j||_\infty,$$

$$||\mathbf{u}_{j+1}||_\infty \leq ||\mathbf{u}_j||_\infty \cdots \leq ||\mathbf{u}_0||_\infty.$$

Therefore, the Explicit Euler Method is conditionally stable, as condition (2.3.15) is satisfied (with $K = 1$) under Assumption (2.3.19).

The *convergence* considers the error that arises when the analytical solution of a partial differential equation is compared with the corresponding solution of the finite difference equation. Denote with U the solution of an initial-boundary value problem for a partial differential equation and suppose that this is well posed, i.e., the solution exists and depends continuously on data. In addition, denote with u_i^j the corresponding approximating solution. The difference of the two solutions, evaluated in the mesh point (i, j), is the *discretization error* in that point

$$E_i^j = U_i^j - u_i^j.$$

The convergence depends on the behavior of E_i^j. A finite difference method is named *convergent* at time t if $||\mathbf{E}_j|| = ||\mathbf{U}_j - \mathbf{u}_j||$ goes to zero when Δx, $\Delta t \to 0$ and $j\Delta t \to t$. It can be proved that the Explicit Euler Method converges under Hypothesis (2.3.19).

The *consistency* considers the error that arises when the partial differential equation is approximated by the finite difference equation. If the analytical solution U of a partial differential equation is substituted in the corresponding finite difference equation, it returns a residual, since the analytical solution generally does not satisfy the finite difference equation. The residual is named a *local truncation error*. For example, for the Explicit Euler Method, it is

$$\frac{U_i^{j+1} - U_i^j}{\Delta t} - \alpha \frac{U_{i+1}^j - 2U_i^j + U_{i-1}^j}{(\Delta x)^2} - f_i^j = \tau_i^j, \qquad (2.3.20)$$

where τ_i^j is the mentioned error. A finite difference method is named *consistent* or *compatible* with the partial differential equation that it approximates if τ_i^j goes to zero when the mesh is refined, i.e., Δx, $\Delta t \to 0$. Explicit Euler Method is consistent with the heat equation. Indeed, consider the formulas of the forward and central approximations

$$(U_t)_i^j = \frac{U_i^{j+1} - U_i^j}{\Delta t} + O(\Delta t), \quad (U_{xx})_i^j = \frac{U_{i+1}^j - 2U_i^j + U_{i-1}^j}{(\Delta x)^2} + O((\Delta x)^2),$$

and substitute them in (2.3.20)

$$\tau_i^j = (U_t - \alpha U_{xx})_i^j - f_i^j + O(\Delta t) + O((\Delta x)^2). \qquad (2.3.21)$$

Since $(U_t - \alpha U_{xx})_i^j = f_i^j$, from (2.3.21), it follows the desired result.

2.3.3 *Boundary Value Problems*

Consider the one-dimensional heat equation

$$U_t - \alpha U_{xx} = F, \quad 0 < x < L, \quad 0 < t \le T, \qquad (2.3.22)$$

with the initial condition

$$U(x, 0) = \varphi(x), \quad 0 \le x \le L, \qquad (2.3.23)$$

and Neumann boundary conditions

$$-U_x(0, t) = g_1(t), \quad U_x(L, t) = g_2(t), \quad 0 < t \le T. \qquad (2.3.24)$$

Different from the Dirichlet problem, the boundary values are now unknowns to be determined, like the other values. Consequently, the unknowns are two more. The Explicit Euler Method

$$u_i^{j+1} = r(u_{i+1}^j + u_{i-1}^j) + (1 - 2r)u_i^j + \Delta t f_i^j, \quad i = 1, \ldots, n - 1, \quad (2.3.25)$$

provides $n - 1$ equations, which are insufficient to determine all $n + 1$ unknowns. The simplest way to get two more equations is to use the forward and backward approximations for the two boundary Conditions (2.3.24)

$$-\frac{u_1^j - u_0^j}{\Delta x} = g_1^j, \quad \frac{u_n^j - u_{n-1}^j}{\Delta x} = g_2^j, \tag{2.3.26}$$

where $g_1^j = g_1(j\Delta t)$ and $g_2^j = g_2(j\Delta t)$. Now, the Neumann problem can be solved. For any j, the unknowns $u_1^{j+1}, \ldots, u_{n-1}^{j+1}$ are obtained from (2.3.25) and the unknowns u_0^{j+1}, u_n^{j+1} from (2.3.26)

$$u_0^{j+1} = u_1^{j+1} + g_1^{j+1}\Delta x, \quad u_n^{j+1} = u_{n-1}^{j+1} + g_2^{j+1}\Delta x. \tag{2.3.27}$$

Formulas (2.3.27) are accurate of order Δx. More accurate formulas, of order $(\Delta x)^2$, can be obtained if the central approximations are used for boundary Conditions (2.3.24):

$$-\frac{u_1^j - u_{-1}^j}{2\Delta x} = g_1^j, \quad \frac{u_{n+1}^j - u_{n-1}^j}{2\Delta x} = g_2^j. \tag{2.3.28}$$

These formulas present the ghost values u_{-1}^j, u_{n+1}^j and cannot be directly applied. The ghost values can be eliminated. Assume that Eq. (2.3.25) also holds on the boundaries, for $i = 0$ and $i = n$,

$$u_0^{j+1} = (1 - 2r)u_0^j + r(u_1^j + u_{-1}^j) + \Delta t f_0^j, \tag{2.3.29}$$

$$u_n^{j+1} = (1 - 2r)u_n^j + r(u_{n+1}^j + u_{n-1}^j) + \Delta t f_n^j. \tag{2.3.30}$$

Solving $(2.3.28)_1$ with respect to u_{-1}^j and substituting the result in (2.3.29) yields

$$u_0^{j+1} = (1 - 2r)u_0^j + 2r(u_1^j + g_1^j\Delta x) + \Delta t f_0^j. \tag{2.3.31}$$

Similarly, from $(2.3.28)_2$ and (2.3.30), it follows

$$u_n^{j+1} = (1 - 2r)u_n^j + 2r(u_{n-1}^j + g_2^j\Delta x) + \Delta t f_n^j. \tag{2.3.32}$$

Equations (2.3.31)–(2.3.32) are the two new equations we need, together with Eqs. (2.3.25), to solve the Neumann problem with greater accuracy. Lastly, all equations can be arranged in compact matrix form as follows

$$\begin{bmatrix} u_0^{j+1} \\ u_1^{j+1} \\ \cdot \\ u_n^{j+1} \end{bmatrix} = \begin{bmatrix} 1-2r & 2r & & \\ r & 1-2r & r & \\ & \cdot & \cdot & \cdot \\ & & 2r & 1-2r \end{bmatrix} \begin{bmatrix} u_0^j \\ u_1^j \\ \cdot \\ u_n^j \end{bmatrix} + \begin{bmatrix} 2rg_1^j\Delta x + \Delta t f_0^j \\ \Delta t f_1^j \\ \cdots \\ 2rg_2^j\Delta x + \Delta t f_n^j \end{bmatrix}.$$

Equivalently, with the apparent definition of the symbols,

$$\mathbf{u}_{j+1} = A\mathbf{u}_j + \mathbf{a}_j. \tag{2.3.33}$$

A function that solves the Neumann Problem (2.3.22)–(2.3.24) in the simplest case (2.3.27) can be obtained from the euler_e function in Example 2.3.1 with few modifications.

Example 2.3.6   A function is presented that applies Explicit Euler Method (2.3.5) to solve the Neumann Problem (2.3.22)–(2.3.24).

```
% function u = euler_en(alpha, L, T, nx, phi, g1, g2)
% This is the function file euler_en.m.
% Explicit Euler Method is applied to solve the Neumann problem:
% Ut - alpha Uxx = 0, U(x,0) = phi(x), -Ux(0,t) = G1(t), Ux(L,t) = G2(t).
% The input arguments are: thermal diffusivity, length of the solid, final time,
% number of points on the space grid, initial-boundary conditions.
% The function returns a vector with the solution at the final time.

% Check data
if alpha <= 0 || L <= 0 || T <= 0 || nx <= 2
    error('Check alpha, L, T, nx')
end

% Stability
dx = L/nx; nt = 10;
st=.5;
while alpha*T/nt/dx² > st
    nt = nt+1;
end

% Initialization
dt = T/nt; r = alpha*dt/dx²;
x = linspace(0,L,nx+1); t = linspace(0,T,nt+1);
u = feval(phi,x);

% Explicit Euler Method
for j=2:nt+1
    u(2:nx) = (1-2*r)*u(2:nx) + r*(u(1:nx-1)+u(3:nx+1));
    u(1) = u(2) + dx*g1(t(j)); u(nx+1) = u(nx) + dx*g2(t(j));
    % Neumann boundary conditions
end
end
```

Some modifications to the euler_en function are suggested in Exercises 2.4.21 and 2.4.22. A way to call euler_en is illustrated in the following example.

Example 2.3.7

```
function u = euler_en_ex1
% This is the function file euler_en_ex1.m.
% Euler_en function is called to solve the special Neumann problem:
% Ut - Uxx = 0, U(x,0) = sin(pi*x), G1(t) = G2(t) = -pi/L*exp(-(pi/L)²*t).
% The approximating solution is plotted together with the exact solution:
% U = sin(pi*x)*exp(-pi²*T). The error is evaluated.

alpha = 1; L = 1.5; T = .3; nx = 30;
phi = @(x)        sin(pi*x/L);
g1 = @(t)        -pi/L*exp(-(pi/L)²*t);
g2 = @(t)        -pi/L*exp(-(pi/L)²*t);
u = euler_en(alpha, L, T, nx, phi, g1, g2);
x = linspace(0,L,nx+1);
U = sin(pi*x/L)*exp(-(pi/L)²*T);
fprintf( 'Maximum error = %g\n',max(abs(U-u)))
plot(x,u,'k',x,U,'r*:');
xlabel('x'); ylabel('u'); title(['t = ',num2str( T )]);
legend('Euler-N,'Exact','Location','NorthWest');
end
```

Example 2.3.8 A function is presented that applies the matrix form (2.3.33) of the Explicit Euler Method and solves the Neumann Problem (2.3.22)–(2.3.24).

```
% function u = euler_enm(alpha, L, T, nx, phi, g1, g2)
% This is the function file euler_enm.m.
% Explicit Euler Method in matrix form is applied to solve the
% Neumann problem:
% Ut - alpha Uxx = 0, U(x,0) = phi(x), -Ux(0,t) = G1(t), Ux(L,t) = G2(t).
% The input arguments are: thermal diffusivity, length of the solid, final time,
% number of points on the space grid, initial-boundary conditions.
% The function returns a vector with the solution at the final time.
```

```
% Check data
if alpha <= 0 || L <= 0 || T <= 0 || nx <= 2
    error('Check alpha, L, T, nx')
end

% Stability
dx = L/nx; nt = 10; st=.5;
while alpha*T/nt/dx² > st
    nt = nt+1;
end

% Initialization
dt = T/nt; r = alpha*dt/dx²;
x = linspace(0,L,nx+1); t = linspace(0,T,nt+1);
u = feval(phi,x');
B = [[r*ones(nx-1,1);2*r;0] (1-2*r)*ones(nx+1,1) [0;2*r;r*ones(nx-1,1)]];
A = spdiags(B, -1:1, nx+1, nx+1);
b = zeros(nx+1,1);

% Explicit Euler Method in matrix form
for j=2:nt+1
    b(1) = 2*r*dx*g1(t(j-1)); b(nx+1) = 2*r*dx*g2(t(j-1));
    u = A*u + b;
end
u = u';
end
```

A way to call **euler_enm** is suggested in Exercise 2.4.23.

Consider the Dirchlet–Neumann problem

$$U_t - \alpha U_{xx} = F, \quad 0 < x < L, \quad 0 < t \leq T, \tag{2.3.34}$$

$$U(x,0) = \varphi(x), \quad 0 \leq x \leq L, \tag{2.3.35}$$

$$U(0,t) = g_1(t), \ U_x(L,t) = g_2(t), \ 0 < t \leq T. \tag{2.3.36}$$

A function for Problem (2.3.34)–(2.3.36) is provided in the following example and a way to call it is illustrated in Example 2.3.10.

Example 2.3.9

```
function u = euler_edn(alpha, L, T, nx, phi, g1, g2)
% This is the function file euler_edn.m.
```

```
% Explicit Euler Method is applied to solve the Dirichlet–Neumann
% problem:
% Ut - alpha Uxx = 0, U(x,0) = phi(x), U(0,t) = G1(t), Ux(L,t) = G2(t).
% The input arguments are: thermal diffusivity, length of the solid, final time,
% number of points on the space grid, initial-boundary conditions.
% The function returns a vector with the solution at the final time.

% Check data
if alpha <= 0 || L <= 0 || T <= 0 || nx <= 2
    error('Check alpha, L, T, nx')
end

% Stability
dx = L/nx; nt = 10;
st=.5;
while alpha*T/nt/dx² > st
    nt = nt+1;
end

% Initialization
dt = T/nt; r = alpha*dt/dx²;
x = linspace(0,L,nx+1); t = linspace(0,T,nt+1);
u = feval(phi,x);

% Explicit Euler Method
for j=2:nt+1
    u(2:nx) = (1-2*r)*u(2:nx) + r*(u(1:nx-1)+u(3:nx+1));
    u(1) = g1(t(j)); % Dirichlet boundary condition
    u(nx+1) = u(nx) + dx*g2(t(j)); % Neumann boundary condition
end
end
```

Example 2.3.10 Consider Problem (2.3.34)–(2.3.36) where

$$U(x,0) = \varphi(x) = \bar{\varphi} = 0.5, \quad 0 \le x \le L,$$

$$U(0,t) = g_1(t) = 0, \ U_x(L,t) = \frac{\bar{\varphi}}{\sqrt{\pi \alpha t}} \exp\left(-\frac{L^2}{4\alpha t}\right), \ 0 < t \le T.$$

The following listing calls euler_edn, solves the the previous problem and
plots the solution (Fig. 2.3.5).

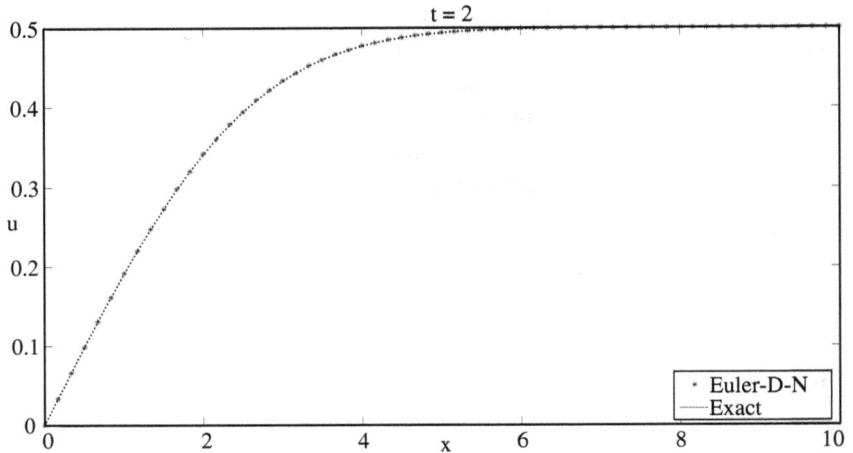

Figure 2.3.5.   Graph of the solution.

```
function u = euler_edn_ex1
% This is the function file euler_edn_ex1.m.
% Euler_edn function is called to solve the special Dirichlet-Neumann problem:
% Ut - Uxx = 0, U(x,0) = phibar, G1(t) = 0,
% G2(t) = phibar*exp(-L^2/(4*alpha*t))/sqrt(pi*alpha*t).
% The approximating solution is plotted together with the exact solution:
% U = phibar*erf(x/sqrt(4*alpha*T)). The error is evaluated.

alpha = 1; L = 10; T = 1; nx = 60;
phibar = .5;
phi = @(x)          phibar*ones(1,nx+1);
g1 = @(t)           0;
g2 = @(t)           phibar*exp(-L^2/(4*alpha*t))/sqrt(pi*alpha*t);
u = euler_edn(alpha, L, T, nx, phi, g1, g2);
x = linspace(0,L,nx+1);
U = phibar*erf(x/sqrt(4*alpha*T));
fprintf( 'Maximum error = %g\n',max(abs(U-u)))
plot(x,u,'k*',x,U,'r:');
xlabel('x'); ylabel('u'); title(['t = ',num2str( T )]);
legend('Euler-D-N','Exact','Location','SouthEast');
end
```

Other applications are suggested in Exercises 2.4.24–2.4.26.

In the previous example, the error function erf(y) was used. It is defined by the formula

$$\text{erf}(y) = \frac{2}{\sqrt{\pi}} \int_0^y \exp(-\eta^2)\, d\eta \quad \Rightarrow \quad \text{erf}(\infty) = \frac{2}{\sqrt{\pi}} \int_0^\infty \exp(-\eta^2)\, d\eta = 1.$$

In addition, the complimentary error function erfc(y) is expressed as

$$\text{erfc}(y) = \frac{2}{\sqrt{\pi}} \int_y^\infty \exp(-\eta^2)\, d\eta \quad \Rightarrow \quad \text{erfc}(y) = 1 - \text{erf}(y).$$

Consider the Robin problem

$$U_t - \alpha U_{xx} = F, \quad 0 < x < L, \quad 0 < t \le T, \tag{2.3.37}$$

$$U(x,0) = \varphi(x), \quad 0 \le x \le L, \tag{2.3.38}$$

$$-U_x(0,t) + h_1 U(0,t) = g_1(t), \ U_x(L,t) + h_2 U(L,t) = g_2(t), \ 0 < t \le T. \tag{2.3.39}$$

The Robin boundary conditions can be approximated exactly as the Neumann conditions. Indeed, we can apply forward and backward approximations to (2.3.39) and get the two following equations

$$-\frac{u_1^j - u_0^j}{\Delta x} + h_1 u_0^j = g_1^j, \quad \frac{u_n^j - u_{n-1}^j}{\Delta x} + h_2 u_n^j = g_2^j. \tag{2.3.40}$$

Solving Formulas (2.3.40) with respect to u_0^{j+1} and u_n^{j+1}, respectively, yields

$$u_0^{j+1} = \frac{u_1^{j+1} + g_1^{j+1}\Delta x}{1 + h_1\Delta x}, \quad u_n^{j+1} = \frac{u_{n-1}^{j+1} + g_2^{j+1}\Delta x}{1 + h_2\Delta x}. \tag{2.3.41}$$

Formulas more accurate than (2.3.41) are obtained by considering the central approximations for boundary Conditions (2.3.39):

$$-\frac{u_1^j - u_{-1}^j}{2\Delta x} + h_1 u_0^j = g_1^j, \quad \frac{u_{n+1}^j - u_{n-1}^j}{2\Delta x} + h_2 u_n^j = g_2^j. \tag{2.3.42}$$

The ghost values u_{-1}^j and u_{n+1}^j are eliminated as in the Neumann problem and we get

$$u_0^{j+1} = [1 - 2r(1 + h_1\Delta x)]u_0^j + 2r(u_1^j + \Delta x g_1^j) + \Delta t f_0^j, \tag{2.3.43}$$

$$u_n^{j+1} = [1 - 2r(1 + h_2\Delta x)]u_n^j + 2r(u_{n-1}^j + \Delta x g_2^j) + \Delta t f_n^j. \tag{2.3.44}$$

The previous equations and the Explicit Euler Method, rewritten below,

$$u_i^{j+1} = r(u_{i+1}^j + u_{i-1}^j) + (1 - 2r)u_i^j + \Delta t f_i^j, \ i = 1, ..., n-1, \tag{2.3.45}$$

provide a system of $n+1$ equations for the $n+1$ unknowns $u_0^{j+1}, \ldots, u_n^{j+1}$. System (2.3.43)–(2.3.45) solves the Robin problem. It can be arranged in a compact matrix form that incorporates the boundary conditions,

$$
\begin{bmatrix} u_0^{j+1} \\ u_1^{j+1} \\ \cdot \\ u_{n-1}^{j+1} \\ u_n^{j+1} \end{bmatrix} = \begin{bmatrix} H_1 & 2r & & & \\ r & 1-2r & r & & \\ & \cdot & \cdot & \cdot & \\ & & r & 1-2r & r \\ & & & 2r & H_2 \end{bmatrix} \begin{bmatrix} u_0^{j} \\ u_1^{j} \\ \cdot \\ u_{n-1}^{j} \\ u_n^{j} \end{bmatrix} + \begin{bmatrix} 2rg_1^j\Delta x + \Delta t f_0^j \\ \Delta t f_1^j \\ \cdots \\ \Delta t f_{n-1}^j \\ 2rg_2^j\Delta x + \Delta t f_n^j \end{bmatrix},
$$

where $H_1 = 1 - 2r(1 + h_1\Delta x)$ and $H_2 = 1 - 2r(1 + h_2\Delta x)$. Equivalently, with the apparent definition of the symbols,

$$\mathbf{u}_{j+1} = A\mathbf{u}_j + \mathbf{a}_j. \tag{2.3.46}$$

2.3.4 *Diffusion in a Multi-layer Medium*

Diffusion, or heat conduction, in a multi-layer medium is involved in many engineering problems. Moreover, the situation of several more or less permeable layers in aquifers is frequent and the analysis of the pore pressure in multi-layer aquifers leads to consider similar mathematical models. A two-layer system is discussed in the following (Fig. 2.3.6). The generalization to more layers is straightforward. In the mentioned case, the physical process is governed by the following equations

$$U_t - \alpha_1 U_{xx} = F, \quad 0 < x < x_h, \quad 0 < t \le T, \tag{2.3.47}$$

$$U_t - \alpha_2 U_{xx} = F, \quad x_h < x < L, \quad 0 < t \le T, \tag{2.3.48}$$

$$U(x_h^-, t) = U(x_h^+, t), \quad 0 < t \le T, \tag{2.3.49}$$

$$k_1 U_x(x_h^-, t) = k_2 U_x(x_h^+, t), \quad 0 < t \le T, \tag{2.3.50}$$

where $x = x_h$ indicates the interphase position. Equations (2.3.49) and (2.3.50) express the continuity of temperature and heat flux at

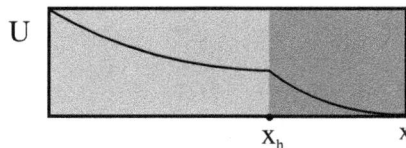

Figure 2.3.6.   Multi-layer medium.

the interphase. Of course, the initial-boundary conditions, depending on special problem, must be associated to Eqs. (2.3.47)–(2.3.50). Since U_x is discontinuous for $x = x_h$ (Fig. 2.3.6), the function U cannot be a solution to the heat equation (in the classical sense).

Consider the Explicit Euler Method. If a mesh is used where x_h is a node $x_h = h\Delta x$, then we get the following discrete version of Problem (2.3.47)–(2.3.50)

$$u_i^{j+1} = r_1(u_{i+1}^j + u_{i-1}^j) + (1 - 2r_1)u_i^j + \Delta t f_i^j, \quad 0 < i < h, \qquad (2.3.51)$$

$$u_i^{j+1} = r_2(u_{i+1}^j + u_{i-1}^j) + (1 - 2r_2)u_i^j + \Delta t f_i^j, \quad h < i < n, \qquad (2.3.52)$$

$$k_1(-4u_{h-1}^{j+1} + 3u_h^{j+1} + u_{h-2}^{j+1}) = k_2(4u_{h+1}^{j+1} - 3u_h^{j+1} - u_{h+2}^{j+1}), \qquad (2.3.53)$$

where $u_h^j = U(x_h^-, t_j) = U(x_h^+, t_j)$ and $r_i = \alpha_i \Delta t/(\Delta x)^2$, $i = 1, 2$. Equation (2.3.53) was obtained from (2.3.50) by applying three-point approximations. Equations (2.3.51) and (2.3.52) explicitly provide the values $u_2^{j+1}, \ldots, u_{h-1}^{j+1}$ and $u_{h+1}^{j+1}, \ldots, u_{n-1}^{j+1}$, respectively. Considering these results in Eq. (2.3.53), we find

$$u_h^{j+1} = (4k_1 u_{h-1}^{j+1} - k_1 u_{h-2}^{j+1} - k_2 u_{h+2}^{j+1} + 4k_2 u_{h+1}^{j+1})/3/(k_1 + k_2). \qquad (2.3.54)$$

Lastly, the values u_0^{j+1} and u_n^{j+1} follow from the boundary conditions.

Example 2.3.11 A function is presented that applies Formulas (2.3.51)–(2.3.54) to solve Problem (2.3.47) to (2.3.50).

```
function u = layers(alpha1, alpha2, k1, k2, T, L, h, nx, phi1, phi2, g1, g2)
% This is the function file layers.m.
% Explicit Euler Method is applied to solve the problem:
% Ut = alpha1 Uxx = 0, x < xh, U(x,0) = phi1(x), U(0,t) = G1(t),
% Ut = alpha2 Uxx = 0, x > xh, U(L,0) = phi2(x), U(L,t) = G2(t),
% U(xh-,t) = U(xh+,t), k1*Ux(xh-,t) = k2*Ux(xh+,t),
% The input arguments are: thermal diffusivity coefficients, thermal
% conductivity coefficients,, final time, length of the solid, interphase
% position, number of points on the space grid, initial-boundary conditions.
% The function returns a vector with the solution at the final time.

% Check data
if any ([alpha1 alpha2 k1 k2 L T nx-2]<= 0)
    error('Check alpha1, alpha2, k1, k2, L, T, nx')
end
```

```
% Stability
dx = L/nx; nt = 10; st = .5;
while max([alpha1 alpha2])*T/nt/dx² > st
    nt = nt+1;
end

% Initialization
dt = T/nt; r1 = alpha1*dt/dx²; r2 = alpha2*dt/dx²;
x = linspace(0,L,nx+1); t = linspace(0,T,nt+1);
u = [phi1(x(1:h))  phi2(x(h+1:nx+1))];

% Explicit Euler Method
for j=2:nt+1
    u(2:h-1) = (1-2*r1)*u(2:h-1) + r1*(u(1:h-2) + u(3:h));
    u(h+1:nx) = (1-2*r2)*u(h+1:nx) + r2*(u(h:end-2) + u(h+2:nx+1));
    u(h) = (4*k1*u(h-1) - k1*u(h-2) + 4*k2*u(h+1)...
    - k2*u(h+2))/(k1+k2)/3;
    u(1) = g1(t(j)); u(nx+1) = g2(t(j));
end
end
```

Example 2.3.12   A way to call layers is illustrated below by considering the simple initial-boundary conditions

$$U(x,0) = \begin{cases} a_1x^2 + c_1x + V, & x < x_h, \\ a_2(x - L)^2 + c_2(x - L), & x > x_h, \end{cases} \qquad (2.3.55)$$

$$U(0,t) = 2t + V, \quad U(0,t) = 2t, \quad t > 0, \qquad (2.3.56)$$

where a_1, a_2, c_1, c_2 and V, are constants specified in the listing. The solution is plotted in Fig. 2.3.7. The error is evaluated.

```
function u = layers_ex1
% This is the function file layers_ex1.m.
% Layers function is called to solve the special problem:
% Ut = alpha1 Uxx = 0,  x < xh,  U(x,0) = a1*x² + c1*x + V,
% Ut = alpha2 Uxx = 0,  x > xh,  U(x,0) = a2*(x - L)² + c2*(x - L),
% U(xh-,t) = U(xh+,t),   k1*Ux(xh-,t) = k2*Ux(xh+,t),
% U(0,t) = 2*t + V,   U(L,t) = 2*t.
% The approximating solution is plotted together with the exact solution:
% U = 2*t + a1*x² + c1*x + V, x < xh,
```

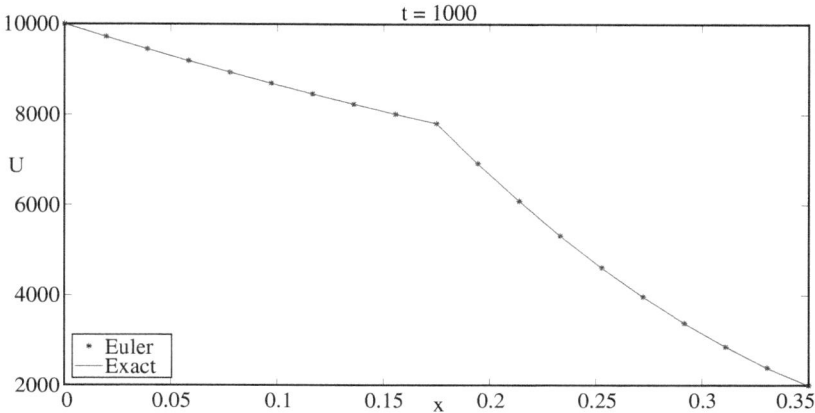

Figure 2.3.7.   Graph of the solution to Problem (2.3.51)–(2.3.56).

```
% U = 2*T + a2*(x - L)² + c2*(x - L), x > xh.
% The error is evaluated.

alpha1 = 85.9*10^( − 6); alpha2=12.4*10^( − 6); k1 = 202.4; k2 = 45;
T = 10³; L = .35; nx = 18; h = 10;
x = linspace(0,L,nx+1); xh = x(h);
V = 8000; a1 = 1/alpha1; a2 = 1/alpha2;
c1 = (-a1*k2*xh²-a2*k2*(xh-L)²+2*k1*a1*xh*(xh-L)-k2*V )/...
    (k2*xh-k1*(xh-L));
c2 = (a2*k1*(xh-L)²+a1*k1*xh²-2*a2*k2*xh*(xh-L)-k1*V )/...
    (k2*xh-k1*(xh-L));
phi1 = @(x)        a1*x.² + c1*x + V;
phi2 = @(x)        a2*(x - L).² + c2*(x - L);
g1 = @(t)        2*t + V;
g2 = @(t)        2*t;
u = strati(alpha1, alpha2, k1, k2, T, L, h, nx, phi1, phi2, g1, g2);
U(1:h) = 2*T + a1*x(1:h).² + c1*x(1:h) + V;
U(h+1:nx+1) = 2*T + a2*(x(h+1:nx+1) - L).² + c2*(x(h+1:nx+1) - L);
    % Exact solution
fprintf( 'Maximum error = %g\n',max( abs(U-u) ) )
plot(x,u,'k*',x,U,'r');
title(['t = ',num2str(T)]);
xlabel('x'); ylabel('U');
legend('Euler','Exact','Location','SouthWest');
end
```

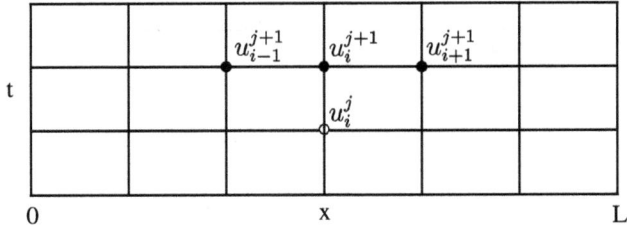

Figure 2.3.8. Implicit Euler Method.

2.3.5 *Implicit Euler Method*

This section presents another numerical method for the heat equation. Let us begin to discuss the Dirichlet problem

$$U_t - \alpha U_{xx} = F, \quad 0 < x < L, \quad 0 < t \le T, \tag{2.3.57}$$

$$U(x,0) = \varphi(x), \quad 0 \le x \le L, \tag{2.3.58}$$

$$U(0,t) = g_1(t), \quad U(L,t) = g_2(t), \quad 0 < t \le T. \tag{2.3.59}$$

Consider the point (x_i, t_{j+1}) of the space-time grid (see Fig. 2.3.8), and discretize Eq. (2.3.57) by using the backward approximation for the time derivative and the central approximation for the space derivative,

$$\frac{u_i^{j+1} - u_i^j}{\Delta t} - \alpha \frac{u_{i+1}^{j+1} - 2u_i^{j+1} + u_{i-1}^{j+1}}{(\Delta x)^2} = f_i^{j+1}. \tag{2.3.60}$$

Hence,

$$-ru_{i-1}^{j+1} + (1+2r)u_i^{j+1} - ru_{i+1}^{j+1} = u_i^j + f_i^{j+1}\Delta t, \quad i = 1, \dots, n-1, \tag{2.3.61}$$

where the usual parameter $r = \alpha \Delta t/(\Delta x)^2$ was introduced. Formula (2.3.61) is named the *Implicit Euler Method*. The method is characterized by the backward approximation of the time derivative. The adjective implicit outlines that equation (2.3.61) alone is unable to determine the three unknowns on the left-hand side when the right-hand side is known. However, when all values u_i^j, $i = 0, \dots, n$, are known, Formula (2.3.61) provides a system of $n-1$ equations in the $n-1$ unknowns u_i^{j+1}, $i = 1, \dots, n-1$. Solving such a system yields the unknowns $u_1^{j+1}, \dots, u_{n-1}^{j+1}$. The method works, as the unknowns are obtained implicitly from Formula (2.3.61). In addition, the values of the first and last points are obtained from the boundary conditions

$$u_0^j = g_1^j = g_1(t_j), \quad u_n^j = g_2^j = g_2(t_j), \quad j = 1, \dots, m, \quad m\Delta t = T. \tag{2.3.62}$$

In conclusion, the method provides the following system for the unknowns

$$\begin{bmatrix} 1+2r & -r & & \\ -r & 1+2r & -r & \\ & \cdot & \cdot & \cdot \\ & & -r & 1+2r \end{bmatrix} \begin{bmatrix} u_1^{j+1} \\ u_2^{j+1} \\ \vdots \\ u_{n-1}^{j+1} \end{bmatrix} = \begin{bmatrix} u_1^{j} \\ u_2^{j} \\ \vdots \\ u_{n-1}^{j} \end{bmatrix} + \begin{bmatrix} rg_1^{j+1} + f_1^{j+1}\Delta t \\ f_2^{j+1}\Delta t \\ \vdots \\ rg_2^{j+1} + f_{n-1}^{j+1}\Delta t \end{bmatrix}.$$

Hence, in compact matrix notation

$$B\mathbf{u}_{j+1} = \mathbf{u}_j + \mathbf{b}_{j+1},$$

$$\mathbf{u}_{j+1} = A\mathbf{u}_j + \mathbf{a}_{j+1}, \quad (A = B^{-1}, \quad \mathbf{a}_{j+1} = B^{-1}\mathbf{b}_{j+1}). \tag{2.3.63}$$

The Implicit Euler Method is unconditionally stable, i.e., being stable without any restriction on the parameter r. This result will be proved in Sec. 2.3.7.

Example 2.3.13 This example presents a function that applies the Implicit Euler Method to the Dirichlet problem for the heat equation.

```
function u = euler_i(alpha, L, T, nx, nt, phi, g1, g2)
% This is the function file euler_i.m.
% Implicit Euler Method is applied to solve the Dirichlet problem:
% Ut = alpha*Uxx, U(x,0) = phi(x), U(0,t) = G1(t), U(L,t) = G2(t).
% The input arguments are: thermal diffusivity, length of the solid, final time,
% number of points on the space-time grid, initial-boundary conditions.
% The function returns a vector with the solution at the final time.

% Check data
if any([alpha L T nx-2 nt-2] <= 0)
     error('Check alpha, L, T, nx, nt')
end

% Initialization
dx = L/nx; dt = T/nt; r = alpha*dt/dx^2;
x = linspace(0,L,nx+1); t = linspace(0,T,nt+1);
u = phi(x');
B = [-r*ones(nx-1,1) (1+2*r)*ones(nx-1,1) -r*ones(nx-1,1)];
A = spdiags(B, -1:1, nx-1, nx-1);
b = zeros(nx-1,1);

% Implicit Euler Method
for j=2:nt+1
```

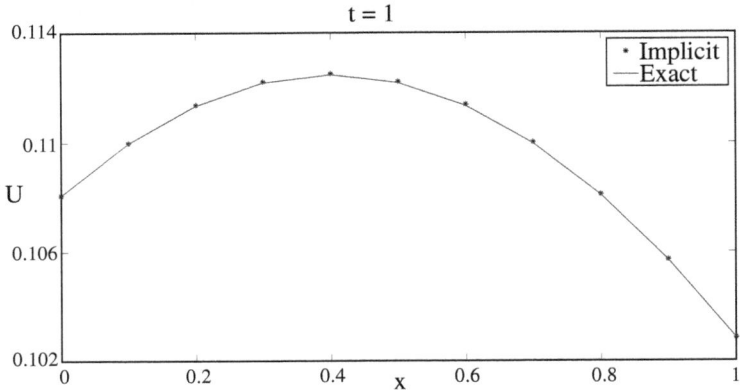

Figure 2.3.9.   Graph of solution.

```
    u(1) = g1(t(j));  u(nx+1) = g2(t(j));
    b(1) = r*u(1); b(nx-1) = r*u(nx+1);
    u(2:nx) = A\(u(2:nx)+b);
end
u = u';
end
```

Example 2.3.14   A way to apply euler_i is suggested in the following
listing, where euler_i is called with the initial-boundary conditions

$$U(x,0) = \begin{cases} \bar{\varphi} = \text{constant if } x \in]x_1, x_2[\subset [0, L], \\ 0 \qquad\qquad \text{if } x \in [0, L]-]x_1, x_2[, \end{cases}$$

$$U(0,t) = \frac{\bar{\varphi}}{2}\left[\text{erf}\left(\frac{-x_1}{\sqrt{4\alpha t}}\right) + \text{erf}\left(\frac{x_2}{\sqrt{4\alpha t}}\right)\right], \quad t > 0,$$

$$U(L,t) = \frac{\bar{\varphi}}{2}\left[\text{erf}\left(\frac{L-x_1}{\sqrt{4\alpha t}}\right) + \text{erf}\left(\frac{x_2-L}{\sqrt{4\alpha t}}\right)\right], \quad t > 0.$$

The graph of the solution is shown in Fig. 2.3.9.

```
function u = euler_i_ex1
% This is the function file euler_i_ex1.m.
% Euler_i function is called to solve the special Dirichlet problem:
% Ut = alpha*Uxx,
% U(x,0) = 0, x <= x1; U(x,0) = phibar, x1 < x < x2; U(x,0) = 0, x >= x2;
```

```
% U(0,t) = .5*phibar* (erf(-x1/sqrt(4*alpha*t)) + erf(x2/sqrt(4*alpha*t)) ),
% U(L,t) = .5*phibar*(erf((L-x1)/sqrt(4*alpha*t))
% + erf((x2-L)/sqrt(4*alpha*t))).
% The approximating solution is plotted together with the exact solution:
% U=.5*phibar*(erf((x-x1)/sqrt(4*alpha*T))+erf((x2-x)/sqrt(4*alpha*T))).
% The error is evaluated.

alpha = 1; L = 1; T = 1; nx = 10; nt = 200;
x = linspace(0,L,nx+1); x1=x(3); x2=x(7); phibar = 1;
phi = @(x)  phibar*((x2-x)>0).*((x-x1)>0);
g1=@(t)  phibar/2*(erf(-x1/sqrt(4*alpha*t)) + erf(x2/sqrt(4*alpha*t)));
g2=@(t)  phibar/2*(erf((L-x1)/sqrt(4*alpha*t))+ erf((x2-L)/sqrt(4*alpha*t)));
u = euler_i(alpha, L, T, nx, nt, phi, g1, g2);
U = .5*phibar*(erf((x-x1)/sqrt(4*alpha*T)) + erf((x2-x)/sqrt(4*alpha*T)));
fprintf( 'Maximum error = %g\n',max(abs(U-u)))
plot(x,u,'k*',x,U,'r');
xlabel('x'); ylabel('U');
title(['t = ',num2str( T )]); legend('Implicit','Exact');
end
```

Consider the Robin problem for Eq. (2.3.57) with initial Condition (2.3.58) and Robin boundary conditions

$$-U_x(0,t) + h_1 U(0,t) = g_1(t), \quad U_x(L,t) + h_2 U(L,t) = g_2(t). \quad (2.3.64)$$

These conditions simplify to the Neumann conditions for $h_1 = h_2 = 0$. Applying the central approximations to the derivatives in (2.3.64) yields

$$-\frac{u_1^{j+1} - u_{-1}^{j+1}}{2\Delta x} + h_1 u_0^{j+1} = g_1^{j+1}, \quad \frac{u_{n+1}^{j+1} - u_{n-1}^{j+1}}{2\Delta x} + h_2 u_n^{j+1} = g_2^{j+1}. \quad (2.3.65)$$

Formulas (2.3.65) cannot be used because of the ghost terms u_{-1}^{j+1} and u_{n+1}^{j+1}. These terms can be eliminated. Indeed, consider Eq. (2.3.61) for $i = 0$ and $i = n$

$$-ru_{-1}^{j+1} + (1 + 2r)u_0^{j+1} - ru_1^{j+1} = u_0^j + f_0^{j+1}\Delta t, \quad (2.3.66)$$

$$-ru_{n-1}^{j+1} + (1 + 2r)u_n^{j+1} - ru_{n+1}^{j+1} = u_n^j + f_n^{j+1}\Delta t. \quad (2.3.67)$$

Solving Eq. $(2.3.65)_1$ with respect to u_{-1}^{j+1} and the result substituted into (2.3.66) yields

$$[1 + 2r(1 + h_1\Delta x)]u_0^{j+1} - 2ru_1^{j+1} = u_0^j + 2r\Delta x g_1^{j+1} + f_0^{j+1}\Delta t. \quad (2.3.68)$$

A similar reasoning for the other ghost value leads to

$$-2ru_{n-1}^{j+1} + [1 + 2r(1 + h_2\Delta x)]u_n^{j+1} = u_n^j + 2r\Delta x g_2^{j+1} + f_n^{j+1}\Delta t. \quad (2.3.69)$$

Equations (2.3.68), (2.3.69) and (2.3.61) are a system of $n+1$ equations in $n+1$ unknowns, written in matrix form as

$$\begin{bmatrix} H_1 & -2r & & \\ -r & 1+2r & -r & \\ & & \cdot & \cdot \\ & & -2r & H_2 \end{bmatrix} \begin{bmatrix} u_0^{j+1} \\ u_1^{j+1} \\ \cdot \\ u_n^{j+1} \end{bmatrix} = \begin{bmatrix} u_0^j \\ u_1^j \\ \cdot \\ u_n^j \end{bmatrix} + \begin{bmatrix} 2r\Delta x g_1^{j+1} + f_0^{j+1}\Delta t \\ f_1^{j+1}\Delta t \\ \cdot \\ 2r\Delta x g_2^{j+1} + f_n^{j+1}\Delta t \end{bmatrix},$$

where

$$H_1 = 1 + 2r(1 + h_1\Delta x), \quad H_2 = 1 + 2r(1 + h_2\Delta x).$$

The previous system solves the Robin problem. In matrix notation, it is written as

$$B\mathbf{u}_{j+1} = \mathbf{u}_j + \mathbf{b}_{j+1},$$

$$\mathbf{u}_{j+1} = A\mathbf{u}_j + \mathbf{a}_{j+1}, \quad (A = B^{-1}, \quad \mathbf{a}_{j+1} = B^{-1}\mathbf{b}_{j+1}). \quad (2.3.70)$$

An application for this problem is suggested in Exercise 2.4.27.

Alternatively, the Robin problem can be solved by using the forward and backward approximations for the derivatives in Formulas (2.3.64). We get two equations

$$-\frac{u_1^j - u_0^j}{\Delta x} + h_1 u_0^j = g_1^j, \quad \frac{u_n^j - u_{n-1}^j}{\Delta x} + h_2 u_n^j = g_2^j, \quad (2.3.71)$$

that solve the Robin problem with less accuracy.

2.3.6   *Crank–Nicolson Method*

Consider the Explicit Euler Method and Implicit Euler Method for the heat equation

$$U_t - \alpha U_{xx} = F, \quad 0 < x < L, \quad 0 < t \le T, \quad (2.3.72)$$

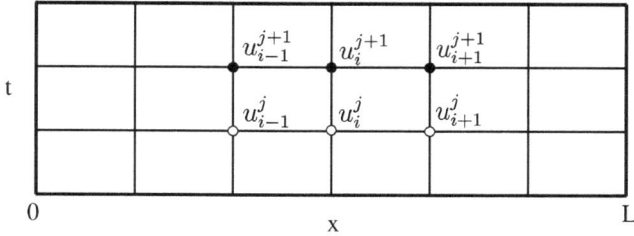

Figure 2.3.10.   Crank–Nicolson Method.

rewritten for convenience

$$\frac{u_i^{j+1} - u_i^j}{\Delta t} - \alpha \frac{u_{i+1}^j - 2u_i^j + u_{i-1}^j}{(\Delta x)^2} = f_i^j,$$

$$\frac{u_i^{j+1} - u_i^j}{\Delta t} - \alpha \frac{u_{i+1}^{j+1} - 2u_i^{j+1} + u_{i-1}^{j+1}}{(\Delta x)^2} = f_i^{j+1}.$$

Summing the two, the previous equations yields

$$\frac{u_i^{j+1} - u_i^j}{\Delta t} = \frac{\alpha(u_{i+1}^{j+1} - 2u_i^{j+1} + u_{i-1}^{j+1} + u_{i+1}^j - 2u_i^j + u_{i-1}^j)}{2(\Delta x)^2} + \frac{1}{2}(f_i^j + f_i^{j+1}).$$

$$(2.3.73)$$

This is a new method, named the *Crank*[13]*–Nicolson*[14] *Method.* (2.3.73). This method is even more accurate than the other two, since the time derivative approximation can be considered a central approximation. To understand this, consider Eq. (2.3.72) on the point $(i, j + 1/2)$ of the space-time mesh

$$(U_t)_i^{j+1/2} - \alpha(U_{xx})_i^{j+1/2} = F_i^{j+1/2},$$

and replace the space derivative and source term with the average on the points (i, j) and $(i, j + 1)$, see Fig. 2.3.10,

$$(U_t)_i^{j+1/2} - \frac{\alpha}{2}[(U_{xx})_i^j + (U_{xx})_i^{j+1}] = \frac{1}{2}(F_i^j + F_i^{j+1}).$$

---

[13]John Crank, a British scientist, 1916–2006. He was also a mathematical physicist. His scientific research centered on numerical solutions of partial differential equations. He worked at the Courtaulds Fundamental Research Laboratory.
[14]Phyllis Lockett Nicolson, a British scientist, 1917–1968. She worked on numerical solutions of partial differential equations, with special attention paid to stability analysis.

Using the central approximation for all derivatives, we get Formula (2.3.73).

Lastly,

$$-ru_{i-1}^{j+1} + 2(1+r)u_i^{j+1} - ru_{i+1}^{j+1} = ru_{i-1}^j + 2(1-r)u_i^j + ru_{i+1}^j + G_i^j, \quad (2.3.74)$$

where $r = \alpha \Delta t/(\Delta x)^2$ and $G_i^j = \Delta t(f_i^j + f_i^{j+1})$. The Crank–Nicolson Method is implicit. Furthermore, it is unconditionally stable, as shown in Sec. 2.3.7.

Consider the Dirichlet problem for Eq. (2.3.72) and assign the following boundary conditions

$$U(0,t) = g_1(t), \quad U(L,t) = g_2(t), \quad t > 0. \tag{2.3.75}$$

In this case, Formula (2.3.74) can be arranged in matrix form as follows

$$\begin{bmatrix} 2(1+r) & -r & & & \\ -r & 2(1+r) & -r & & \\ & & \cdot & \cdot & \cdot \\ & & & -r & 2(1+r) \end{bmatrix} \begin{bmatrix} u_1^{j+1} \\ u_2^{j+1} \\ \vdots \\ u_{n-1}^{j+1} \end{bmatrix}$$

$$= \begin{bmatrix} 2(1-r) & r & & & \\ r & 2(1-r) & r & & \\ & & \cdot & \cdot & \cdot \\ & & & r & 2(1-r) \end{bmatrix} \begin{bmatrix} u_1^j \\ u_2^j \\ \vdots \\ u_{n-1}^j \end{bmatrix} + \begin{bmatrix} r(g_1^j + g_1^{j+1}) + G_1^j \\ G_2^j \\ \vdots \\ r(g_2^j + g_2^{j+1}) + G_{n-1}^j \end{bmatrix},$$

$$B\mathbf{u}_{j+1} = C\mathbf{u}_j + \mathbf{b}_j,$$

$$\mathbf{u}_{j+1} = A\mathbf{u}_j + \mathbf{a}_j, \quad (A = B^{-1}C, \quad \mathbf{a}_j = B^{-1}\mathbf{b}_j). \tag{2.3.76}$$

Formula (2.3.76) solves the Dirichlet problem. Indeed, $\mathbf{a}_j$ is a known vector depending on boundary conditions, a source term and matrix B. When the initial conditions are given, the computational process can start and Eq. (2.3.76) provides the solution $\mathbf{u}_j \ \forall \ j$.

Example 2.3.15   The example presents a function that applies the Crank–Nicolson Method to solve the Dirichlet problem for the heat equation. A way to call the function is suggested in Exercise 2.4.28.

```
function u = crank(alpha, L, T, nx, nt, phi, g1, g2)
% This is the function file crank.m.
% Crank–Nicolson Method is applied to solve the Dirichlet problem:
```

```
% Ut = alpha*Uxx, U(x,0) = phi(x), U(0,t) = G1(t), U(L,t) = G2(t).
% The input arguments are: thermal diffusivity, length of the solid, final time,
% number of points on the space-time grid, initial-boundary conditions.
% The function returns a vector with the solution at the final time.

% Check data
if any([alpha L T nx-2] <= 0)
    error('Check alpha, L, T, nx, nt')
end
 % Initialization

dx = L/nx; dt = T/nt; r = alpha*dt/dx²;
x = linspace(0,L,nx+1); t = linspace(0,T,nt+1);
u = phi(x');
BB = [-r*ones(nx-1,1)  2*(1+r)*ones(nx-1,1)  -r*ones(nx-1,1)];
B = spdiags(BB, -1:1, nx-1, nx-1);
CC = [r*ones(nx-1,1)  2*(1-r)*ones(nx-1,1)  r*ones(nx-1,1)];
C = spdiags(CC, -1:1, nx-1,nx-1);
b = zeros(nx-1,1);

% Crank–Nicolson Method
for j = 2:nt+1
    b(1) = r*(g1(t(j-1)) + g1(t(j)));  b(nx-1) = r*(g2(t(j-1)) + g2(t(j)));
    u(2:nx) = B\(C*u(2:nx)+b);
    u(1) = g1(t(j)); u(nx+1) = g2(t(j));
end
u = u';
end
```

Let us apply the Crank–Nicolson Method to the Robin problem where the following boundary conditions are assigned

$$-U_x(0,t) + h_1 U(0,t) = g_1(t), \quad U_x(L,t) + h_2 U(L,t) = g_2(t). \quad (2.3.77)$$

Consider the first condition and apply the central approximation for $t = t_j$ and $t = t_{j+1}$

$$-\frac{u_1^j - u_{-1}^j}{2\Delta x} + h_1 u_0^j = g_1^j, \qquad -\frac{u_1^{j+1} - u_{-1}^{j+1}}{2\Delta x} + h_1 u_0^{j+1} = g_1^{j+1}.$$

Summing these formulas yields

$$u_{-1}^j + u_{-1}^{j+1} - u_1^j - u_1^{j+1} + 2\Delta x h_1 (u_0^j + u_0^{j+1}) = 2\Delta x (g_1^j + g_1^{j+1}). \quad (2.3.78)$$

The ghost sum $u_{-1}^j + u_{-1}^{j+1}$ can be eliminated by considering (2.3.74) for $i = 0$

$$r(u_{-1}^j + u_{-1}^{j+1}) = 2(1+r)u_0^{j+1} - 2(1-r)u_0^j - r(u_1^j + u_1^{j+1}) - G_0^j. \quad (2.3.79)$$

Solving (2.3.78) with respect to $u_{-1}^j + u_{-1}^{j+1}$ and the result substituted into (2.3.79) yields

$$(1 + r + r\Delta x h_1)u_0^{j+1} - ru_1^{j+1}$$

$$= ru_1^j + (1 - r - r\Delta x h_1)u_0^j + r\Delta x(g_1^j + g_1^{j+1}) + G_0^j/2. \quad (2.3.80)$$

A similar reasoning for the second boundary condition leads to

$$(1 + r + r\Delta x h_2)u_n^{j+1} - ru_{n-1}^{j+1}$$

$$= ru_{n-1}^j + (1 - r - r\Delta x h_2)u_n^j + r\Delta x(g_2^j + g_2^{j+1}) + G_n^j/2. \quad (2.3.81)$$

From Formulas (2.3.80)–(2.3.81), it follows that

$$(1+H_1)u_0^{j+1} - ru_1^{j+1} = ru_1^j + (1-H_1)u_0^j + r\Delta x(g_1^j + g_1^{j+1}) + G_0^j/2, \quad (2.3.82)$$

$$(1 + H_2)u_n^{j+1} - ru_{n-1}^{j+1} = ru_{n-1}^j + (1 - H_2)u_n^j + r\Delta x(g_2^j + g_2^{j+1}) + G_n^j/2, \quad (2.3.83)$$

where

$$H_i = r + r\Delta x h_i, \quad i = 1, 2.$$

Equations (2.3.82), (2.3.83) and (2.3.74) can be arranged in matrix form as follows

$$\begin{bmatrix} 1+H_1 & -r & & \\ -r & 2(1+r) & -r & \\ & \cdot & \cdot & \cdot \\ & & -r & 1+H_2 \end{bmatrix} \begin{bmatrix} u_0^{j+1} \\ u_1^{j+1} \\ \cdot \\ u_n^{j+1} \end{bmatrix}$$

$$= \begin{bmatrix} 1-H_1 & r & & \\ r & 2(1-r) & r & \\ & \cdot & \cdot & \cdot \\ & & r & 1-H_2 \end{bmatrix} \begin{bmatrix} u_0^j \\ u_1^j \\ \cdot \\ u_n^j \end{bmatrix} + \begin{bmatrix} r\Delta x(g_1^j + g_1^{j+1}) + G_0^j/2 \\ G_1^j \\ \cdot \\ r\Delta x(g_2^j + g_2^{j+1}) + G_n^j/2 \end{bmatrix},$$

$$B\mathbf{u}_{j+1} = C\mathbf{u}_j + \mathbf{b}_{j,j+1},$$

$$\mathbf{u}_{j+1} = A\mathbf{u}_j + \mathbf{a}_{j,j+1}, \quad (A = B^{-1}C, \quad \mathbf{a}_{j,j+1} = B^{-1}\mathbf{b}_{j,j+1}). \quad (2.3.84)$$

Formula (2.3.84) solves the Robin problem. Finally, note that boundary Conditions (2.3.77) can be also approximated with the less accurate formulas

$$-\frac{u_1^j - u_0^j}{\Delta x} + h_1 u_0^j = g_1^j, \qquad \frac{u_n^j - u_{n-1}^j}{\Delta x} + h_2 u_n^j = g_2^j. \qquad (2.3.85)$$

2.3.7 Von Neumann Stability Criterium

Von Neumann[15] approach to stability analysis is based on the assumption that the solution of a numerical method can be expressed by a finite Fourier series

$$u_i^j = \sum_{p=-(n-1)}^{n-1} b_p \xi_p^j e^{I p \pi i \Delta x}, \quad (I = \sqrt{-1}). \qquad (2.3.86)$$

For example, this result for the Explicit Euler Method is shown in Exercise 2.4.29. If each term of the Fourier series is controlled by the initial condition, the numerical solution is controlled as well and the method is stable. Therefore, following the Von Neumann approach, it is sufficient to consider only one term of the series

$$u_i^j = \xi^j e^{I \beta i \Delta x}, \qquad (2.3.87)$$

and investigate its stability. The error behavior on time depends only on the factor ξ. If it is

$$|\xi| \le 1, \qquad (2.3.88)$$

the error does not amplify and the method is stable. For this reason, ξ is named the *amplification factor*. Formula (2.3.88) expresses the *Von Neumann criterium* for the numerical stability. Its application is relatively easy — it is probably the most used tool for the stability analysis of numerical methods for partial differential equations. Condition (2.3.88) is very strong. A less restrictive condition is the following

$$|\xi| \le 1 + C \Delta t, \qquad (2.3.89)$$

where the positive constant C is independent of Δt and Δx.

---

[15] János Von Neumann, a Hungarian-American scientist, 1903–1957. He was Professor at Princeton University. He worked on Numerical Analysis and Quantum Mechanics.

Finally, let us recall the Euler formulas that are frequently applied in the following examples

$$e^{Iz} = \cos z + I \sin z, \ e^{-Iz} = \cos z - I \sin z, \quad (2.3.90)$$

$$\cos z = (e^{Iz} + e^{-Iz})/2, \ \sin z = (e^{Iz} - e^{-Iz})/(2I). \quad (2.3.91)$$

Example 2.3.16   As the first application, consider the Explicit Euler Method

$$u_i^{j+1} = r(u_{i+1}^j + u_{i-1}^j) + (1 - 2r)u_i^j.$$

Substituting (2.3.87) into the the previous equation yields

$$\xi^{j+1} e^{I\beta i \Delta x} = \xi^j e^{I\beta i \Delta x} [1 - 2r + r(e^{-I\beta \Delta x} + e^{I\beta \Delta x})],$$

$$\xi = 1 - 2r + 2r \cos(\beta \Delta x) = 1 - 2r(1 - \cos \beta \Delta x) = 1 - 4r \sin^2(\beta \Delta x/2).$$

Therefore, the Von Neumann condition $|\xi| \leq 1$ is written

$$-1 \leq 1 - 4r \sin^2(\beta \Delta x/2) \leq 1.$$

The second inequality is always satisfied. The first is equivalent to

$$2r \sin^2(\beta \Delta x/2) \leq 1,$$

that is satisfied if

$$r \leq 1/2.$$

Of course, this condition is the same as that found in Sec. 2.3.2.

Example 2.3.17   Consider the Implicit Euler Method

$$-ru_{i-1}^{j+1} + (1 + 2r)u_i^{j+1} - ru_{i+1}^{j+1} = u_i^j.$$

Substituting $u_i^j = \xi^j e^{I\beta i \Delta x}$ in the previous equation yields

$$-\xi r e^{-I\beta \Delta x} + \xi(1 + 2r) - \xi r e^{I\beta \Delta x} = 1,$$

$$\xi[1 + 2r - 2r \cos(\beta \Delta x)] = 1,$$

$$\xi = 1/(1 + 4r \sin^2(\beta \Delta x/2)).$$

The Von Neumann condition $|\xi| \leq 1$ is satisfied for any value of r. The Implicit Euler Method is unconditionally stable, i.e., stable for any value of r. However, since the method is accurate as $O((\Delta x)^2) + O(\Delta t)$, the error grows with Δx and much more with Δt.

Example 2.3.18 Consider the Crank–Nicolson Method

$$-ru_{i-1}^{j+1} + 2(1+r)u_i^{j+1} - ru_{i+1}^{j+1} = ru_{i-1}^j + 2(1-r)u_i^j + ru_{i+1}^j.$$

Substituting $u_i^j = \xi^j e^{I\beta i \Delta x}$ into the previous equation yields

$$\xi[1 + r - r\cos(\beta\Delta x)] = 1 - r + r\cos(\beta\Delta x),$$

$$\xi = [1 - 2r\sin^2(\beta\Delta x/2)]/[1 + 2r\sin^2(\beta\Delta x/2)].$$

The Von Neumann condition $|\xi| \leq 1$ is satisfied for any value of r. The Crank–Nicolson Method is unconditionally stable.

Example 2.3.19 Consider the equation

$$U_t + \lambda U - \alpha U_{xx} = 0, \quad \lambda = \text{constant}. \tag{2.3.92}$$

Following the Explicit Euler Method, we get

$$\frac{u_i^{j+1} - u_i^j}{\Delta t} + \lambda u_i^j - \frac{\alpha}{(\Delta x)^2}(u_{i+1}^j - 2u_i^j + u_{i-1}^j) = 0,$$

$$u_i^{j+1} = (1 - 2r)u_i^j - \lambda\Delta t u_i^j + r(u_{i+1}^j + u_{i-1}^j), \tag{2.3.93}$$

where $r = \alpha\Delta t/(\Delta x)^2$. Use the Von Neumann criterium for the stability analysis. Substituting $u_i^j = \xi^j e^{I\beta i \Delta x}$ into the previous equation yields

$$\xi = 1 - 2r + 2r\cos(\beta\Delta x) - \lambda\Delta t = 1 - 2r(1 - \cos\beta\Delta x) - \lambda\Delta t,$$

$$\xi = 1 - 4r\sin^2(\beta\Delta x/2) - \lambda\Delta t.$$

If $r \leq 1/2$, then it is $|1 - 4r\sin^2(\beta\Delta x/2)| \leq 1$, as shown in Example 2.3.16. Therefore, from the last formula, we get

$$|\xi| \leq 1 + |\lambda|\Delta t.$$

Condition (2.3.89) is satisfied and Method (2.3.92) is conditionally stable.

Note that the change of unknown function

$$W = U \exp(\lambda t) \quad \Leftrightarrow \quad U = W \exp(-\lambda t), \qquad (2.3.94)$$

simplifies Eq. (2.3.92) to the heat equation

$$W_t - \alpha W_{xx} = 0. \qquad (2.3.95)$$

Example 2.3.20 Formulas (2.3.94)–(2.3.95) suggest that a function for Eq. (2.3.92) can be obtained easily from euler_e in Example 2.3.1.

```
function u = euler_el(alpha, L, T, nx, phi, g1, g2, lambda)
% This is the function file euler_el.m.
% Explicit Euler Method is applied to solve the Dirichlet problem:
% Ut+lambda*U=alpha*Uxx, U(x,0) = phi(x), U(0,t) = G1(t), U(L,t)=G2(t).
% The input arguments are: thermal diffusivity, length of the solid, final time,
% number of points on the space grid, initial-boundary conditions, parameter
% lambda. The function returns a vector with the solution at the final time.

Same code as euler_e

% Explicit Euler Method
for j=2:nt+1
    u(2:nx)=(1-2*r)*u(2:nx)+r*(u(1:nx-1)+u(3:nx+1));
    u(1) = g1(t(j))*exp(lambda*t(j)); u(nx+1) = g2(t(j))*exp(lambda*t(j));
end
u = u*exp(-lambda*T);
end
```

A way to apply euler_el is illustrated in the following listing where the simple initial-boundary conditions are considered

$$U(x,0) = x^2, U(0,t) = 2t \exp(-\lambda t), U(L,t) = (L^2+2t) \exp(-\lambda t). \quad (2.3.96)$$

```
function u = euler_el_ex1
% This is the function file euler_el_ex1.m
% Euler_el function is called to solve the special Dirichlet problem:
% Ut + lambda*U = Uxx, U(x,0) = x^2,
% U(0,t) = 2*t*exp(-lambda*t), U(L,t) = (L^2+2*t)*exp(-lambda*t).
% The approximating solution is plotted together with the exact solution:
% U = (x^2+2*T)*exp(-lambda*T). The error is evaluated.
alpha = 1; L = 1; T = 1; nx = 10; lambda = -2;
phi = @(x)        x.^2;
```

```
g1 = @(t)         2*t*exp(-lambda*t);
g2 = @(t)         (L²+2*t)*exp(-lambda*t);
u = euler_el(alpha, L, T, nx, phi, g1, g2, lambda);
x = linspace(0,L,nx+1); U = (x.²+2*T)*exp(-lambda*T);
fprintf( 'Maximum error = %g\n',max(abs(U-u)))
plot(x,u,'k*',x,U,'r');
xlabel('x'); ylabel('U');
legend('Euler','Exact');
title(['time = ',num2str(T)]);
end
```

Another application is suggested in Exercise 2.4.30.

2.4 Exercises

Exercise 2.4.1 Execute forward_ex1.m twice. First, with nx = 32 and then with nx = 64, which is double the amount of the previous number. Consider the ratio of the second error to the first and explain the result.

Exercise 2.4.2 Derive Formula (2.1.7) related to the backward approximation.

Hint. Consider the Taylor series for $f(x_i - \Delta x)$.

Exercise 2.4.3 Write a function that returns the backward approximation of derivatives.

Answer.

```
function y = backward(u,h)
% This is the function file backward.m.
% It returns the backward approximation of derivatives. Since the backward
% approximation cannot be applied in the first point, the vector length
% returned by the backward function is equal to that of vector u minus 1.
% Example
% a = 0; b = 1; nx = 40; x = linspace(a,b,nx+1); dx = (b - a)/nx;
% u = x.^2;
% dbu = backward(u,dx)

n = length(u) - 1;
y = (u(2:n+1) - u(1:n))/h;
end
```

Exercise 2.4.4 Write a listing, say backward_ex1.m, that calls and applies the backward function.

Exercise 2.4.5 Execute central_ex1.m twice. First, with nx = 32 and then with nx = 64, which is double the amount of the previous number. Consider the ratio of the second central error on the first and explain the result.

Exercise 2.4.6 Derive three-point approximation Formulas (2.1.9) and (2.1.10).

Answer. Consider the Taylor series

$$f_{i+1} - f_i = f_i'h + f_i''h^2/2 + O(h^3), \tag{2.4.1}$$

$$f_{i+2} - f_i = f_i'2h + f_i''2h^2 + O(h^3). \tag{2.4.2}$$

Subtract (2.4.2) from (2.4.1)×4 and solve the result with respect to f_i'

$$f_i' = \frac{4f_{i+1} - 3f_i - f_{i+2}}{2h} + O(h^2).$$

Hence, Formula (2.1.9). Similarly, Formula (2.1.10) is derived.

Exercise 2.4.7 Derive the four-point forward and backward approximation formulas

$$f_i' \approx \frac{-f_{i+2} + 6f_{i+1} - 3f_i - 2f_{i-1}}{6h}, \quad f_i' \approx \frac{f_{i-2} - 6f_{i-1} + 3f_i + 2f_{i+1}}{6h},$$

with an error of order h^3.

Exercise 2.4.8 Obtain the numerical values of **gradient** in the first and last points from central_ex2.m. Try to understand the approximating derivatives used to get those values.

Exercise 2.4.9 Derive Formula $(2.1.12)_1$ related to the forward approximation of u_t.

Answer. From Taylor's series

$$u(x_i, t_j + \Delta t) = u(x_i, t_j) + u_t(x_i, t_j)\Delta t + O((\Delta t)^2),$$

$$u_i^{j+1} = u_i^j + (u_t)_i^j \Delta t + O((\Delta t)^2),$$

$$(u_t)_i^j = \frac{u_i^{j+1} - u_i^j}{\Delta t} + O(\Delta t).$$

Exercise 2.4.10   Derive forward and backward three-point approximation formulas for u_x and u_t.

Exercise 2.4.11   Derive Formula (2.1.19) for the central approximation of u_{xx}.

Exercise 2.4.12   Derive Formulas (2.1.21)-(2.1.22) for the backward and central approximations of u_{xt}.

Exercise 2.4.13   Calculate the gradient of the function

$$u(x_1, x_2, x_3) = x_3. \tag{2.4.3}$$

Answer. $\nabla u = (0, 0, 1)$. ∇u is perpendicular to the surface $u = 0$ and directed towards the region where $u > 0$.

Exercise 2.4.14   Calculate the gradient of the function

$$u = x_1^2 + x_2^2 + x_3^2 - R^2. \tag{2.4.4}$$

Answer. $\nabla u = (2x_1, 2x_2, 2x_3)$. ∇u is perpendicular to the surface $u = 0$. In addition, it is directed towards the region where $u > 0$.

Exercise 2.4.15   The function $\sin(x)$ is continuous on $B = [0, \ 2\pi]$ and it is

$$\int_0^{2\pi} \sin(x) \ dx = 0.$$

Explain why the previous result does not imply $\sin(x) = 0$ on $[0, \ 2\pi]$.

Exercise 2.4.16   Explain why $\partial u / \partial n$ is related to the heat flux.

Exercise 2.4.17   Show the identity $\nabla \cdot (u\mathbf{v}) = \mathbf{v} \cdot \nabla u + u \nabla \cdot \mathbf{v}$ used to derive Formula (2.2.19) from (2.2.18).

Answer. $$\nabla \cdot (u\mathbf{v}) = \frac{\partial(uv_1)}{\partial x_1} + \frac{\partial(uv_2)}{\partial x_2} + \frac{\partial(uv_3)}{\partial x_3}$$

$$= \frac{\partial u}{\partial x_1} v_1 + \frac{\partial u}{\partial x_2} v_2 + \frac{\partial u}{\partial x_3} v_3 + u \frac{\partial v_1}{\partial x_1} + u \frac{\partial v_2}{\partial x_2} + u \frac{\partial v_3}{\partial x_3} = \nabla u \cdot \mathbf{v} + u \nabla \cdot \mathbf{v}.$$

Exercise 2.4.18   Use the logical function any to write the following code line of the euler_e function in a compact manner.

if alpha <= 0 || L <= 0 || T <= 0 || nx <= 2

Exercise 2.4.19   Write a function, say euler_e_ex2, that calls euler_e to solve the Dirichlet problem

$$U_t - U_{xx} = 2t - 6x, \tag{2.4.5}$$

$$U(x,0) = x^3, \quad U(0,t) = t^2, \quad U(L,t) = L^3 + t^2. \tag{2.4.6}$$

Compare the results with the exact solution: $U = x^3 + t^2$.

Exercise 2.4.20   The norm $\|\mathbf{v}\|_\infty$ is calculated in Matlab with the norm(v,Inf) command. Use it after creating a vector. Also, try these two commands: norm(v,2) and norm(v,1).

Exercise 2.4.21   Modify the euler_en function by considering the source term.

Exercise 2.4.22   Modify the euler_en function so that it can return a matrix with the approximating solution during the whole computational process.

Exercise 2.4.23   Call the euler_enm function presented in Example 2.3.8.

Hint. Consider the euler_en_ex1 function in Example 2.3.7.

Exercise 2.4.24   Consider Example 2.3.10 and the graph of the solution in Fig. 2.3.5. Explain why the solution near $x = L$ behaves according to the Neumann boundary condition.

Exercise 2.4.25   Write a function, say euler_end, for the Neumann–Dirichlet problem and apply it.

Exercise 2.4.26   Write the matrix form for the Dirichlet–Neumann problem and Neumann–Dirichlet problem.

Exercise 2.4.27   Write a function that solves, the Robin Problem (2.3.70) for the Implicit Euler Method.

Exercise 2.4.28   Write a listing that calls and applies the crank function presented in Example 2.3.15.

Hint. Consider Example 2.3.14.

Exercise 2.4.29 Show that the solution of the following Dirichlet problem for the Explicit Euler Method

$$u_i^{j+1} = r(u_{i+1}^j + u_{i-1}^j) + (1 - 2r)u_i^j, \quad i = 1, \ldots, n - 1, \tag{2.4.7}$$

$$u_i^0 = \varphi(x_i) = \varphi_i, \quad i = 0, \ldots, n, \quad n\Delta x = L = 1, \tag{2.4.8}$$

$$u_0^j = 0, \quad u_n^j = 0, \quad j = 1, \ldots, m, \tag{2.4.9}$$

can be expressed by a finite Fourier series.

Answer. Consider the solution of System (2.4.7)–(2.4.9) in the following form

$$u_i^j = \sum_{p=1}^{n-1} a_p(\xi_p)^j \sin(p\pi i\Delta x), \quad i = 0, \ldots, n, \tag{2.4.10}$$

where a_p, $p = 1, \ldots, n - 1$, are unknown coefficients, and

$$\xi_p = 1 - 4r \sin^2 \frac{p\pi\Delta x}{2}. \tag{2.4.11}$$

Imposing that (2.4.10) satisfies initial conditions, (2.4.8) yields

$$\sum_{p=1}^{n-1} a_p \sin(p\pi i\Delta x) = \varphi_i, \quad i = 1, \ldots, n - 1.$$

Solving the previous system gives the coefficients a_p. In addition, boundary Conditions (2.4.9) are satisfied; as for $i = 0$ and $i = n$ from (2.4.10), it follows $u_0^j = u_n^j = 0$. Now, use the Euler Formulas (2.3.90)–(2.3.91) and get

$$\sin(p\pi i\Delta x) = \left[e^{Ip\pi i\Delta x} - e^{-Ip\pi i\Delta x} \right] / (2I).$$

Substituting the previous result into (2.4.10) yields

$$u_i^j = \sum_{p=1}^{n-1} (a_p/2I)\xi_p^j e^{Ip\pi i\Delta x} - \sum_{p=1}^{n-1} (a_p/2I)\xi_p^j e^{-Ip\pi i\Delta x},$$

$$u_i^j = \sum_{p=1}^{n-1} (a_p/2I)\xi_p^j e^{Ip\pi i\Delta x} - \sum_{p=-(n-1)}^{-1} (a_{-p}/2I)\xi_p^j e^{Ip\pi i\Delta x}.$$

Hence,

$$u_i^j = \sum_{p=-(n-1)}^{n-1} b_p \xi_p^j e^{Ip\pi i\Delta x}, \tag{2.4.12}$$

where

$$b_p = a_p/2I \ \text{ if } \ p > 0, \quad b_0 = 0, \quad b_p = -a_{-p}/2I \ \text{ if } \ p < 0.$$

While Formula (2.4.12) shows that u_i^j is expressed by a finite Fourier series, u_i^j as the solution of (2.4.7) remains to be shown. Preliminary, note that

$$\xi_p = 1 - 4r \sin^2 \frac{p\pi\Delta x}{2} = 1 - 2r[1 - \cos(p\pi\Delta x)]$$

$$\xi_p = 1 - 2r + 2r \cos(p\pi\Delta x) = 1 - 2r + r \left[e^{Ip\pi\Delta x} + e^{-Ip\pi\Delta x} \right]. \quad (2.4.13)$$

Now, from (2.4.12)–(2.4.13), we get

$$u_i^{j+1} = \sum_{p=-(n-1)}^{n-1} b_p \xi_p \xi_p^j e^{Ip\pi i \Delta x}$$

$$= \sum_{p=-(n-1)}^{n-1} b_p \xi_p^j e^{Ip\pi i \Delta x} \left\{ 1 - 2r + r \left[e^{Ip\pi\Delta x} + e^{-Ip\pi\Delta x} \right] \right\}$$

$$= (1 - 2r) u_i^j + r(u_{i+1}^j + u_{i-1}^j),$$

that is the desired result.

Exercise 2.4.30   Write a function for Method (2.3.93).

Hint. Consider the euler_e function, see Example 2.3.1.

Chapter 3

Diffusion and Convection

The convection-diffusion physical process is very important since it is involved in many engineering problems. The convection-diffusion equation, named the advection-diffusion equation in some contexts, was derived in Chap. 2. This equation models two physical processes: the diffusive process, governed by the diffusion equation, and the convection (or advection) process, governed by the convection (or advection) equation Crank (1979). The Finite Difference Method for the convection-diffusion equation is presented in the first section of this chapter Lapidus and Pinter (1982); Mitchell and Griffiths (1995); Necati Ozisik (1994).

The second section introduces the Method of Lines. This method is a semi-discrete numerical method for the integration of partial differential equations, where only some variables are discretrized. The Method of Lines is illustrated with engineering applications.

Finally, some sections are devoted to Matlab functions that can help to save and load data and figures.

3.1 Convection-diffusion Equation

3.1.1 *Upwind Method*

Consider the convection-diffusion equation

$$U_t + vU_x - \alpha U_{xx} = 0, \quad 0 < x < L, \quad 0 < t \leq T, \qquad (3.1.1)$$

introduced in Sec. 2.2.1. The same equation is named the advection-diffusion equation, depending on the context. Equation (3.1.1) models two physical processes: diffusion and advection. The first is governed by the parabolic partial differential equation

$$U_t - \alpha U_{xx} = 0,$$

and the second by the hyperbolic partial differential equation

$$U_t + vU_x = 0.$$

The last equation, the *advection equation*, or *convection equation*, governs the process of the transport of matter in the absence of diffusion. In this context, the parameter v represents velocity.

The prevailing process depends on the parameters v and α. More precisely, it depends on a suitable ratio of the two parameters: the *Péclet*[1] *number*

$$P = vL/\alpha.$$

To understand why, a change of variables

$$\begin{array}{ll} \xi = \xi(x,t), & x = x(\xi,\tau), \\ \tau = \tau(x,t), & t = t(\xi,\tau), \end{array} \quad \Leftrightarrow$$

must be performed that converts Eq. (3.1.1) to a non-dimensional form. As shown in the worked Exercise 3.4.1, a special change of variables converts Eq. (3.1.1) to

$$W_\tau + P\, W_\xi = W_{\xi\xi}.$$

This equation emphasizes that the prevailing process depends on $|P|$. A high value of $|P|$ indicates that advection is the prevailing process. On the contrary, a low value of $|P|$ indicates that diffusion is the prevailing process.

We now introduce the *Upwind Method* for Eq. (3.1.1). If the parameter v is positive, then by using the forward approximation for U_t, the backward for U_x and the central for U_{xx} yields

$$\frac{U_i^{j+1} - U_i^j}{\Delta t} + v\frac{U_i^j - U_{i-1}^j}{\Delta x} = \alpha\frac{U_{i+1}^j - 2U_i^j + U_{i-1}^j}{(\Delta x)^2} + O(\Delta t + \Delta x).$$

Therefore, the finite difference method

$$\frac{u_i^{j+1} - u_i^j}{\Delta t} + v\frac{u_i^j - u_{i-1}^j}{\Delta x} = \alpha\frac{u_{i+1}^j - 2u_i^j + u_{i-1}^j}{(\Delta x)^2} \qquad (3.1.2)$$

is consistent with Eq. (3.1.1). With the positions $r = \Delta t/(\Delta x)^2$ and $s = \Delta t/\Delta x$, it follows from (3.1.2) that

$$u_i^{j+1} = (r\alpha + sv)u_{i-1}^j + (1 - 2r\alpha - sv)u_i^j + r\alpha u_{i+1}^j. \qquad (3.1.3)$$

---

[1] Jean Claude Eugène Péclet, a French scientist, 1793–1857. He was Professor at the Collège de Marseille and at the École Normale Supérièure in Paris.

Method (3.1.3) is conditionally stable. Indeed, it is stable under the hypothesis

$$2r\alpha + sv \leq 1, \tag{3.1.4}$$

as proved later on. If the parameter v in (3.1.1) is negative, then using the forward approximation for U_t and U_x, and the central for U_{xx} yields

$$\frac{u_i^{j+1} - u_i^j}{\Delta t} + v \frac{u_{i+1}^j - u_i^j}{\Delta x} = \alpha \frac{u_{i+1}^j - 2u_i^j + u_{i-1}^j}{(\Delta x)^2},$$

$$u_i^{j+1} = r\alpha u_{i-1}^j + (1 - 2r\alpha + sv)u_i^j + (r\alpha - sv)u_{i+1}^j. \tag{3.1.5}$$

Method (3.1.5) is conditionally stable, as it is stable under the hypothesis

$$2r\alpha - sv \leq 1. \tag{3.1.6}$$

Equations (3.1.3) and (3.1.5) are named the *Upwind Method*. Setting

$$p = \begin{cases} r\alpha + s|v| & \text{if } v \geq 0, \\ r\alpha & \text{if } v < 0, \end{cases} \qquad q = \begin{cases} r\alpha & \text{if } v \geq 0, \\ r\alpha + s|v| & \text{if } v < 0, \end{cases} \tag{3.1.7}$$

Eqs. (3.1.3) and (3.1.5) are grouped into a single equation

$$u_i^{j+1} = pu_{i-1}^j + (1 - p - q)u_i^j + qu_{i+1}^j. \tag{3.1.8}$$

Let us show that the Upwind Method (3.1.8) is stable under the hypothesis

$$p + q \leq 1 \quad \Leftrightarrow \quad 2r\alpha + s|v| \leq 1. \tag{3.1.9}$$

Indeed,

$$|u_i^{j+1}| \leq p\|\mathbf{u}_j\|_\infty + (1 - p - q)\|\mathbf{u}_j\|_\infty + q\|\mathbf{u}_j\|_\infty \leq \|\mathbf{u}_j\|_\infty,$$

and therefore,

$$\|\mathbf{u}_{j+1}\|_\infty \leq \|\mathbf{u}_j\|_\infty \cdots \leq \|\mathbf{u}_0\|_\infty.$$

The stability Condition (3.1.9) is the same as (3.1.4) if $v \geq 0$ and (3.1.6) if $v \leq 0$. See Exercise 3.4.2.

Example 3.1.1 A function is presented that applies Method (3.1.8) to solve the Dirichlet problem for Eq. (3.1.1).

```
function u = upwind(alpha, v, L, T, nx, phi, g1, g2)
% This is the function file upwind.m.
% Upwind Method is applied to solve the Dirichlet problem:
```

```
% Ut +v Ux = alpha Uxx, U(x,0) = phi(x), U(0,t) = G1(t), U(L,t) = G2(t).
% The input arguments are: thermal diffusivity, velocity v, length of the solid,
% final time, number of points on the space grid, initial-boundary conditions.
% The function returns a vector with the solution at the final time.

% Check data
if any([alpha L T nx-2] <= 0)
    error('Check alpha, L, T, nx')
end

% Stability
nt = 150; st = 1; dx = L/nx;
while 2*alpha*T/nt/dx² + T/nt/dx*abs(v) > st
    nt = nt + 1;
end

% Initialization
dt = T/nt; r = dt/dx²; s = dt/dx;
x = linspace(0,L,nx+1); t = linspace(0,T,nt+1);
u = feval(phi,x);
if v >= 0
    p = alpha*r+s*v; q = alpha*r;
else
    p = alpha*r; q = alpha*r-s*v;
end

% Upwind Method
for j = 2:nt + 1
    u(2:end-1) = p*u(1:end-2) + (1 - p - q)*u(2:end-1) + q*u(3:end);
    u(1) = g1(t(j)); u(end) = g2(t(j));
end
end
```

Example 3.1.2   The listing below illustrates a way to call the upwind function. It considers the special Dirichlet problem

$$U_t + vU_x - \alpha U_{xx} = 0, \quad 0 < x < L, \quad 0 < t \leq T, \qquad (3.1.10)$$

$$U(x,0) = \begin{cases} 0 \text{ if } x \in [0, x_1[\, \cup \,]x_2, L], \\ k \text{ if } x \in [x_1, x_2], \end{cases} \qquad (3.1.11)$$

$$U(0,t) = 0, \quad U(L,t) = 0, \quad 0 < t \leq T. \qquad (3.1.12)$$

The graph of the numerical solution is shown in Fig. 3.1.1.

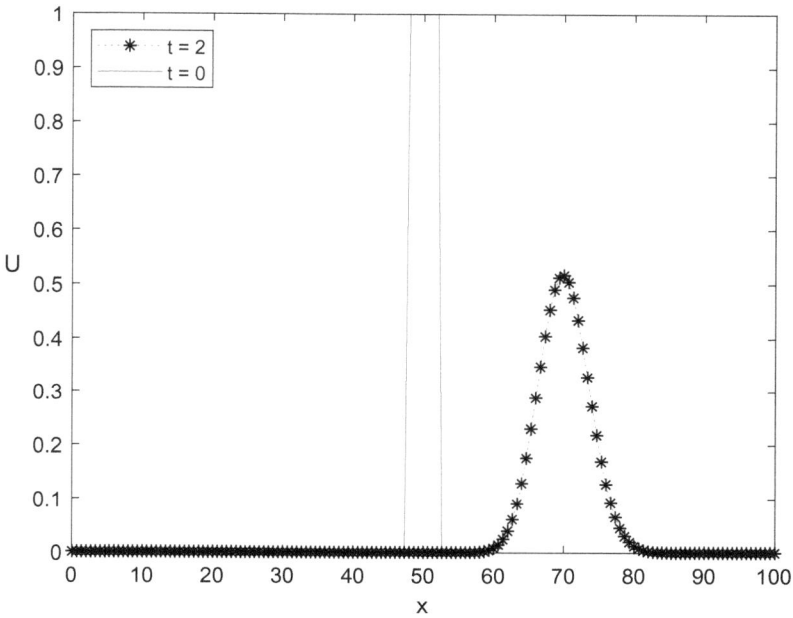

Figure 3.1.1.   Graph of the numerical solution to Problem (3.1.10) to (3.1.12).

```
function u = upwind_ex1
% This is the function file upwind_ex1.m.
% Upwind function is called to solve the special Dirichlet problem:
% Ut + v Ux = alpha Uxx, U(0,t) = 0, U(L,t) = 0,
% U(x,0) = 0, if x < x1,
% U(x,0) = k, if x1 <= x <= x2,
% U(x,0) = 0, if x > x2.
% The approximating solution is plotted.

alpha = .1; v = 10; L = 100; T = 2; nx = 150;
x = linspace(0,L,nx+1);
i1 = 73; i2 = 79; x1 = x(i1); x2 = x(i2); k = 1;
phi = @(x)   k*(x >= x1).*(x <= x2);
g1 = @(t)   0*t;
g2 = @(t)   0*t;
U = feval(phi,x);
nt = 30; t = linspace(0,T,nt+1);
for j = 2:nt+1
```

```
u = upwind(alpha, v, L, t(j), nx, phi, g1, g2);
plot(x,u,'k*:',x,U,'k','LineWidth',.1);
xlabel('x'); ylabel('U');
legend(['t = ',num2str(t(j))],'t = 0','Location','NorthWest');
pause(.01);
end
end
```

Example 3.1.3 Another application is provided below that considers the Dirichlet problem

$$U_t + vU_x - \alpha U_{xx} = 0, \quad 0 < x < L, \quad 0 < t \le T, \tag{3.1.13}$$

$$U(x,0) = \sin(\pi x), \quad 0 \le x \le L, \tag{3.1.14}$$

$$\begin{cases} U(0,t) = \sin(-\pi vt)\exp(-\alpha\pi^2 t), \\ U(L,t) = \sin(\pi(L-vt))\exp(-\alpha\pi^2 t), \end{cases} \quad 0 < t \le T. \tag{3.1.15}$$

The Upwind function is called and Problem (3.1.13) to (3.1.15) is solved. The graph of the numerical solution is shown in Fig. 3.1.2, together with the analytical solution

$$U(x,t) = \sin(\pi(x-vt))\exp(-\alpha\pi^2 t).$$

```
function u = upwind_ex2
% This is the function file upwind_ex2.m.
% Upwind function is called to solve the special Dirichlet problem:
% Ut +v Ux = alpha Uxx, U(x,0) = sin(pi*x),
% U(0,t) = sin(-pi*v*t)*exp(-pi²*t),
% U(L,t) = sin(pi*(L-v*t))*exp(-pi²*t).
% The approximating solution is plotted. The error is evaluated.

alpha = 1; v = .2; L = 1; T = 1; nx = 40;
phi = @(x)        sin(pi*x);
g1 = @(t)        sin(-pi*v*t)*exp(-pi²*t);
g2 = @(t)        sin(pi*(L-v*t))*exp(-pi²*t);
u = upwind(alpha, v, L, T, nx, phi, g1, g2);
x = linspace(0,L,nx+1);
U = sin(pi*(x - v*T))*exp(-pi²*T);
plot(x,u,'k',x,U,'r*','LineWidth',.1);
xlabel('x'); ylabel('U'); title(['t = ',num2str(T)]);
```

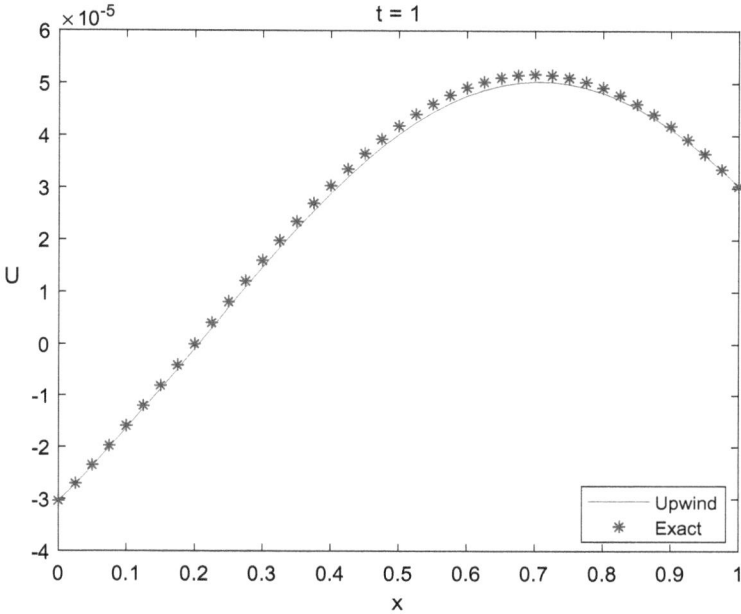

Figure 3.1.2. Graph of the numerical solution of Problem (3.1.13) to (3.1.15).

```
legend('Upwind','Exact','Location','Best');
fprintf( 'Maximum error = %g\n',max(abs(U - u)))
end
```

Remark 3.1.1 *The Upwind Method simplifies to the Explicit Euler Method when $v = 0$. The stability Condition (3.1.9) simplifies to the stability condition of that method as well. In addition, the* upwind *function simplifies to the* euler_e *function (Sec. 2.3.1) with the consequence that the* upwind *function works even when it is called with $v = 0$, as the reader is invited to try. Now, let us investigate what happens when α goes to zero. Equation (3.1.1) simplifies to the advection equation*

$$U_t + vU_x = 0,$$

and stability Condition (3.1.9) to

$$s|v| \leq 1.$$

In Sec. 3.1.3, it will be proved that the Upwind Method for the advection equation is stable under the condition above. Nevertheless, the situation is completely different. Indeed, the advection equation is a first-order equation that requires one boundary condition, whereas the advection-diffusion equation is a second-order equation that requires two boundary conditions.

As noted, the Upwind Method simplifies to the Explicit Euler Method when $v = 0$. Now, the upwind version of the Implicit Euler Method is presented. If $v > 0$, using the backward approximation for U_t and U_x, and the central for U_{xx} in Eq. (3.1.1) then yields

$$\frac{u_i^{j+1} - u_i^j}{\Delta t} + v \frac{u_i^{j+1} - u_{i-1}^{j+1}}{\Delta x} = \alpha \frac{u_{i+1}^{j+1} - 2u_i^{j+1} + u_{i-1}^{j+1}}{(\Delta x)^2}.$$

Hence, with the usual positions $r = \Delta t/(\Delta x)^2$, $s = \Delta t/\Delta x$, we get

$$-(r\alpha + vs)u_{i-1}^{j+1} + (1 + 2r\alpha + vs)u_i^{j+1} - r\alpha u_{i+1}^{j+1} = u_i^j. \qquad (3.1.16)$$

If $v < 0$, using the forward approximation for U_x then yields

$$\frac{u_i^{j+1} - u_i^j}{\Delta t} + v \frac{u_{i+1}^{j+1} - u_i^{j+1}}{\Delta x} = \alpha \frac{u_{i+1}^{j+1} - 2u_i^{j+1} + u_{i-1}^{j+1}}{(\Delta x)^2},$$

$$-r\alpha u_{i-1}^{j+1} + (1 + 2r\alpha - vs)u_i^{j+1} - (r\alpha - sv)u_{i+1}^{j+1} = u_i^j. \qquad (3.1.17)$$

Methods (3.1.16) and (3.1.17) are unconditionally stable. See Exercise 3.4.3. Moreover, they can be combined into a single formula. See Exercise 3.4.4.

Let us present the upwind version of the Crank–Nicolson Method. Of course, the method depends on the sign of v. If $v > 0$, we get

$$\frac{u_i^{j+1} - u_i^j}{\Delta t} + \frac{v}{2}\left[\frac{u_i^{j+1} - u_{i-1}^{j+1}}{\Delta x} + \frac{u_i^j - u_{i-1}^j}{\Delta x}\right]$$

$$= \frac{\alpha}{2}\left[\frac{u_{i+1}^{j+1} - 2u_i^{j+1} + u_{i-1}^{j+1}}{(\Delta x)^2} + \frac{u_{i+1}^j - 2u_i^j + u_{i-1}^j}{(\Delta x)^2}\right].$$

Hence,

$$-(r\alpha + sv)u_{i-1}^{j+1} + (2 + 2r\alpha + sv)u_i^{j+1} - r\alpha u_{i+1}^{j+1}$$

$$= (r\alpha + sv)u_{i-1}^j + (2 - 2r\alpha - sv)u_i^j + r\alpha u_{i+1}^j. \qquad (3.1.18)$$

Method (3.1.18) is unconditionally stable. See Exercise 3.4.5. Similarly, when $v < 0$, we get

$$-r\alpha u_{i-1}^{j+1} + (2 + 2r\alpha - sv)u_i^{j+1} + (vs - r\alpha)u_{i+1}^{j+1}$$

$$= r\alpha u_{i-1}^j + (2 - 2r\alpha + sv)u_i^j + (r\alpha - vs)u_{i+1}^j. \tag{3.1.19}$$

Method (3.1.19) is unconditionally stable. See Exercise 3.4.6.

3.1.2 *Other Finite Difference Methods for the Convection-Diffusion Equation*

This section presents some finite difference methods that use a different approach for the approximation of the derivative U_x in Eq. (3.1.1). The first method is derived from the Explicit Euler Method and uses the central approximation for U_x. We get

$$\frac{u_i^{j+1} - u_i^j}{\Delta t} + v\frac{u_{i+1}^j - u_{i-1}^j}{2\Delta x} = \alpha\frac{u_{i+1}^j - 2u_i^j + u_{i-1}^j}{(\Delta x)^2},$$

$$u_i^{j+1} = (r\alpha + sv/2)u_{i-1}^j + (1 - 2r\alpha)u_i^j + (r\alpha - sv/2)u_{i+1}^j, \tag{3.1.20}$$

where

$$r = \Delta t/(\Delta x)^2, \quad s = \Delta t/\Delta x. \tag{3.1.21}$$

Method (3.1.20) could be named the *Central Explicit Euler Method*. It is stable under the hypothesis

$$s|v| \le 2r\alpha, \quad 2r\alpha \le 1. \tag{3.1.22}$$

Indeed, if conditions in (3.1.22) are satisfied, the terms in parentheses are non-negative and

$$|u_i^{j+1}| \le (r\alpha+sv/2)||\mathbf{u}_j||_\infty+(1-2r\alpha)||\mathbf{u}_j||_\infty+(r\alpha-sv/2)||\mathbf{u}_j||_\infty \le ||\mathbf{u}_j||_\infty,$$
$$||\mathbf{u}_{j+1}||_\infty \le ||\mathbf{u}_j||_\infty \cdots \le ||\mathbf{u}_0||_\infty.$$

The second method is derived from the Implicit Euler Method and again uses the central approximation for U_x. We get

$$\frac{u_i^{j+1} - u_i^j}{\Delta t} + v\frac{u_{i+1}^{j+1} - u_{i-1}^{j+1}}{2\Delta x} = \alpha\frac{u_{i+1}^{j+1} - 2u_i^{j+1} + u_{i-1}^{j+1}}{(\Delta x)^2},$$

$$-(r\alpha + vs/2)u_{i-1}^{j+1} + (1 + 2r\alpha)u_i^{j+1} - (r\alpha - vs/2)u_{i+1}^{j+1} = u_i^j, \tag{3.1.23}$$

where Notations (3.1.21) were used. Method (3.1.23) could be named the *Central Implicit Euler Method*. It is unconditionally stable, as proved by the Von Neumann criterium. Indeed, substituting $u_i^j = \xi^j e^{I\beta i \Delta x}$ in (3.1.23) yields

$$\xi[1 + 2r\alpha - 2r\alpha \cos(\beta \Delta x) + vsI \sin(\beta \Delta x)] = 1,$$

$$|\xi|^2 = 1/\{[1 + 4r\alpha \sin^2(\beta \Delta x/2)]^2 + v^2 s^2 \sin^2(\beta \Delta x)\} < 1.$$

Setting $p = r\alpha + sv/2$ and $q = r\alpha - sv/2$, Method (3.1.23) is written as

$$-pu_{i-1}^{j+1} + (1 + p + q)u_i^{j+1} - qu_{i+1}^{j+1} = u_i^j,$$

and in compact matrix form

$$\begin{bmatrix} 1+p+q & -q & & \\ -p & 1+p+q & -q & \\ & & \cdot & \\ & & \cdot & \\ & & -p\ 1+p+q \end{bmatrix} \begin{bmatrix} u_1^{j+1} \\ u_2^{j+1} \\ \cdot \\ u_{n-1}^{j+1} \end{bmatrix} = \begin{bmatrix} u_1^j \\ u_2^j \\ \cdot \\ u_{n-1}^j \end{bmatrix} + \begin{bmatrix} pu_0^{j+1} \\ 0 \\ \cdot \\ qu_n^{j+1} \end{bmatrix},$$

$$B\mathbf{u}_{j+1} = \mathbf{u}_j + \mathbf{b}_{j+1},$$

$$\mathbf{u}_{j+1} = A\mathbf{u}_j + \mathbf{a}_{j+1}, \quad (A = B^{-1}, \quad \mathbf{a}_{j+1} = B^{-1}\mathbf{b}_{j+1}). \tag{3.1.24}$$

Example 3.1.4   A function is provided that applies Method (3.1.24) to solve the Dirichlet problem for Eq. (3.1.1).

```
function u = central_implicit(alpha, v, L, T, nx, phi, g1, g2)
% This is the function file central_implicit.m.
% Central Implicit Euler Method is applied to solve the Dirichlet problem:
% Ut + v Ux = alpha Uxx, U(x,0) = phi(x), U(0,t) = G1(t), U(L,t) = G2(t).
% The input arguments are: thermal diffusivity, velocity v, length of the solid,
% final time, number of points on the space grid, initial-boundary conditions.
% The function returns a vector with the solution at the final time.

% Check data
if any([alpha L T nx-2] <= 0)
    error('Check alpha, L, T, nx')
end

% Initialization
nt = 1000; dx = L/nx; dt = T/nt; r = dt/dx^2; s = dt/dx;
p = alpha*r+v*s/2; q = alpha*r-v*s/2;
x = linspace(0,L,nx+1); t = linspace(0,T,nt+1);
b = zeros(nx-1,1);
```

```
BB = [-p*ones(nx-1,1)  (1+p+q)*ones(nx-1,1)  -q*ones(nx-1,1)];
B = spdiags(BB, -1:1, nx-1, nx-1);
u = feval(phi,x');

% Central Implicit Euler Method
for j = 2:nt + 1
    b(1) = p*g1(t(j)); b(end) = q*g2(t(j));
    u(2:nx) = B\(u(2:nx) + b);
    u(1) = g1(t(j)); u(end) = g2(t(j));
end
end
```

Example 3.1.5   Consider the Dirichlet Problem (3.1.10) to (3.1.12), already discussed with the Upwind Method in Example 3.1.2. The listing below illustrates a way to call the central_implicit function to solve Problem (3.1.10) to (3.1.12). The graph of the numerical solution is shown in Fig. 3.1.3. The oscillating behavior is due to the central approximation of U_x. Oscillations do not grow without bounds as the method is stable. However, they are undesirable.

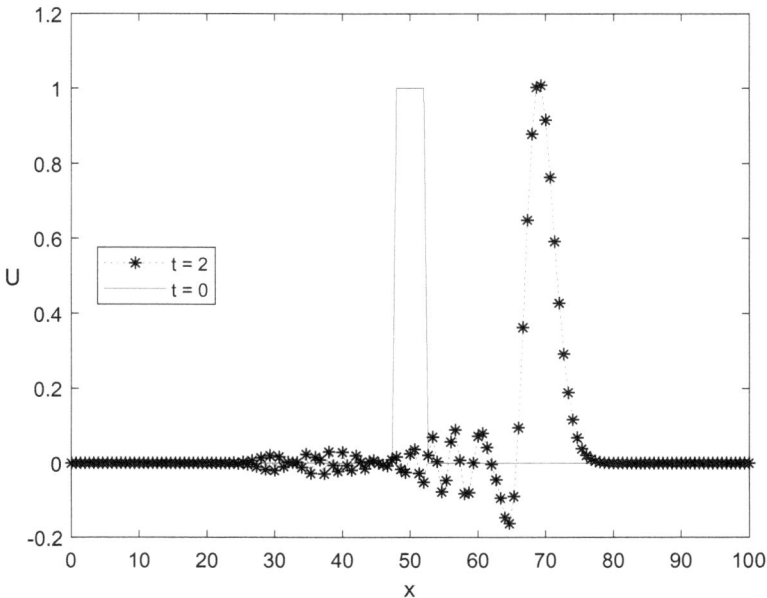

Figure 3.1.3.   Numerical solution of Problem (3.1.10) to (3.1.12).

```
function central_implicit_ex1
% This is the function file central_implicit_ex1.m.
% Central_implicit function is called to solve the special Dirichlet problem:
% Ut +v Ux = alpha Uxx, U(0,t) = 0, U(L,t) = 0,
% U(x,0) = 0, if x < x1,
% U(x,0) = k, if x1 <= x <= x2,
% U(x,0) = 0, if x > x2.
% The approximating solution is plotted.

alpha = 0.1; v = 10; L = 100; T = 2; nx = 150;
x = linspace(0,L,nx+1);
i1 = 73; i2 = 79; x1 = x(i1); x2 = x(i2); k = 1;
phi = @(x)        k*(x >= x1).*(x <= x2);
g1 = @(t)         0*t;
g2 = @(t)         0*t;
U = feval(phi,x');
u = central_implicit(alpha, v, L, T, nx, phi, g1, g2);
plot(x,u,'k*:',x,U,'k','LineWidth',.1);
xlabel('x'); ylabel('U');
legend(['t = ',num2str(T)],'t = 0','Location','Best');
end
```

The last method of this section is the central version of the Crank–Nicolson Method for Eq. (3.1.1)

$$\frac{u_i^{j+1} - u_i^j}{\Delta t} + \frac{v}{4\Delta x}(u_{i+1}^{j+1} - u_{i-1}^{j+1} + u_{i+1}^j - u_{i-1}^j)$$

$$= \frac{\alpha}{2(\Delta x)^2}(u_{i+1}^{j+1} - 2u_i^{j+1} + u_{i-1}^{j+1} + u_{i+1}^j - 2u_i^j + u_{i-1}^j).$$

Hence, by using Notations (3.1.21),

$$-(r\alpha + sv/2)u_{i-1}^{j+1} + 2(1 + r\alpha)u_i^{j+1} - (r\alpha - sv/2)u_{i+1}^{j+1}$$

$$= (r\alpha + sv/2)u_{i-1}^j + 2(1 - r\alpha)u_i^j + (r\alpha - sv/2)u_{i+1}^j. \tag{3.1.25}$$

Method (3.1.25) could be named the *Central Crank–Nicolson Method*. It is unconditionally stable, as proved by the Von Neumann criterium. Indeed, substituting $u_i^j = \xi^j e^{I\beta i \Delta x}$ in (3.1.25) yields

$$\xi\{2 + 2r\alpha[1 - \cos(\beta\Delta x)] + vsI\sin(\beta\Delta x)\}$$
$$= 2 - 2r\alpha[1 - \cos(\beta\Delta x)] - vsI\sin(\beta\Delta x),$$
$$|\xi|^2 = \frac{4[1 - 2r\alpha\sin^2(\beta\Delta x/2)]^2 + v^2 s^2 \sin^2(\beta\Delta x)}{4[1 + 2r\alpha\sin^2(\beta\Delta x/2)]^2 + v^2 s^2 \sin^2(\beta\Delta x)} < 1.$$

Moreover, setting

$$p = r\alpha + sv/2, \quad q = r\alpha - sv/2, \tag{3.1.26}$$

Method (3.1.25) is written as

$$-pu_{i-1}^{j+1} + 2(1 + r\alpha)u_i^{j+1} - qu_{i+1}^{j+1} = pu_{i-1}^j + 2(1 - r\alpha)u_i^j + qu_{i+1}^j,$$

and in compact matrix form

$$\begin{bmatrix} 2(1+r\alpha) & -q & & \\ -p & 2(1+r\alpha) & -q & \\ & \cdot & \cdot & \\ & & -p & 2(1+r\alpha) \end{bmatrix} \begin{bmatrix} u_1^{j+1} \\ u_2^{j+1} \\ \cdot \\ u_{n-1}^{j+1} \end{bmatrix}$$

$$= \begin{bmatrix} 2(1-r\alpha) & q & & \\ p & 2(1-r\alpha) & q & \\ & \cdot & \cdot & \\ & & p & 2(1-r\alpha) \end{bmatrix} \begin{bmatrix} u_1^j \\ u_2^j \\ \cdot \\ u_{n-1}^j \end{bmatrix} + \begin{bmatrix} p(u_0^{j+1} + u_0^j) \\ 0 \\ \cdot \\ q(u_n^{j+1} + u_n^j) \end{bmatrix},$$

$$B\mathbf{u}_{j+1} = C\mathbf{u}_j + \mathbf{b}_j,$$
$$\mathbf{u}_{j+1} = A\mathbf{u}_j + \mathbf{a}_j, \quad (A = B^{-1}C, \quad \mathbf{a}_j = B^{-1}\mathbf{b}_j).$$

3.1.3 Advection Equation

Consider the *advection equation*

$$U_t + vU_x = 0, \tag{3.1.27}$$

also named the *convection equation*. The Upwind Method for Eq. (3.1.27) can be derived from that related to Eq. (3.1.1) as a special case. Therefore, for $v > 0$, we get (Fig. 3.1.4)

$$\frac{u_i^{j+1} - u_i^j}{\Delta t} + v\frac{u_i^j - u_{i-1}^j}{\Delta x} = 0,$$

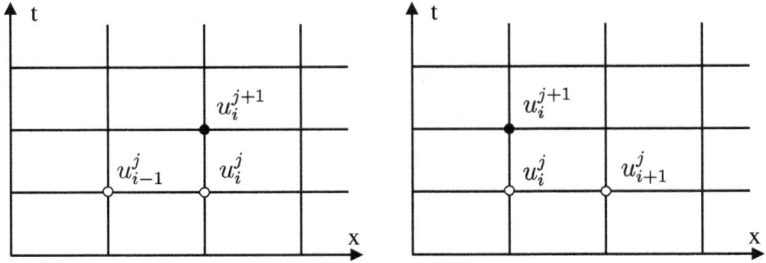

Figure 3.1.4.   Upwind Method for $v > 0$ (left) and $v < 0$ (right).

and for $v < 0$, it is

$$\frac{u_i^{j+1} - u_i^j}{\Delta t} + v\frac{u_{i+1}^j - u_i^j}{\Delta x} = 0.$$

Setting $s = \Delta t/\Delta x$, the equations above are written as

$$u_i^{j+1} = (1 - sv)u_i^j + svu_{i-1}^j. \tag{3.1.28}$$

$$u_i^{j+1} = (1 + sv)u_i^j - svu_{i+1}^j. \tag{3.1.29}$$

Methods (3.1.28) and (3.1.29) are named the *FTBS Method* (forward in time, backward in space) and *FTFS Method* (forward in time, forward in space), respectively. They are stable under the hypothesis

$$|v|s \leq 1. \tag{3.1.30}$$

Condition (3.1.30) can be proved easily, as suggested in Exercise 3.4.7. Of course, it follows from the stability condition of the Upwind Method for the advection-diffusion as a special case.

Consider the following initial-boundary value problem

$$U_t + vU_x = 0, \quad x > 0, \quad 0 < t \leq T, \quad v > 0, \tag{3.1.31}$$

$$U(x,0) = \varphi(x), \quad x \geq 0, \tag{3.1.32}$$

$$U(0,t) = g(t), \quad 0 < t \leq T. \tag{3.1.33}$$

The analytical solution of Problem (3.1.31) to (3.1.33) is suggested in Exercise 3.4.8.

Example 3.1.6   This example presents a function that applies Method (3.1.28) to solve Problem (3.1.31) to (3.1.33).

```
function [u, nt] = ftbs(v, L, T, nx, phi, g)
% This is the function file ftbs.m.
% FTBS Method is applied to solve the Dirichlet problem:
% Ut +v Ux = 0, U(x,0) = phi(x), U(0,t) = G(t).
% The input arguments are: velocity v, length, final time,
% number of points on the space grid, initial-boundary conditions.
% It returns a matrix with the approximating solutions at tⱼ, j = 1, ..., nt + 1.

% Check data
if any([v L T nx-2] <= 0)
    error('Check v, L, T, nx')
end

% Stability
nt = 5; dx = L/nx; st = 1;
while v*T/nt/dx > st
    nt = nt+1;
end

% Initialization
dt = T/nt; s = dt/dx;
x = linspace(0,L,nx+1); t = linspace(0,T,nt+1);
u(:,1) = feval(phi,x);

% FTBS Method
for j = 2:nt + 1
    u(2:nx+1,j) = (1 - s*v)*u(2:nx+1,j-1) + s*v*u(1:nx,j-1);
    u(1,j) = g(t(j));
end
end
```

Example 3.1.7 Consider the following initial-boundary value problem

$$U_t + vU_x = 0, \quad 0 < x < L, \quad 0 < t \le T, \quad v > 0, \qquad (3.1.34)$$

$$U(x,0) = 0, \quad 0 \le x \le L, \quad U(0,t) = \begin{cases} K \text{ if } 0 < t < t_0, \\ 0 \text{ if } t_0 \le t \le T. \end{cases} \qquad (3.1.35)$$

Problem (3.1.34) and (3.1.35) has the following analytical solution

$$U(x,t) = \begin{cases} K, \text{ if } v(t-t_0) < x < vt, \\ 0, \text{ if } x \ge vt. \end{cases}$$

The listing below calls the ftbs function and solves Problem (3.1.34) and (3.1.35). The graph of the numerical solution is shown in Fig. 3.1.5.

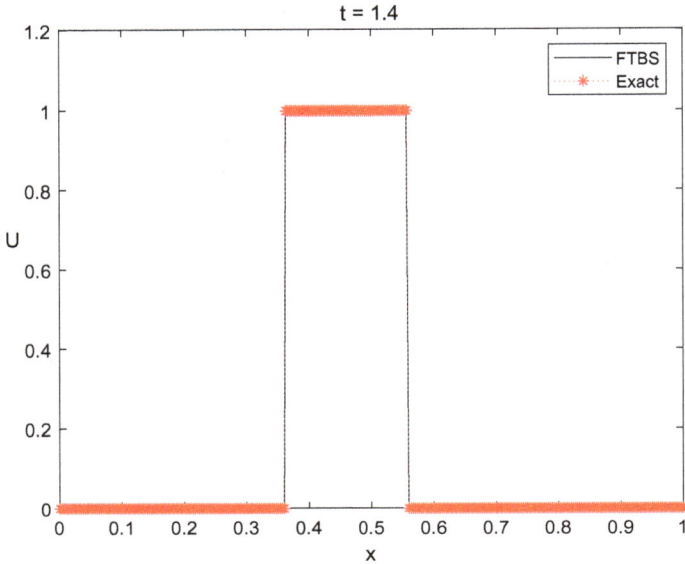

Figure 3.1.5. Graph of the numerical solution of Problem (3.1.34) and (3.1.35).

```
function u = ftbs_ex1
% This is the function file ftbs_ex1.m.
% Ftbs function is called to solve the special Dirichlet problem:
% Ut +v Ux = 0, U(x,0) = 0,
% U(0,t) = K, if t < t0, U(0,t) = 0, if t >= t0.
% The approximating solution is plotted.
% Exact solution:
% U = K, if v*(t-t0) < x < v*t,  U = 0, if x >= v*t.

v = .4; L = 1; T = 1.4; nx = 300;
t0 = .5; K = 1;
phi = @(x)        0*x;
g = @(t)          K*(t0 - t > 0);
[u, nt] = ftbs(v, L, T, nx, phi, g);
x = linspace(0,L,nx+1); t = linspace(0,T,nt+1);
U(:,1) = K*(x - v*(T - t0) > 0).*(x - v*T < 0);
for j = 2:nt + 1
    plot(x,u(:,j),'k',x,U,'r*:');
    axis([0 L 0 K+.201]);
    xlabel('x'); ylabel('U');
```

```
    title(['t = ',num2str(t(j))]);
    legend('FTBS','Exact');
    pause(.01);
end
fprintf( 'Maximum error = %g\n',max(abs(U - u(:,nt+1))))
end
```

Other applications are suggested in Exercises 3.4.9 and 3.4.10.
Consider the following initial-boundary value problem

$$U_t + vU_x = 0, \quad x < L, \quad 0 < t \le T, \quad v < 0, \tag{3.1.36}$$

$$U(x,0) = \varphi(x), \quad x \le L, \tag{3.1.37}$$

$$U(L,t) = g(t), \quad 0 < t \le T. \tag{3.1.38}$$

The analytical solution of Problem (3.1.36) to (3.1.38) is suggested in
Exercise 3.4.11.

Example 3.1.8 A function is provided that applies Method (3.1.29) to
solve Problem (3.1.36) to (3.1.38).

```
function [u, nt] = ftfs(v, L, T, nx, phi, g)
% This is the function file ftfs.m.
% FTFS Method is applied to solve the Dirichlet problem:
% Ut +v Ux = 0, U(x,0) = phi(x), U(L,t) = G(t).
% The input arguments are: velocity v, length, final time,
% number of points on the space grid, initial-boundary conditions.
% It returns a matrix with the approximating solutions at $t_j, j = 1, ..., nt + 1$.

% Check data
if any([-v L T nx-2] <= 0)
    error('Check v, L, T, nx')
end

% Stability
nt = 5; dx = L/nx; st = 1;
while abs(v)*T/nt/dx > st
    nt = nt+1;
end

% Initialization
dt = T/nt; s = dt/dx;
```

```
x = linspace(0,L,nx+1); t = linspace(0,T,nt+1);
u(:,1) = feval(phi,x);

% FTFS Method
for j = 2:nt + 1
    u(1:nx,j) = (1 + s*v)*u(1:nx,j-1) - s*v*u(2:nx+1,j-1);
    u(nx+1,j) = g(t(j));
end
end
```

Example 3.1.9 Let us illustrate a way to call the ftfs function. Consider the special initial-boundary value problem

$$U_t + vU_x = 0, \quad 0 < x < L, \quad 0 < t \le T, \quad v < 0, \tag{3.1.39}$$

$$U(x,0) = 0, \quad 0 \le x \le L, \quad U(L,t) = \sin(\omega t), \quad 0 < t \le T, \tag{3.1.40}$$

that has the following analytical solution

$$U(x,t) = \begin{cases} \sin(\omega(t - x/v + L/v)), & \text{if} \quad x > vt + L, \\ 0, & \text{if} \quad x \le vt + L. \end{cases}$$

The listing below calls the ftfs function and solves Problem (3.1.39 and 3.1.40). The graph of the numerical solution is shown in Fig. 3.1.6.

```
function u = ftfs_ex1
% This is the function file ftfs_ex1.m.
% Ftfs function is called to solve the special Dirichlet problem:
% Ut +v Ux = 0, U(x,0) = 0, U(L,t) = sin(om*t).
% The approximating solution is plotted.
% Exact solution:
% U(x,t) = sin(om*(t-x/v+L/v)), if x > vt + L,
% U(x,t) = 0, if x <= vt + L.

v = -.4; L = 1; T = 2.5; nx = 50;
om = 5;
phi = @(x)         0*x;
g = @(t)          sin(om*t);
[u, nt] = ftfs(v, L, T, nx, phi, g);
x = linspace(0,L,nx+1); t = linspace(0,T,nt+1);
U(:,1) = sin(om*(- x/v + L/v)).*(T - x/v + L/v > 0);
fprintf( 'Maximum error = %g\n',max(abs(U - u(:,nt+1))))
for j = 2:nt + 1
```

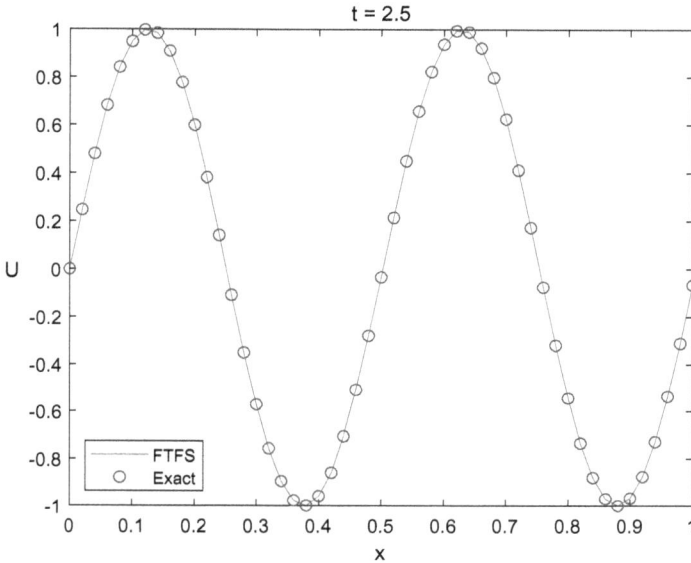

Figure 3.1.6.   Graph of the numerical solution of Problem (3.1.39 and 3.1.40).

```
plot(x,u(:,j),'k',x,U,'ro','LineWidth',.1);
xlabel('x'); ylabel('U');
title(['t = ',num2str(t(j))]);
legend('FTFS','Exact','Location','SouthWest');
pause(.001);
end
end
```

Consider the *advection equation with decay*

$$U_t + vU_x = \lambda U, \tag{3.1.41}$$

where λ is the decay coefficient. The change of unknown function

$$W = U \exp(-\lambda t), \tag{3.1.42}$$

converts Eq. (3.1.41) to the advection equation

$$W_t + vW_x = 0.$$

Consequently, all Matlab functions, introduced for the advection equation, can be used for Eq. (3.1.41) as well, after small modifications. See Exercise 3.4.12.

Consider the *advection equation with a variable coefficient*

$$U_t + v(x,t)U_x = 0, \qquad (3.1.43)$$

where $v(x,t)$ is a given function. If $v(x,t) > 0$, applying the FTBS Method yields

$$u_i^{j+1} = (1 - sv_i^j)u_i^j + sv_i^j u_{i-1}^j, \qquad (3.1.44)$$

where $v_i^j = v(x_i, t_j)$. Method (3.1.44) is stable under the hypothesis

$$s|v_i^j| \le 1, \quad \forall\, i, j, \qquad (3.1.45)$$

as it is

$$|u_i^{j+1}| \le (1 - sv_i^j)|u_i^j| + sv_i^j|u_{i-1}^j| \le ||\mathbf{u}^j||_\infty \quad \Rightarrow \quad ||\mathbf{u}^{j+1}||_\infty \le ||\mathbf{u}^j||_\infty.$$

If $v(x,t) < 0$, applying the FTFS Method yields

$$u_i^{j+1} = (1 + sv_i^j)u_i^j - sv_i^j u_{i+1}^j, \qquad (3.1.46)$$

that is stable under the same Hypothesis (3.1.45).

3.2 Method of Lines

3.2.1 *Heat Equation*

The Method of Lines is a semi-discrete finite difference method for the numerical integration of partial differential equations. Only the space variables are discretized, whereas the time variable remains continuous. We begin to illustrate the method by considering the one-dimensional heat equation

$$U_t = \alpha U_{xx}, \quad 0 < x < L, \quad t > 0. \qquad (3.2.1)$$

The variable x is discretized and the variable t is not. Using the central approximation for U_{xx} yields

$$\dot{u}_i = p(u_{i+1} - 2u_i + u_{i-1}), \quad i = 1, \ldots, n-1, \quad n\Delta x = L, \qquad (3.2.2)$$

where $p = \alpha/(\Delta x)^2$, and the following notations have been introduced

$$u_i = u(x_i, t), \quad \dot{u}_i = u_t(x_i, t).$$

Equations (3.2.2) are a system of $n - 1$ ordinary differential equations to be integrated along the lines $x = x_i$ (Fig. 3.2.1). System (3.2.2) has $n + 1$ unknown functions: $u_0(t), \ldots, u_n(t)$, two more than equations. In the case of

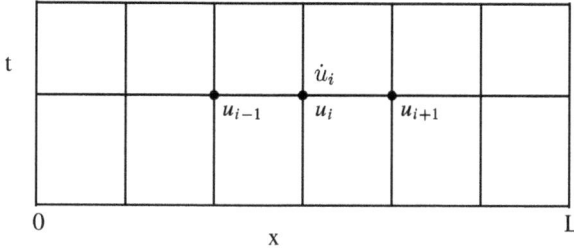

Figure 3.2.1. Method of Lines.

the Dirichlet problem, two functions are known, as given by the boundary conditions $u_0 = u(0, t) = g_1(t)$ and $u_n = u(L, t) = g_2(t)$, and the differential system can be solved. In addition, System (3.2.2) can be arranged in matrix form as follows

$$
\begin{bmatrix} \dot{u}_1 \\ \dot{u}_2 \\ \cdot \\ \dot{u}_{n-1} \end{bmatrix} = p \begin{bmatrix} -2 & 1 & & \\ 1 & -2 & 1 & \\ & \cdot & \cdot & \cdot \\ & & 1 & -2 \end{bmatrix} \begin{bmatrix} u_1 \\ u_2 \\ \cdot \\ u_{n-1} \end{bmatrix} + p \begin{bmatrix} u_0 \\ 0 \\ \cdot \\ u_n \end{bmatrix},
$$

$$
\dot{\mathbf{u}} = A\mathbf{u} + \mathbf{a}, \tag{3.2.3}
$$

where $\mathbf{a}$ is a known term. System (3.2.3) solves the Dirichlet problem.

Consider the Neumann problem for Eq. (3.2.1) with the following boundary conditions

$$
-U_x(0, t) = g_1(t), \quad U_x(L, t) = g_2(t). \tag{3.2.4}
$$

Apply the central approximations to the derivatives in (3.2.4)

$$
-\frac{u_1 - u_{-1}}{2\Delta x} = g_1, \quad \frac{u_{n+1} - u_{n-1}}{2\Delta x} = g_2. \tag{3.2.5}
$$

Formulas (3.2.5) are not applicable because of the ghost terms u_{-1} and u_{n+1}. Therefore, consider the equations obtained from (3.2.2) for $i = 0$ and $i = n$:

$$
\dot{u}_0 = p(u_1 - 2u_0 + u_{-1}), \quad \dot{u}_n = p(u_{n+1} - 2u_n + u_{n-1}). \tag{3.2.6}
$$

Solving the algebraic Equations (3.2.5) with respect to the ghost terms and the result substituted into the differential Eqs. (3.2.6) yields

$$
\dot{u}_0 = 2p(u_1 - u_0) + 2p\Delta x g_1, \quad \dot{u}_n = 2p(u_{n-1} - u_n) + 2p\Delta x g_2. \tag{3.2.7}
$$

Equations (3.2.2) and (3.2.7) are the system that solves the Neumann problem. In matrix notation, it is written as

$$
\begin{bmatrix} \dot{u}_0 \\ \dot{u}_1 \\ \cdot \\ \dot{u}_n \end{bmatrix} = p \begin{bmatrix} -2 & 2 & & \\ 1 & -2 & 1 & \\ & & \cdot & \cdot & \cdot \\ & & & 2 & -2 \end{bmatrix} \begin{bmatrix} u_0 \\ u_1 \\ \cdot \\ u_n \end{bmatrix} + 2p\Delta x \begin{bmatrix} g_1 \\ 0 \\ \cdot \\ g_2 \end{bmatrix},
$$

$$
\dot{\mathbf{u}} = A\mathbf{u} + \mathbf{a}. \tag{3.2.8}
$$

Another way to approximate boundary Conditions (3.2.4) is the following

$$
-\frac{u_1 - u_0}{\Delta x} = g_1, \qquad \frac{u_n - u_{n-1}}{\Delta x} = g_2,
$$

where forward and backward approximations were applied. Conditions above do not present ghost terms, but are less accurate. The Robin problem is discussed similarly.

Example 3.2.1   Consider the process of consolidation in a clay layer due to a uniform load q (Fig. 3.2.2). The evolution of the overpressure, excess pore pressure U, is governed by the following Dirichlet problem

$$
U_t - c_\nu U_{zz} = 0, \quad 0 < z < L, \quad 0 < t \le T, \tag{3.2.9}
$$

$$
U(z,0) = q, \quad 0 \le z \le L, \tag{3.2.10}
$$

$$
U(0,t) = 0, \quad U(L,t) = 0, \quad 0 < t \le T. \tag{3.2.11}
$$

where $c_\nu = 10^{-7} m^2/s$ is the consolidation coefficient. A function that applies the Method of Lines to solve Problem (3.2.9) to (3.2.11) is provided below. For $L = 8\ m$, $T = 3\ years$, the graph of the numerical solution is shown in Fig. 3.2.3.

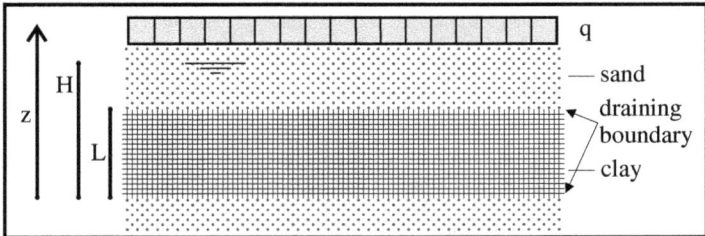

Figure 3.2.2.   Consolidation.

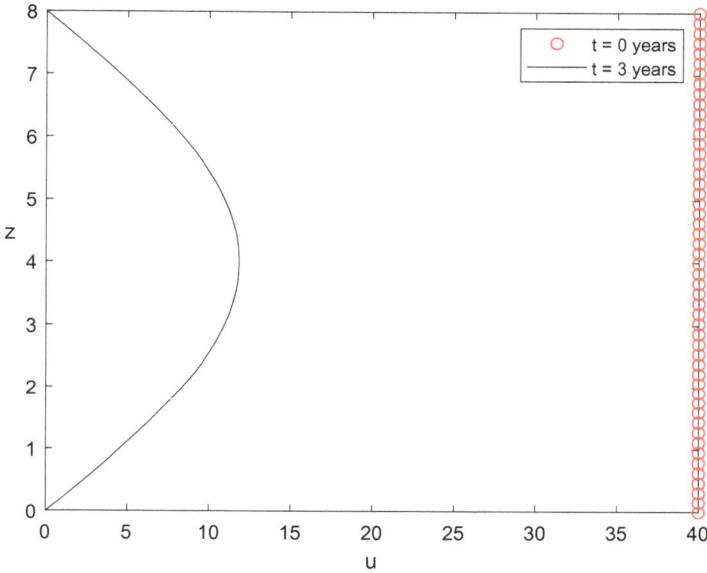

Figure 3.2.3.   Graph of the numerical solution to Problem (3.2.9) to (3.2.11).

```
function u = consolidation1
% This is the function file consolidation1.m.
% Method of Lines is applied to solve the Dirichlet problem
% Ut - cv Uzz = 0, U(z,0) = q, U(0,t) = U(L,t) = 0.

% Initialization
cv = 10^(-7); L = 8; T = 3*365*24*3600; % 1 year = 365*24*3600 seconds
q = 40;
n = 50; dz = L/n; p = cv/dz^2; z = linspace(0,L,n+1);
AA = [ones(n-1,1)  -2*ones(n-1,1)  ones(n-1,1)];
A = p*spdiags(AA, -1:1, n-1,n-1);
u = q*ones(n+1,1);% Vector u is initialized with the initial condition.
plot(u,z,'ro'); % The initial condition is plotted.
hold on;% Retains plots. New plots do not delete previous plots.

% Method of Lines
[~,y] = ode45(@system, [0 T ], u(2:end-1), [ ], A);
    % [t,y] = ode45(@fun, ti, ic, options, p1, p2,...)
    % This function solves the systems of ordinary differential equations;
```

```
% fun is the local function where the differential system is defined;
% ti is the vector containing the initial and final times;
% ic is the vector containing the initial values;
% The symbol [ ] replaces the structure 'options' that is not used in
% this case;
% p1, p2,... are other parameters passed to ode45, in this example A.
% The function returns the column vector t and the matrix y that has the
% same number of rows as t. In this example, t has been replaced by the
% symbol ~ as not used. The first row in y contains the solution
% at the initial time and the last at the final time. The other rows in y
% contain the solution at the time specified by the corresponding row
% in t.
u(2:n) = y(end,:);% The final solution is copied in u.
u(1) = 0; u(n+1) = 0;% Boundary conditions.
plot(u,z,'k');
xlabel('u'); ylabel('z');
year = 365*24*3600; legend('t = 0 years',['t = ',num2str(T/year),' years']);
hold off;% The default behavior is restored.
end

%———— Local function ————
function Du = system(~, u, B)
    % The symbol ~ replaces the variable t that is not used in this case.
Du = B*u;
end
```

When the boundary conditions depend on time, the application of the Method of Lines may present some difficulty, as outlined in the example below.

Example 3.2.2   Consider the following Dirichlet problem

$$U_t - \alpha U_{xx} = 0, \quad 0 < x < L, \quad 0 < t \le T, \qquad (3.2.12)$$

$$U(x,0) = 0, \quad 0 \le x \le L, \qquad (3.2.13)$$

$$U(0,t) = \begin{cases} t/t_0 - (t/t_0)^2 & \text{if } t \le t_0, \\ 0 & \text{if } t > t_0, \end{cases} \quad U(L,t) = 0, \quad 0 < t \le T. \quad (3.2.14)$$

A way to apply the Method of Lines to Problem (3.2.12) to (3.2.14) is illustrated below. The graph of the numerical solution is shown in Fig. 3.2.4.

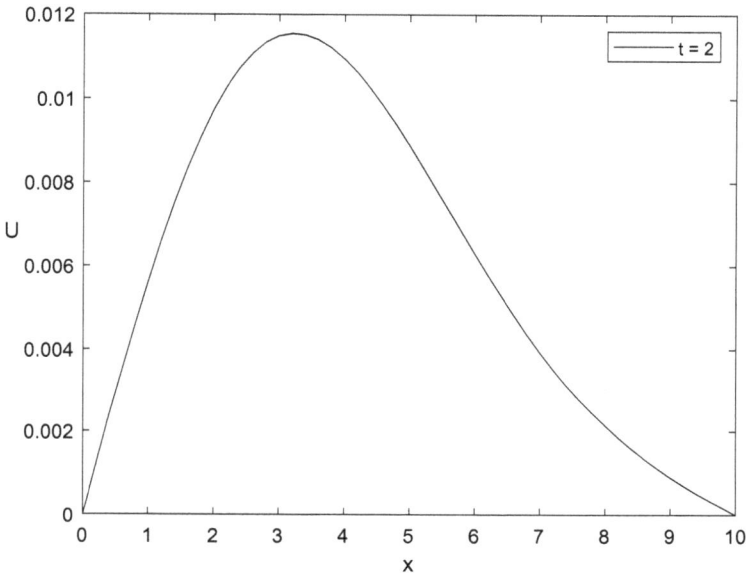

Figure 3.2.4. Graph of the numerical solution to Problem (3.2.12) to (3.2.14).

```
function u = lines_heat1
% This is the function file lines_heat1.m.
% Method of Lines is applied to solve the Dirichlet problem:
% Ut = alpha Uxx, U(x,0) = 0, U(L,t) = 0.
% U(0,t) = t/t0 - t^2/t0^2, if t <= t0,
% U(0,t) = 0, if t > t0.

% Initialization
alpha = 3; L = 10; T = 2; nx = 50;
dx = L/nx; p = alpha/dx^2; x = linspace(0,L,nx+1);
AA = [ones(nx-1,1)  -2*ones(nx-1,1)  ones(nx-1,1)];
A = p*spdiags(AA, -1:1, nx-1,nx-1);
a = zeros(nx-3,1);
u = zeros(nx+1,1); % Initial condition.

% Method of Lines
tic % tic and toc functions measure the time elapsed between the two.
[~,y] = ode45(@system, [0 T], u(2:end-1), [ ], A, a, p);
u(2:nx) = y(end,:);
u(1) = g1(T); u(nx+1) = 0;
```

```
toc % See tic
```

```
% Plot
plot(x,u,'k');
xlabel('x'); ylabel('U');
legend(['t = ',num2str(T)],'Location','NorthEast');
end
```

```
% ———— Local functions ————
function f = g1(t)
t0 = .5;
f = (t/t0 - t²/t0²)*(t <= t0);
end
function Du = system(t, u, A, a, p)
Du = A*u + [p*g1(t); a; 0];
end
```

Other applications are suggested in Exercises 3.4.13–3.4.17.

Consider the Neumann–Dirichlet problem

$$-U_x(0,t) = g_1(t), \quad U(L,t) = g_2(t). \tag{3.2.15}$$

The equations for the Method of Lines are derived from Formulas $(3.2.7)_1$ and $(3.2.2)$, rewritten for convenience, are

$$\dot{u}_0 = 2p(u_1 - u_0) + 2p\Delta x g_1,$$

$$\dot{u}_i = p(u_{i+1} - 2u_i + u_{i-1}), \quad i = 1,\ldots,n-2, \quad \dot{u}_{n-1} = p(u_{n-2} - 2u_{n-1}) + pg_2.$$

Equivalently, in matrix form, we have

$$\begin{bmatrix} \dot{u}_0 \\ \dot{u}_1 \\ \cdot \\ \dot{u}_{n-1} \end{bmatrix} = p \begin{bmatrix} -2 & 2 & & \\ 1 & -2 & 1 & \\ & \cdot & \cdot & \cdot \\ & & 1 & -2 \end{bmatrix} \begin{bmatrix} u_0 \\ u_1 \\ \cdot \\ u_{n-1} \end{bmatrix} + p \begin{bmatrix} 2\Delta x g_1 \\ 0 \\ \cdot \\ g_2 \end{bmatrix},$$

$$\dot{\mathbf{u}} = A\mathbf{u} + \mathbf{a}. \tag{3.2.16}$$

Equation (3.2.16) will be applied in the example below. The Dirichlet–Neumann problem

$$-U_x(0,t) = g_1(t), \quad U(L,t) = g_2(t), \tag{3.2.17}$$

is discussed similarly. See Exercise 3.4.18.

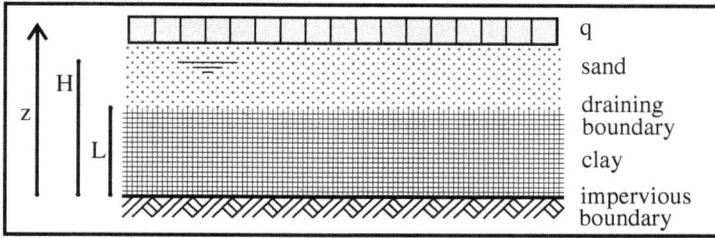

Figure 3.2.5. Consolidation.

Example 3.2.3 Consider again the consolidation process due to a uniform load q, as in Example 3.2.1. Let us discuss a different mechanical situation: a layer of clay between an upper draining boundary and a lower impervious boundary (Fig. 3.2.5). The evolution of the pore overpressure U is governed by the following Neumann-Dirichlet problem

$$U_t - c_v U_{zz} = 0, \quad 0 < z < L, \quad 0 < t \le T, \tag{3.2.18}$$

$$U(z,0) = q, \quad 0 \le z \le L, \tag{3.2.19}$$

$$U_z(0,t) = 0, \quad U(L,t) = 0, \quad 0 < t \le T. \tag{3.2.20}$$

A function that applies the Method of Lines to solve Problem (3.2.18 to 3.2.20) is presented below. The graph of the numerical solution is shown in Fig. 3.2.6.

```
function u = consolidation2
% This is the function file consolidation2.m.
% Method of Lines is applied to solve the Neumann–Dirichlet problem:
% Ut - cv*Uzz = 0, U(z,0) = q, Ux(0,t) = 0, U(L,t) = 0.

% Initialization
cv = 10^(-7); L = 8; T = 3*365*24*3600;
n = 50; dz = L/n; p = cv/dz^2; z = linspace(0,L,n+1);
q = 40; z = linspace(0,L,n+1);
AA = [ones(n,1)  -2*ones(n,1) [0; 2; ones(n-2,1)]];
A = p*spdiags(AA, -1:1, n,n);
a = zeros(n-2,1);
u = q*ones(n+1,1);
plot(u,z,'r*');
hold on;
```

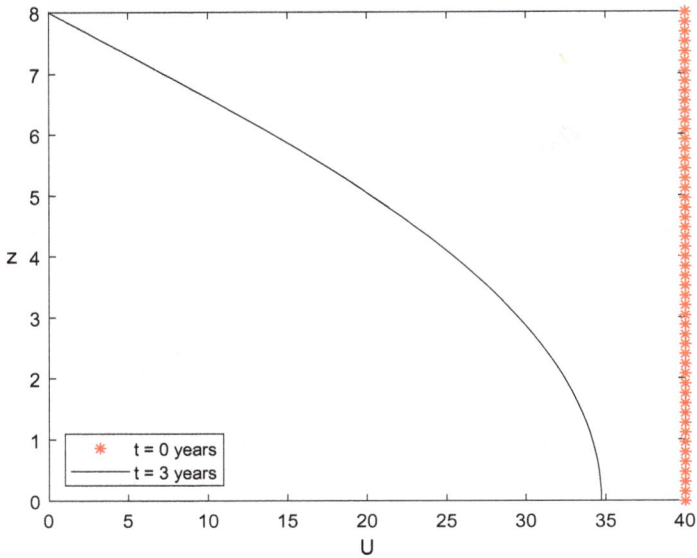

Figure 3.2.6.   Graph of the numerical solution to Problem (3.2.18) to (3.2.20).

```
% Method of Lines
[~,y] = ode45(@system, [0 T ], u(1:end-1), [ ], A, a);
u(1:n) = y(end,:);
u(n+1) = 0;
plot(u,z,'k');
xlabel('u'); ylabel('z');
year = 365*24*3600;
legend('t = 0 year',['t = ',num2str(T/year),' years'],'Location','SouthWest');
hold off;
end
%——— Local function ———
function Du = system(~, u, B, b)
Du = B*u + [0; b; 0];
end
```

Other applications are suggested in Exercises 3.4.19 and 3.4.20.

3.2.2 *Nonlinear Equations*

Consider the equation

$$U_t = \alpha U_{xx} + F(U, U_x), \tag{3.2.21}$$

where F can depend nonlinearly on U and U_x. For example, if $F = -UU_x$, Eq. (3.2.21) simplifies to the Burgers[2] equation

$$U_t + UU_x = \alpha U_{xx}, \tag{3.2.22}$$

and when $F = U_x^2/(1 + U)$, Eq. (3.2.21) simplifies to

$$U_t = U_x^2/(1 + U) + \alpha U_{xx}. \tag{3.2.23}$$

Let us apply the Method of Lines to Eq. (3.2.21). Using the central approximations for the space derivatives, we get the following system of ordinary differential equations

$$\dot{u}_i = p(u_{i+1} - 2u_i + u_{i-1}) + F(u_i, (u_{i+1} - u_{i-1})/(2\Delta x)), \tag{3.2.24}$$

where $p = \alpha/(\Delta x)^2$.

For the Burgers Eq. (3.2.22), System (3.2.24) is written as

$$\dot{u}_i = p(u_{i+1} - 2u_i + u_{i-1}) - qu_i(u_{i+1} - u_{i-1}), \quad i = 1, \ldots, n - 1,$$

where $q = 1/2/\Delta x$. More explicitly,

$$\begin{cases} \dot{u}_1 = p(u_2 - 2u_1 + u_0) - qu_1(u_2 - u_0), \\ \dot{u}_i = p(u_{i+1} - 2u_i + u_{i-1}) - qu_i(u_{i+1} - u_{i-1}), \\ \quad i = 2, \ldots, n - 2, \\ \dot{u}_{n-1} = p(u_n - 2u_{n-1} + u_{n-2}) - qu_{n-1}(u_n - u_{n-2}). \end{cases} \tag{3.2.25}$$

Consider the Dirichlet problem with the boundary conditions

$$U(0, t) = g_1(t), \quad U(L, t) = g_2(t). \tag{3.2.26}$$

In this case, Eqs. (3.2.25) are a system of $n - 1$ equations with the $n - 1$ unknown functions $u_1, \ldots, u_{n-1}$, since the functions u_0, u_n are known and

---

[2]Johannes Martinus Burgers, a Dutch scientist, 1895–1981. He explored the equation that bears his name. The Burgers equation occurs in gas and fluid mechanics, and traffic flow.

given by

$$u_0(t) = g_1(t), \quad u_n(t) = g_2(t).$$

Example 3.2.4   Consider the Burgers Eq. (3.2.22) with the special initial-boundary conditions

$$U(x, 0) = 2\alpha[1 - \tanh(x)], \quad 0 \le x \le L, \tag{3.2.27}$$

$$\begin{cases} U(0, t) = 2\alpha[1 - \tanh(-2\alpha t)], \\ U(L, t) = 2\alpha[1 - \tanh(L - 2\alpha t)], \end{cases} \quad 0 < t < T. \tag{3.2.28}$$

A function that applies the Method of Lines to solve Problem (3.2.27 and 3.2.28) is presented below. Since the index 0 cannot be used in Matlab, all indices in System (3.2.25) will be rescaled, $u_0 \to u_1$, $u_n \to u_{n+1}$. Next, the following notations

$$w(1) = u(2), ..., w(n - 1) = u(n), \ g1 = u(1), \ g2 = u(n + 1),$$

will be introduced. Considering this, System (3.2.25) is written in Matlab notation as

$$Dw(1) = p * (w(2) - 2 * w(1) + g1(t)) - q * w(1) * (w(2) - g1(t)),$$

$$Dw(2 : n - 2) = p * (w(3 : n - 1) - 2 * w(2 : n - 2) + w(1 : n - 3))$$

$$-q * w(2 : n - 2). * (w(3 : n - 1) - w(1 : n - 3)),$$

$$Dw(n-1) = p*(w(n-2)-2*w(n-1)+g2(t))-q*w(n-1)*(g2(t)-w(n-2)).$$

The graph of the numerical solution provided by the Method of Lines is shown in Fig. 3.2.7, together with the exact solution: $U(x, t) = 2\alpha[1 - \tanh(x - 2\alpha t)]$.

```
function u = burgers
% This is the function file burgers.m.
% Method of Lines is applied to solve the following Dirichlet problem:
% Ut + U Ux = alpha Uxx,  U(x,0) = 2*alpha - 2*alpha*tanh(x)
% U(0,t) = 2*alpha - 2*alpha*tanh(-2*alpha*t),
% U(L,t) = 2*alpha - 2*alpha*tanh(L - 2*alpha*t),
% Analytical solution: U = 2*alpha - 2*alpha*tanh(x - 2*alpha*t).

% Initialization
L = 1; alpha = 1; T = 1; n = 20;
```

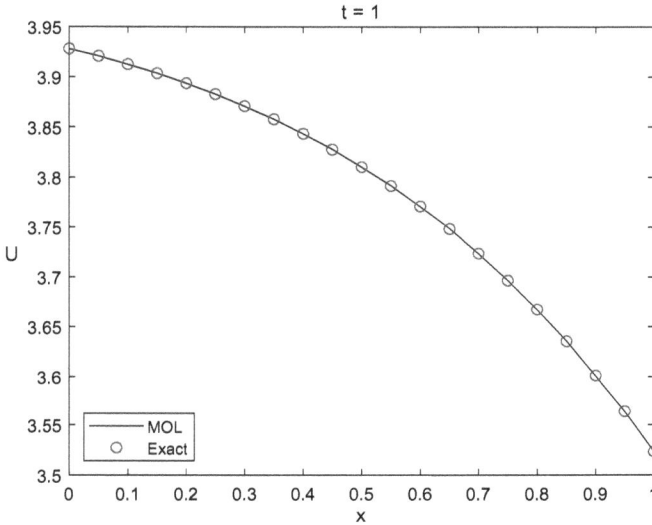

Figure 3.2.7. Graph of the numerical solution to Problem (3.2.27 and 3.2.28).

```
x = linspace(0,L,n+1);
u = 2*alpha - 2*alpha*tanh(x');

% Method of Lines
[~,y] = ode45(@system,[0 T],u(2:end-1),[ ],L,n,alpha);
u(2:n) = y(end,:);
u(1) = g1(T, alpha);
u(n+1) = g2(T, alpha, L);
U = 2*alpha - 2*alpha*tanh(x' - 2*alpha*T);
plot(x,u,'k',x,U,'ro');
title(['t = ',num2str(T)]); xlabel('x');  ylabel('U');
fprintf( 'Maximum error = %g\n',max(abs(U - u)))
legend('MOL','Exact','Location','SouthWest');
end

%——— Local functions ———
function f = g1(t, alpha)
f = 2*alpha - 2*alpha*tanh(-2*alpha*t);
end

function f = g2(t, alpha, L)
f = 2*alpha - 2*alpha*tanh(L-2*alpha*t);
```

end

```
function Dw = system(t, w, L, n, alpha)
dx = L/n; p = alpha/dx²; q = 1/2/dx;
Dw(1,1) = p*(w(2) - 2*w(1) + g1(t,alpha)) - q*w(1)*(w(2) - g1(t,alpha));
Dw(2:n-2,1) = p*(w(3:n-1) - 2*w(2:n-2) + w(1:n-3))...
            -q*w(2:n-2).*(w(3:n-1) - w(1:n-3));
Dw(n-1,1) = p*(w(n-2) - 2*w(n-1) + g2(t,alpha,L))...
            -q*w(n-1)*(g2(t,alpha,L) - w(n-2));
    % Note that w(1) = u(2),...,w(n-1) = u(n), g1 = u(1), g2 = u(n+1).
end
```

Consider the nonlinear Eq. (3.2.23). System (3.2.24) simplifies to

$$\dot{u}_i = p(u_{i+1} - 2u_i + u_{i-1}) + \frac{p(u_{i+1} - u_{i-1})^2}{4(1 + u_i)}, \quad i = 1, \ldots, n - 1, \quad (3.2.29)$$

and, more explicitly,

$$\begin{cases} \dot{u}_1 = p(u_2 - 2u_1 + u_0) + p(u_2 - u_0)^2/4(1 + u_1), \\ \dot{u}_i = p(u_{i+1} - 2u_i + u_{i-1}) + p(u_{i+1} - u_{i-1})^2/4(1 + u_i), \\ i = 2, \ldots, n - 2, \\ \dot{u}_{n-1} = p(u_n - 2u_{n-1} + u_{n-2}) + p(u_n - u_{n-2})^2/4(1 + u_{n-1}). \end{cases} \quad (3.2.30)$$

In the Dirichlet problem with boundary Conditions (3.2.26), the functions u_0 and u_n are known

$$u_0(t) = g_1(t), \quad u_n(t) = g_2(t),$$

and System (3.2.30) contains $n - 1$ unknown functions $u_1, \ldots, u_{n-1}$.

Example 3.2.5   Consider Eq. (3.2.23) with the following initial-boundary conditions

$$U(x, 0) = -1 + \sqrt{1 + 2x^2}, \quad 0 \le x \le L, \quad (3.2.31)$$

$$\begin{cases} U(0, t) = -1 + \sqrt{1 + 4t}, \\ U(L, t) = -1 + \sqrt{1 + 2(L^2 + 2t)}, \end{cases} \quad 0 < t < T. \quad (3.2.32)$$

A function that applies the Method of Lines to solve Problem (3.2.31 and 3.2.32) is presented below. Since the index 0 cannot be used in Matlab, all indices in System (3.2.30) will be rescaled: $u_0 \rightarrow u_1$, $u_n \rightarrow u_{n+1}$. Next, the following notations

$$w(1) = u(2), ..., w(n-1) = u(n), \quad g1 = u(1), \quad g2 = u(n+1),$$

will be introduced. Considering this, System (3.2.30) is written in Matlab notation as

$$Dw(1) = p * (w(2) - 2 * w(1) + g1(t)) + p/4 * (w(2) - g1(t))^2/(1 + w(1))$$

$$Dw(2:n-2, 1) = p * (w(3:n-1) - 2 * w(2:n-2) + w(1:n-3))$$

$$+p/4*(w(3:n-1) - w(1:n-3)).^2./(1 + w(2:n-2))$$

$$Dw(n-1) = p * (w(n-2) - 2 * w(n-1) + g2(t, L))$$

$$+p/4 * (g2(t) - w(n-2))^2/(1 + w(n-1)).$$

The graph of the numerical solution provided by the Method of Lines is shown in Fig. 3.2.8, together with the exact solution: $U(x, t) = -1 + \sqrt{1 + 2(x^2 + 2t)}$.

function u = nonlinear
% This is the function file nonlinear.m.

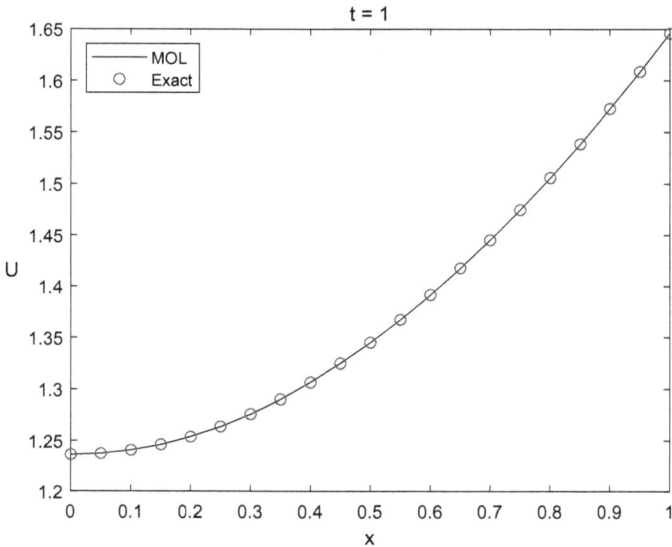

Figure 3.2.8. Graph of the numerical solution to Problem (3.2.31 and 3.2.32).

```
% Method of Lines is applied to solve the following Dirichlet problem:
% Ut = (Ux)²/(1 + U) + Uxx, U(x,0) = -1 + sqrt(1 + 2*x²),
% U(0,t) = -1 + sqrt(1 + 4*t), U(L,t) = -1 + sqrt(1 + 2*(L² + 2*t)).
% Analytical solution: U = -1 + sqrt(1 + 2*(x² + 2*t)),

% Initialization
L = 1; T = 1; n = 20;
x = linspace(0,L,n+1);
u = -1 + sqrt(1 + 2*x'.²);

% Method of Lines
[~,y] = ode15s(@system,[0 T],u(2:end-1),[ ],L,n);
% For stiff problems ode15s can work better than ode45
u(2:n) = y(end,:);
u(1) = g1(T); u(n+1) = g2(T, L);
U = -1 + sqrt(1 + 2*(x'.² + 2*T));
plot(x,u,'k',x,U,'ro');
title(['t = ',num2str(T)]); xlabel('x'); ylabel('U');
legend('MOL','Exact','Location','NorthWest');
fprintf( 'Maximum error = %g\n',max(abs(U - u)))
end

% ———— Local functions ————
function f = g1(t)
f = -1 + sqrt(1 + 4*t);
end
function f = g2(t, L)
f = -1 + sqrt(1 + 2*(L² + 2*t));
end
function Dw = system(t, w, L, n)
dx = L/n; p = 1/dx²;
Dw(1,1) = p*(w(2) - 2*w(1) + g1(t))+ p/4*(w(2) - g1(t))²/(1 + w(1));
Dw(2:n-2,1) = p*(w(3:n-1) - 2*w(2:n-2) + w(1:n-3))...
              + p/4*(w(3:n-1) - w(1:n-3)).²./(1 + w(2:n-2));
Dw(n-1,1) = p*(w(n-2) - 2*w(n-1) + g2(t,L))...
              + p/4*(g2(t, L) - w(n-2))²/(1 + w(n-1));
     % Note that w(1) = u(2),...,w(n-1) = u(n), g1 = u(1), g2 = u(n+1).
end
```

Another application is suggested in Exercise 3.4.21.

Finally, let us present a finite difference method for equation

$$U_t = \alpha U_{xx} + F(U, U_x). \tag{3.2.33}$$

Using the forward approximation for the time derivative and the central approximation for the space derivatives yields

$$u_i^{j+1} = (1 - 2r\alpha)u_i^j + r\alpha(u_{i+1}^j + u_{i-i}^j) + \Delta t F\left(u_i^j, \frac{u_{i+1}^j - u_{i-1}^j}{2\Delta x}\right), \tag{3.2.34}$$

where $r = \Delta t/(\Delta x)^2$. It can be proved that Method (3.2.34) is stable under the hypotheses

$$1 - M\Delta t - 2r\alpha \geq 0, \quad r\alpha - M_1 s/2 \geq 0,$$

where

$$M = \sup_{(x,t)} |F_u|, \quad M_1 = \sup_{(x,t)} |F_{u_x}|, \quad s = \Delta t/\Delta x.$$

3.2.3 *Variable Diffusivity Coefficient*

Consider the one-dimensional heat equation with variable diffusivity coefficient

$$U_t - \alpha(x,t)U_{xx} = 0, \quad 0 < x < L, \quad 0 < t \leq T, \tag{3.2.35}$$

where $\alpha(x,t)$ is a strictly positive function. Let us apply the Method of Lines to Eq. (3.2.35). We get the following system of ordinary differential equations

$$\dot{u}_i = p_i(u_{i+1} - 2u_i + u_{i-1}), \quad i = 1, \ldots, n-1, \tag{3.2.36}$$

where

$$u_i(t) = u(x_i, t), \quad \alpha_i(t) = \alpha(x_i, t), \quad p_i(t) = \alpha_i(t)/(\Delta x)^2, \quad i = 1, \ldots, n-1.$$

In matrix form, System (3.2.36) is written as

$$\begin{bmatrix} \dot{u}_1 \\ \cdot \\ \cdot \\ \dot{u}_{n-1} \end{bmatrix} = \begin{bmatrix} -2p_1 & p_1 & & \\ p_2 & -2p_2 & p_2 & \\ & & \cdot & \cdot & \cdot \\ & & & p_{n-1} & -2p_{n-1} \end{bmatrix} \begin{bmatrix} u_1 \\ \cdot \\ \cdot \\ u_{n-1} \end{bmatrix} + \begin{bmatrix} p_1 u_0 \\ 0 \\ \cdot \\ p_{n-1} u_n \end{bmatrix},$$

$$\dot{\mathbf{u}} = A(\mathbf{p})\mathbf{u} + \mathbf{a}(\mathbf{p}), \tag{3.2.37}$$

where the notations $A(\mathbf{p})$ and $\mathbf{a}(\mathbf{p})$ outline that A and $\mathbf{a}$ depend on $\mathbf{p} = (p_1, ..., p_{n-1})$.

Example 3.2.6   A function that applies the Method of Lines to solve the following Dirichlet problem

$$U_t - (1+t)U_{xx} = 0, \quad 0 < x < L, \quad 0 < t \le T, \tag{3.2.38}$$

$$U(x,0) = \sin(\pi x/L), \quad 0 \le x \le L, \tag{3.2.39}$$

$$U(0,t) = 0, \quad U(L,t) = 0, \quad t > 0, \tag{3.2.40}$$

is provided below. The graph of the numerical solution is shown in Fig. 3.2.9, together with the analytical solution $U(x,t) = \sin(\pi x/L) \exp(-(\pi/L)^2(t + t^2/2))$.

```
function u = variable1
% This is the function file variable1.m.
% Method of Lines is applied to solve the following Dirichlet problem:
% Ut = (1 + t) Uxx, U(x,0) = sin(pi*x/L), U(0,t) = 0, U(L,t) = 0.
% Analytical solution: U = sin(pi*x/L)*exp(-(pi/L)^2*(t+t^2/2)).
```

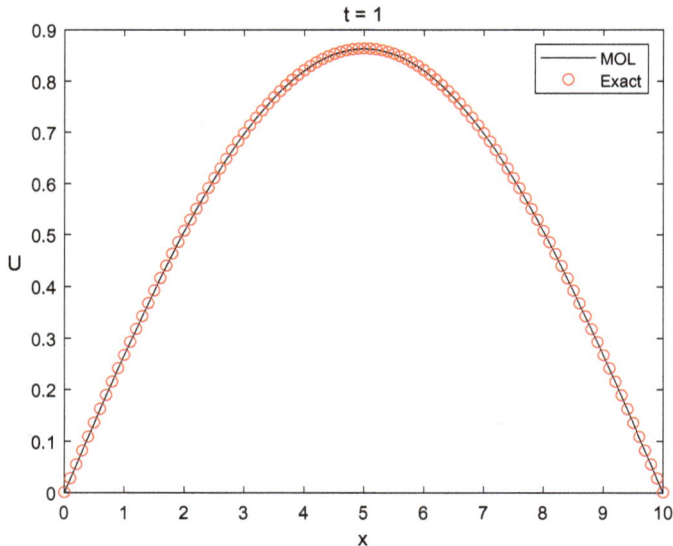

Figure 3.2.9.   Graph of the numerical solution to Problem (3.2.38 and 3.2.40).

```
% Initialization
L = 10; T = 1; n = 100;
x = linspace(0,L,n+1);
u = sin(pi*x'/L);
U = sin(pi*x'/L)*exp(-(pi/L)^2*(T+T^2/2));

% Method of Lines
[~,y] = ode45(@system,[0 T],u(2:end-1),[ ],L,n);
u(2:n) = y(end,:);
plot(x,u,'k',x,U,'ro');
xlabel('x'); ylabel('U');
title(['t = ',num2str(T)]); legend('MOL','Exact');
fprintf( 'Maximum error = %g\n',max(abs(U - u)))
end

% ———— Local function ————
function Du = system(t, u, L, n)
dx = L/n;
p = (1 + t)*ones(n+1,1)/dx^2;
AA = [p(3:n+1) -2*p(2:n) p(1:n-1)];
A = spdiags(AA, -1:1, n-1, n-1);
Du = A*u + [0; zeros(n-3,1); 0];
end
```

Another application is suggested in Exercise 3.4.22.

Lastly, a finite difference method for Eq. (3.2.35) is presented. The method is derived from the Explicit Euler Method and is expressed by

$$\frac{u_i^{j+1} - u_i^j}{\Delta t} = \alpha_i^j \frac{u_{i+1}^j - 2u_i^j + u_{i-1}^j}{\Delta x^2}. \tag{3.2.41}$$

where $\alpha_i^j = \alpha(x_i, t_j)$. Hence, with the usual position $r = \Delta t / \Delta x^2$, we have

$$u_i^{j+1} = (1 - 2r\alpha_i^j)u_i^j + r\alpha_i^j(u_{i+1}^j + u_{i-1}^j). \tag{3.2.42}$$

Method (3.2.42) is stable under the hypothesis

$$2r\alpha_i^j \leq 1, \quad \forall\, i, j. \tag{3.2.43}$$

See Exercises 3.4.23 and 3.4.24.

3.2.4 *Convection-Diffusion Equation*

Consider the convection-diffusion equation

$$U_t + vU_x - \alpha U_{xx} = 0, \quad 0 < x < L, \quad 0 < t \leq T, \tag{3.2.44}$$

discussed with the Upwind Method in Sec. 3.1.1. The Method of Lines for Eq. (3.2.44) leads to different expressions that depend on the approximation of U_x. If $v > 0$, the backward approximation is used and we get

$$\dot{u}_i = -\frac{v}{\Delta x}(u_i - u_{i-1}) + \frac{\alpha}{(\Delta x)^2}(u_{i+1} - 2u_i + u_{i-1}), \quad i = 1, \ldots, n-1. \tag{3.2.45}$$

If $v < 0$, using the forward approximation yields

$$\dot{u}_i = -\frac{v}{\Delta x}(u_{i+1} - u_i) + \frac{\alpha}{(\Delta x)^2}(u_{i+1} - 2u_i + u_{i-1}), \quad i = 1, \ldots, n-1. \tag{3.2.46}$$

From Formulas (3.2.45 and 3.2.46), it follows that

$$\dot{u}_i = \left(\frac{\alpha}{(\Delta x)^2} + \frac{v}{\Delta x}\right)u_{i-1} - \left(\frac{2\alpha}{(\Delta x)^2} + \frac{v}{\Delta x}\right)u_i + \frac{\alpha}{(\Delta x)^2}u_{i+1}, \quad i = 1, \ldots, n-1,$$

$$\dot{u}_i = \frac{\alpha}{(\Delta x)^2}u_{i-1} - \left(\frac{2\alpha}{(\Delta x)^2} - \frac{v}{\Delta x}\right)u_i + \left(\frac{\alpha}{(\Delta x)^2} - \frac{v}{\Delta x}\right)u_{i+1}, \quad i = 1, \ldots, n-1.$$

The equations above can be combined into a single equation

$$\dot{u}_i = pu_{i-1} - (p+q)u_i + qu_{i+1}, \quad i = 1, \ldots, n-1, \tag{3.2.47}$$

where

$$p = \begin{cases} \alpha/(\Delta x)^2 + |v|/\Delta x & \text{if } v > 0, \\ \alpha/(\Delta x)^2 & \text{if } v < 0, \end{cases} \quad q = \begin{cases} \alpha/(\Delta x)^2 & \text{if } v > 0, \\ \alpha/(\Delta x)^2 + |v|/\Delta x & \text{if } v < 0. \end{cases}$$

Equation (3.2.47) is the upwind version of the Method of Lines for Eq. (3.2.44). It can be arranged in the following matrix form

$$\begin{bmatrix} \dot{u}_1 \\ \cdot \\ \cdot \\ \dot{u}_{n-1} \end{bmatrix} = \begin{bmatrix} -(p+q) & q & & \\ p & -(p+q) & q & \\ & \cdot & \cdot & \cdot \\ & & p & -(p+q) \end{bmatrix} \begin{bmatrix} u_1 \\ \cdot \\ \cdot \\ u_{n-1} \end{bmatrix} + \begin{bmatrix} pu_0 \\ 0 \\ \cdot \\ qu_n \end{bmatrix},$$

$$\dot{\mathbf{u}} = A\mathbf{u} + \mathbf{a}. \tag{3.2.48}$$

Of course, the central approximation for U_x can be used as well. In this case the Method of Lines for Eq. (3.2.44) leads to

$$\dot{u}_i = -\frac{v}{2\Delta x}(u_{i+1} - u_{i-1}) + \frac{\alpha}{(\Delta x)^2}(u_{i+1} - 2u_i + u_{i-1}), \quad i = 1, \ldots, n-1.$$

$$(3.2.49)$$

Example 3.2.7   Consider the Dirichlet problem

$$U_t + vU_x - \alpha U_{xx} = 0, \quad v > 0, \quad 0 < x < L, \quad 0 < t \le T, \qquad (3.2.50)$$

$$U(x, 0) = 0, \quad 0 \le x \le L, \qquad (3.2.51)$$

$$U(0, t) = \begin{cases} t/t_0 - (t/t_0)^2 & \text{if } t \le t_0, \\ 0 & \text{if } t > t_0, \end{cases} \quad U(L, t) = 0, \quad t > 0. \qquad (3.2.52)$$

A function that applies Method (3.2.48) to solve Problem (3.2.50 to 3.2.52) is presented below. The graph of the numerical solution is shown in Fig. 3.2.10.

```
function u = lines_c_d_1
% This is the function file lines_c_d_1.m.m.
% Method of Lines is applied to solve the special Dirichlet problem:
```

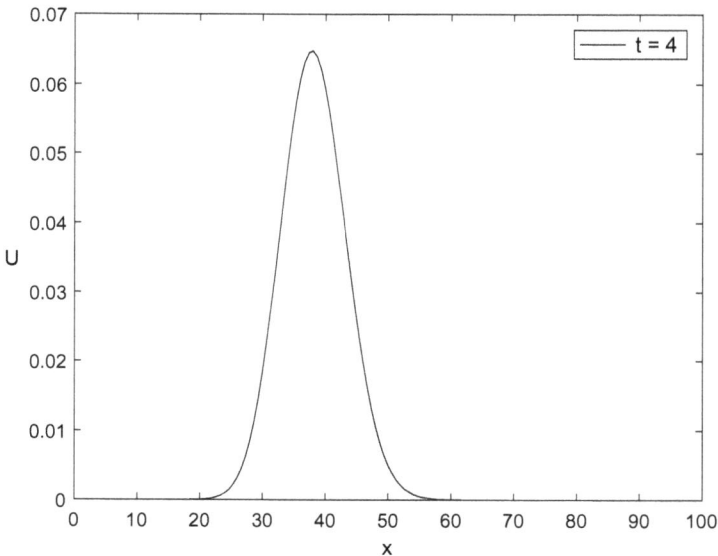

Figure 3.2.10.   Graph of the numerical solution to Problem (3.2.50 to 3.2.52).

```
% Ut +v*Ux = alpha*Uxx,  U(x,0) = 0, U(L,t) = 0.
% U(0,t) = t/t0 - t^2/t0^2, if t < t0,
% U(0,t) = 0, if t >= t0.

% Initialization
alpha = 0.01; v = 10; L = 100; T = 4; n = 150;
if v <= 0
    error('v must be positive')
end
dx = L/n; x = linspace(0,L,n+1);
t0 = .5;
g1 = @(t)        (t/t0 - t²/t0²).*(t < t0);
p = alpha/dx² + v/dx; q = alpha/dx²;
a = zeros(n-3,1);
AA = [p*ones(n-1,1)  -(p + q)*ones(n-1,1)  q*ones(n-1,1)];
A = spdiags(AA, -1:1, n-1,n-1);
u = zeros(n+1,1);

% Method of Lines
[~,y] = ode45(@system, [0 T], u(2:end-1), [ ], A, a, p, g1);
u(2:n) = y(end,:); u(1) = g1(T);
plot(x,u,'k'); xlabel('x'); ylabel('U');
legend(['t = ',num2str(T)]);
end

% ——— Local function ———
function Du = system(t, u, A, a, p, g1)
Du = A*u + [p*g1(t); a; 0];
end
```

Example 3.2.8 Consider the Dirichlet problem

$$U_t + vU_x - \alpha U_{xx} = 0, \quad 0 < x < L, \quad 0 < t \leq T, \tag{3.2.53}$$

$$U(x,0) = \begin{cases} 0 \text{ if } x \in [0, x_1[\,\cup\,]x_2, L], \\ k \text{ if } x \in [x_1, x_2], \end{cases} \tag{3.2.54}$$

$$U(0,t) = 0, \quad U(L,t) = 0, \quad t > 0. \tag{3.2.55}$$

A function that applies Method (3.2.48) to solve Problem (3.2.53 and 3.2.55) is presented below. The graph of the numerical solution is shown in Fig. 3.2.11.

```
function u = lines_c_d_2
```

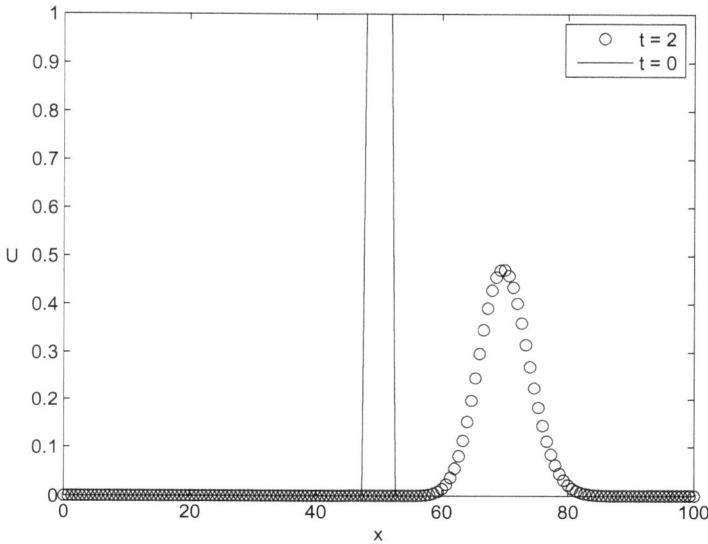

Figure 3.2.11. Graph of the numerical solution to Problem (3.2.53 to 3.2.55).

```
% This is the function file lines_c_d_2.m.m.
% Method of Lines is applied to solve the special Dirichlet problem:
% Ut +v Ux = alpha Uxx, U(0,t) = 0, U(L,t) = 0,
% U(x,0) = 0, if x < x1,
% U(x,0) = k, if x1 <= x <= x2,
% U(x,0) = 0, if x > x2.

% Initialization
alpha = 0.1; v = 10; L = 100; T = 2; n = 150;
dx = L/n; x = linspace(0,L,n+1);
if v >= 0
    p = alpha/dx^2 + v/dx; q = alpha/dx^2;
else
    p = alpha/dx^2; q = alpha/dx^2 - v/dx;
end
i1 = 73; i2 = 79; x1 = x(i1); x2 = x(i2);
k = 1;
phi = @(x)        k*(x >= x1).*(x <= x2);
U = feval(phi,x);
a = zeros(n-3,1);
```

```
AA = [p*ones(n-1,1)  -(p + q)*ones(n-1,1)  q*ones(n-1,1)];
A = spdiags(AA, -1:1, n-1,n-1);

% Method of Lines
[~,y] = ode45(@system, [0 T], U(2:end-1), [ ], A, a);
u(2:n) = y(end,:); u(1) = 0; u(n+1) = 0;
plot(x,u,'ko',x,U,'k','LineWidth',2);
xlabel('x'); ylabel('U');
legend(['t = ',num2str(T)],'t = 0');
end

% ——— Local function ———
function Du = system(~, u, A, a)
Du = A*u + [0; a; 0];
end
```

Other applications are suggested in Exercises 3.4.25 and 3.4.26.

Consider the *convection-diffusion equation with decay*

$$U_t + vU_x = \alpha U_{xx} - \lambda U, \qquad (3.2.56)$$

where λ is the decay coefficient. The change of unknown function

$$W = U \exp(\lambda t), \qquad (3.2.57)$$

simplifies Eq. (3.2.56) to the convection-diffusion equation

$$W_t + vW_x = \alpha W_{xx}. \qquad (3.2.58)$$

Consequently, the Method of Lines can be applied to Eq. (3.2.56) using Transformation (3.2.56), as suggested in Exercise 3.4.27. Of course, the Method of Lines can be applied to Eq. (3.2.56) and gives

$$\dot{u}_i = pu_{i-1} - (p + q + \lambda)u_i + qu_{i+1}, \quad i = 1, \ldots, n - 1. \qquad (3.2.59)$$

Finally, note that the result (3.2.58) holds even when $\lambda = \lambda(t)$, as shown in Exercise 3.4.28.

3.3 Saving Data and Figures

3.3.1 Save *Function*

Variables created in the Command Window can be saved with the **save** function. The simplest syntax is the following

```
save name_of_file var1 var2 ....
```

For example, after creating the variables

    a = 1; M = [1 2; 3 4]; s = 'string';

the command

    save sv_1 a M s

creates the sv_1.mat file in the current directory and saves the variables a, M. s in that file. As noted, the extension mat can be omitted. Variables saved in a file can be inspected with the command

    whos -file name_of_file.

For example, after saving a, M. s in the sv_1 file, the command

    whos -file sv_1

produces

Name	Size	Bytes	Class	Attributes
M	2x2	32	double	
a	1x1	8	double	
s	1x6	12	char	

If the file already exists, then the **save** function overwrites the file and the previous contents are lost. See Exercise 3.4.29. If the user wants to save new variables in an existing file, then the command

    save -append name_of_file n_var1 n_var2 ...

must be used. In the command above, n_var1 n_var2 ... are the new variables that will be appended to those already saved. If some new variables are already in the file, their values are updated. See Exercise 3.4.30. The command

    save name_of_file

saves all variables of the Command Window in the name_of_file file. As noted, the command does not require the variables to be specified. The command above can be shortened to

    save

that saves all variables in the matlab.mat file. See Exercise 3.4.31.

The **save** function saves the variables in binary format. It is possible to save the variables in Ascii format by specifying that as follows

    save name_of_file -ascii.

For example, after creating the variable a = pi;, the command

    save sv_4.txt a -ascii

saves a in Ascii format using 8 digits. Note that the whos command does not work with these files. However, they can be opened with any text editor, or with the Matlab command

    type('sv_4.txt').

See Exercise 3.4.32. The save function can be called by using the typical function syntax as well

    save('name_of_file','var1','var2',...).

Therefore, commands equivalent to those introduced above are the following,

    save('name_of_file','n_var1','n_var2','-append'),

    save('name_of_file','var1','var2','-ascii').

See Exercise 3.4.33. The save function may be useful in applications of real interest. Since executing complex programs can last days, it is important to save final and intermediate results.

3.3.2 Load *Function*

The load function performs the reverse operations of save. It imports the variables saved by the save function in the Command Window. The syntax to call the load function is analogous to that of save

    load name_of_file var1 var2 ....

Equivalently, the load function can be called using the typical function syntax

    load('name_of_file', 'var1', 'var2', ...).

If the load function loads a variable already existing in the Command Window, the value of such a variable is updated and the variable takes the value that it has in the loaded file. For example, create the variable x = pi;. Next, load the variable x from sv_1.mat file

    load sv_1 x.

Note that it is now

    x = 15.

If the variable does not exist in the file indicated in the load function, an error message is sent by Matlab. See Exercise 3.4.34. The following command

    load name_of_file

imports all variables of the name_of_file file in the Command Window. As noted, the variable name was omitted. The above commands work only with binary files, i.e., those with mat extension.

3.3.3 *Saving Figures*

The simplest way to save a figure is to select File/Save as ... in the menu of the figure. Next, specify the name, directory and extension (fig, eps, jpg, pdf, tif, ...). For example, let us produce a figure with the command

   fplot(@sin, [-pi pi])

and save it as figure_1.fig in the current directory. The saved figure can be opened by selecting Open in the Matlab menu and looking for the file. Alternatively, the following command can be used

   open('C:\... \current_directory\figure_1.fig').

The figures produced by Matlab can be easily copied to other programs. Select

   File/Preferences

in the figure menu and follow the indications.

3.4 Exercises

Exercise 3.4.1 Derive the non-dimensional form of Eq. (3.1.1).

Answer. Consider the following change of variables

$$\xi = x/L, \qquad x = L\xi,$$
$$\tau = \alpha t/L^2, \qquad t = L^2\tau/\alpha, \qquad (3.4.1)$$

and notations

$$W(\xi, \tau) = U(x(\xi, \tau), t(\xi, \tau)) \quad \Leftrightarrow \quad U(x, t) = W(\xi(x, t), \tau(x, t)). \quad (3.4.2)$$

From (3.4.1 and 3.4.2), we get

$$U_x = W_\xi/L, \quad U_{xx} = W_{\xi\xi}/L^2, \quad U_t = W_\tau\alpha/L^2.$$

Substituting the formulas above into (3.1.1) leads to the desired equation

$$W_\tau + P\, W_\xi = W_{\xi\xi}, \quad (P = vL/\alpha).$$

Exercise 3.4.2   Derive the stability Condition (3.1.4) directly from (3.1.3) and stability Condition (3.1.6) directly from (3.1.5).

Exercise 3.4.3   Prove the unconditional stability of Methods (3.1.16) and (3.1.17).

Hint. Apply the Von Neumann criterium. Substituting $u_i^j = \xi^j e^{I\beta i \Delta x}$ in (3.1.16) yields

$$\xi\{1 + 2r\alpha + vs - 2r\alpha\cos(\beta\Delta x) - vs\exp(-I\beta\Delta x)\} = 1,$$

$$\xi\{1 + 2r\alpha + vs - 2r\alpha\cos(\beta\Delta x) - vs[\cos(\beta\Delta x) + I\sin(\beta\Delta x)]\} = 1,$$

$$|\xi|^2 = 1/\{[1 + 4r\alpha\sin^2(\beta\Delta x/2) + 2vs\sin^2(\beta\Delta x/2)]^2 + v^2 s^2 \sin^2(\beta\Delta x)\} < 1.$$

Prove that the same result holds for Method (3.1.17).

Exercise 3.4.4   Group Methods (3.1.16) and (3.1.17) in one formula.
Answer.

$$-pu_{i-1}^{j+1} + (1 + p + q)u_i^{j+1} - qu_{i+1}^{j+1} = u_i^j, \qquad (3.4.3)$$

where

$$p = \begin{cases} r\alpha + s|v| & \text{if } v \geq 0, \\ r\alpha & \text{if } v < 0, \end{cases} \qquad q = \begin{cases} r\alpha & \text{if } v \geq 0, \\ r\alpha + s|v| & \text{if } v < 0. \end{cases} \qquad (3.4.4)$$

Exercise 3.4.5   Prove the unconditional stability of Method (3.1.18).
Answer. Apply the Von Neumann criterium and obtain

$$|\xi|^2 = \frac{4[1 - (2r\alpha + vs)\sin^2(\beta\Delta x/2)]^2 + v^2 s^2 \sin^2(\beta\Delta x)}{4[1 + (2r\alpha + vs)\sin^2(\beta\Delta x/2)]^2 + v^2 s^2 \sin^2(\beta\Delta x)} < 1.$$

Exercise 3.4.6   Prove the unconditional stability of Method (3.1.19).

Exercise 3.4.7   Derive the stability Condition (3.1.30).

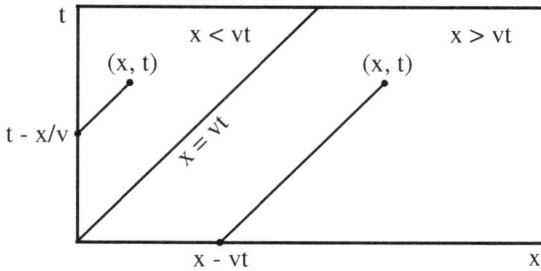

Figure 3.4.1. Regions where $x < vt$ and $x > vt$.

Exercise 3.4.8 Verify that the function

$$U(x,t) = \begin{cases} \varphi(x - vt) & \text{if } x \geq vt, \\ g(t - x/v) & \text{if } x < vt, \end{cases} \tag{3.4.5}$$

is the solution of Problem (3.1.31 to 3.1.33).

Hint. Consider Fig. 3.4.1.

Exercise 3.4.9 Consider the ftbs_ex1 function. Replace $T = 0.4$ with $T = 3.4$ and execute the listing. Explain what happens.

Hint. Consider the error.

Exercise 3.4.10 Consider the Dirichlet problem

$$U_t + vU_x = 0, \quad 0 < x < L, \quad 0 < t \leq T, \quad v > 0, \tag{3.4.6}$$

$$U(x,0) = \begin{cases} 0 & \text{if } x \in [0, x_1[\, \cup \,]x_2, L], \\ K & \text{if } x \in [x_1, x_2], \end{cases} \quad U(0,t) = 0, \quad 0 < t \leq T, \tag{3.4.7}$$

that has the following analytical solution

$$U = \begin{cases} 0 & \text{if } x \in [0, x_1 + vt[\, \cup \,]x_2 + vt, L], \\ K & \text{if } x \in [x_1 + vt, x_2 + vt]. \end{cases}$$

Write a listing, say ftbs_ex2, that applies the ftbs function to solve Problem (3.4.6 and 3.4.7).

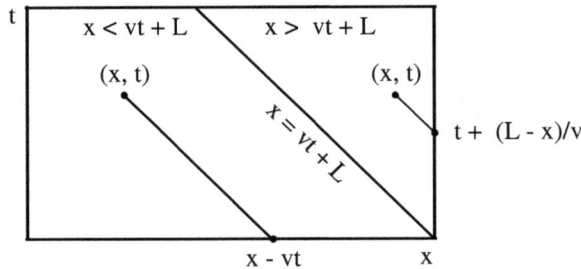

Figure 3.4.2.   Regions where $x < vt + L$ and $x > vt + L$.

Exercise 3.4.11   Verify that the function

$$U(x,t) = \begin{cases} \varphi(x - vt) & \text{if } x \leq vt + L, \\ g(t + (L - x)/v) & \text{if } x > vt + L, \end{cases} \tag{3.4.8}$$

is the solution of Problem (3.1.36 to 3.1.38).

Hint. Consider the Fig. 3.4.2.

Exercise 3.4.12   Write a listing, say ftbs_ex3, that calls the fbfs function and solves the following initial-boundary value problem

$$U_t + vU_x = \lambda U, \quad 0 < x < L, \quad 0 < t \leq T, \quad v > 0, \tag{3.4.9}$$

$$U(x,0) = \begin{cases} \sin(\pi x/x_1), & \text{if } 0 \leq x \leq 2x_1, \\ 0, & \text{if } 2x_1 < x \leq L, \end{cases} \quad U(0,t) = 0, \ 0 < t \leq T. \tag{3.4.10}$$

The analytical solution of Problem (3.4.9 and 3.4.10) is the function

$$U(x,t) = \begin{cases} 0, & \text{if } 0 \leq x \leq vt, \\ \exp(\lambda t)\sin(\pi(x - vt)/x_1), & \text{if } vt < x \leq vt + 2x_1, \\ 0, & \text{if } vt + 2x_1 < x \leq L. \end{cases}$$

Hint. The listing is partially provided below and the graph of the numerical solution is shown in Fig. 3.4.3.

```
function u = ftbs_ex3
v = .5; L = 1; T = 1; nx = 150;
x1 = 0.2; lambda = 1;
phi = @(x)        sin(pi*x/x1).*(2*x1 >= x);
g = @(t)          0*t;
```

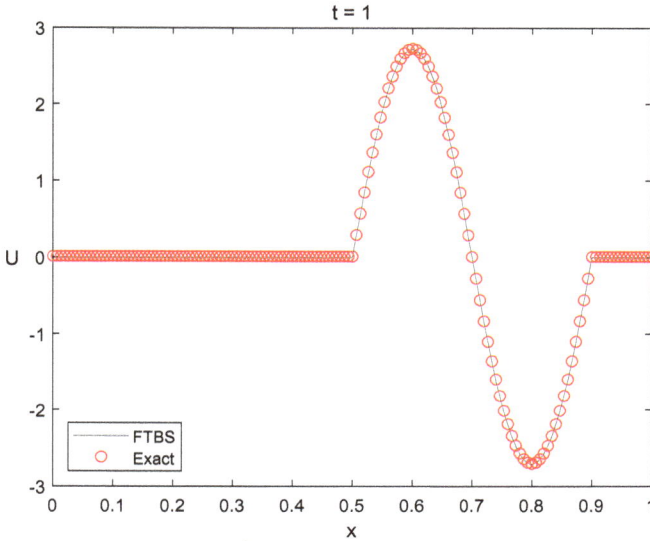

Figure 3.4.3. Graph of the numerical solution of Problem (3.4.9 and 3.4.10).

```
[u, nt] = ftbs(v, L, T, nx, phi, g);
x = linspace(0,L,nx+1); t = linspace(0,T,nt+1);
U(1:nx+1,1)=exp(lambda*T)*sin(pi*(x-
v*T)/x1).*(2*x1+v*T>=x).*(x>=v*T);
for j=1:nt+1
    u(:,j) = u(:,j).*exp(lambda*t(j));
end
...
end
```

Exercise 3.4.13 Write a listing, say consolidation1_ee, that solves the consolidation Problem (3.2.9 to 3.2.11) by calling the euler_e function in Example 2.3.1.

Hint. See Example 2.3.2. A way to introduce the initial condition may be

  phi = @(z)   q*ones(1,nz+1);

Exercise 3.4.14 Write a listing, say consolidation1_ie, where the consolidation Problem the (3.2.9 to 3.2.11) is solved by calling the euler_i function in Example 2.3.13.

Hint. See Example 2.3.14. Another way to introduce the initial condition could be

phi = @(z)   0*z + q;

Exercise 3.4.15   Write a listing, say consolidation1_c, that solves consolidation Problem (3.2.9 to 3.2.11) by calling the crank function in Example 2.3.15.

Exercise 3.4.16   Write a listing, say lines_heat2, that applies the Method of Lines to solve the following problem

$$U_t - U_{xx} = 0, \quad 0 < x < L, \quad 0 < t \le T, \qquad (3.4.11)$$

$$U(x, 0) = x^2, \quad 0 \le x \le L, \qquad (3.4.12)$$

$$U(0, t) = 2t, \quad U(L, t) = 2t + L^2, \quad 0 < t \le T. \qquad (3.4.13)$$

Exercise 3.4.17   Write listings that solve Problem (3.4.11 to 3.4.13) by calling the euler_e, euler_i and crank functions.

Exercise 3.4.18   Consider the Dirichlet–Neumann Problem (3.2.17). Write the equations for the Method of Lines in matrix form similar to (3.2.16).

Exercise 3.4.19   Write a listing, say consolidation2_end, that solves the consolidation Problem (3.2.18 to 3.2.20) by using the Explicit Euler Method.

Exercise 3.4.20   Write a listing that solves the consolidation Problem (3.2.18 to 3.2.20) by using the Implicit Euler Method.

Exercise 3.4.21   Verify that $U = -1 - \sqrt{1 + 2(x^2 + 2t)}$ is a solution of Eq. (3.2.23). Write a Dirichlet problem for (3.2.23) that has U as a solution. Then, provide a function that applies the Method of Lines to solve the Dirichlet problem above.

Exercise 3.4.22   Write a function that applies the Method of Lines to solve the Dirichlet problem

$$U_t - (1 + 2t)U_{xx} = 0, \quad 0 < x < L, \quad 0 < t \le T, \qquad (3.4.14)$$

$$U(x, 0) = x^2, \quad 0 \le x \le L, \qquad (3.4.15)$$

$$U(0, t) = 2t + 2t^2, \quad U(L, t) = L^2 + 2t + 2t^2, \quad t > 0, \qquad (3.4.16)$$

that has the following analytical solution

$$U(x, t) = x^2 + 2t + t^2.$$

Hint. Some code lines are provided below.

```
function u = variable2
...
g1 = @(t)              2*t + t²;
g2 = @(t)              2*t + t² + L²;
% Method of Lines
[~,y] = ode15s(@system,[0 T],u(2:end-1),[ ], L, n, g1, g2);
...
end
% ———— Local function ————
function Du = system(t, u, L, n, g1, g2)
    dx = L/n;
    p = (1 + t)*ones(n+1,1)/dx²;
    AA = [p(3:n+1) -2*p(2:n) p(1:n-1)];
    A = spdiags(AA, -1:1, n-1, n-1);
    Du = A*u + [p(2)*g1(t); zeros(n-3,1); p(n)*g2(t)];
end
```

Exercise 3.4.23 Prove the stability Condition (3.2.43).

Exercise 3.4.24 Write a function that applies Method (3.2.42) to solve Problem (3.2.38 to 3.2.40).

Hint. Note that the stability Condition (3.2.43) is the satisfied if $2rM \leq 1$, where M is the maximum of the function $\alpha(x, t)$.

Exercise 3.4.25 Consider the Dirichlet problem

$$U_t + vU_x - \alpha U_{xx} = 0, \quad v < 0, \quad 0 < x < L, \quad 0 < t \leq T, \qquad (3.4.17)$$

$$U(x, 0) = 0, \quad 0 \leq x \leq L, \qquad (3.4.18)$$

$$U(0, t) = 0, \; U(L, t) = \begin{cases} t/t_0 - (t/t_0)^2 & \text{if } t \leq t_0, \\ 0 & \text{if } t > t_0, \end{cases} \quad t > 0. \qquad (3.4.19)$$

Write a function that applies Method (3.2.48) to solve Problem (3.4.17 to 3.4.19).

Exercise 3.4.26   Consider the Dirichlet problem

$$U_t + vU_x - \alpha U_{xx} = 0, \quad 0 < x < L, \quad 0 < t \le T, \tag{3.4.20}$$

$$U(x,0) = x^2, \quad 0 \le x \le L, \tag{3.4.21}$$

$$U(0,t) = v^2 t^2 + 2t, \quad U(L,t) = (L - vt)^2 + 2t, \quad t > 0. \tag{3.4.22}$$

Write a function that applies Method (3.2.48) to solve the problem above. The analytical solution of Problem (3.4.20 to 3.4.22) is $U = (x - vt)^2 + 2t$.

Exercise 3.4.27   Write a function that applies the Method of Lines (3.2.48) to solve Eq. (3.2.56) using Transformation (3.2.57 and 3.2.58).

Exercise 3.4.28 Simplify Eq. (3.2.56), where $\lambda = \lambda(t)$, to the convection-diffusion equation.

Hint. Use the change of unknown function

$$W = U \exp\left(\int_0^t \lambda(\tau)d\tau\right),$$

and get

$$W_t = U_t \exp\left(\int_0^t \lambda(\tau)d\tau\right) + \lambda(t)U \exp\left(\int_0^t \lambda(\tau)d\tau\right),$$

$$W_x = U_x \exp\left(\int_0^t \lambda(\tau)d\tau\right), \quad W_{xx} = U_{xx} \exp\left(\int_0^t \lambda(\tau)d\tau\right).$$

Exercise 3.4.29   Create the variable x = 15;. Use the save sv_1 x command to save it in the already existing file. Use the whos -file sv_1 command to verify that sv_1 x file has been updated.

Exercise 3.4.30   Create the variable a = pi;. Save it with the save sv_2 a command. Create the vector v = [1 2 3]; and append it in the same file. Inspect the sv_2 file. Modify v into v = 123;. Append v in the sv_2 file and inspect the file again.

Exercise 3.4.31   Use the commands

save sv_3

save

and verify that the two files have the same content.

Exercise 3.4.32 Create the variable a = pi; and save it in the sv_4.txt file. Next, save the same variable with the command

save sv_5.txt a -ascii -double.

Open both files and note the difference.

Exercise 3.4.33 Try the commands

save('name_of_file','var1','var2',...)

save('name_of_file','n_var1','n_var2','-append').

Exercise 3.4.34 Use the load sv_1 z command to load the variable z that does not exist in the sv_1 file. Note the message

Warning: Variable 'z' not found.

Chapter 4

Introduction to the Finite Element Method

The Finite Element Method (FEM) is introduced in this chapter Hutton (2004). Scientific research on the FEM began after the Second World War and was partially supported by the American aerospace industry. The name appeared for the first time in the paper by Clough (1960). The Engineering community immediately understood the power of the new method. The structural engineers made a great contribution to its development. Expressions such as stiffness matrix, load vector, and others initially introduced for applications in Structural Mechanics, are used today for FEM applications in different scientific fields Fenner (2005). Matlab applications are presented in this book by Kwon and Bang (2000).

The concept of a weak solution to differential equations is closely related to the FEM. Examples of engineering applications that lead to consider weak solutions will be discussed. It will be shown that the FEM is able to provide satisfactory answers in these situations too.

The immediate section below is devoted to the numerical integration that often occurs in FEM applications.

4.1 Numerical Integration

Consider the function $f(x)$, defined and integrable on the interval (a, b). Recalling the integral additive property, the interval (a, b) is divided into n subintervals (x_i, x_{i+1}), $x_0 = a$, $x_n = b$, with equal $h = x_{i+1} - x_i$ spacing. The numerical integration is performed on each subinterval separately. The integral on (a, b) is obtained as the sum of all partial integrals.

The simplest method of numerical integration is the *Rectangle Rule*. See Fig. 4.1.1 (left). This method approximates the integral on (x_i, x_{i+1}) by replacing $f(x)$ with the zero-order polynomial, i.e., the constant function

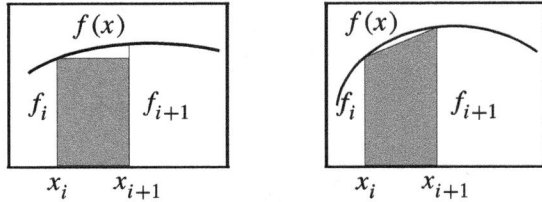

Figure 4.1.1.　Rectangle Rule (left) and Trapezium Rule (right).

$$p_0(x) = f(x_i) = f_i,$$

$$\int_{x_i}^{x_{i+1}} f(x) \, dx \approx \int_{x_i}^{x_{i+1}} p_0(x) \, dx = h f_i.$$

A similar method is *Trapezium Rule*. See Fig. 4.1.1 (right). The integral on (x_i, x_{i+1}) is approximated by replacing $f(x)$ with the first-order polynomial, i.e., the linear function $p_1(x) = (f_{i+1} - f_i)(x - x_i)/h + f_i$,

$$\int_{x_i}^{x_{i+1}} f(x) \, dx \approx \int_{x_i}^{x_{i+1}} p_1(x) \, dx = (f_{i+1} + f_i)h/2.$$

The two previous methods are rarely applied, as they provide unsatisfactory approximations of integrals.

A more accurate method is the *Simpson[1] Rule*. See Fig. 4.1.2. This method approximates the integral on (x_i, x_{i+2}) by replacing $f(x)$ with the second-order polynomial, i.e., the parabola $p_2(x)$ passing through the points (x_i, f_i), (x_{i+1}, f_{i+1}), (x_{i+2}, f_{i+2}),

$$p_2(x) = \frac{f_i + f_{i+2} - 2f_{i+1}}{2h^2}(x - x_{i+1})^2 + \frac{f_{i+2} - f_i}{2h}(x - x_{i+1}) + f_{i+1}. \quad (4.1.1)$$

See Exercise 4.4.1. Therefore, the integral of $f(x)$ on (x_i, x_{i+2}) is approximated with

$$\int_{x_i}^{x_{i+2}} f(x) \, dx \approx \int_{x_i}^{x_{i+2}} p_2(x) \, dx = \frac{f_i + f_{i+2} - 2f_{i+1}}{2h^2} \int_{x_i}^{x_{i+2}} (x - x_{i+1})^2 \, dx$$

$$+ \frac{f_{i+2} - f_i}{2h} \int_{x_i}^{x_{i+2}} (x - x_{i+1}) \, dx + 2h f_{i+1} = \frac{h}{3}(f_i + 4f_{i+1} + f_{i+2}). \quad (4.1.2)$$

The integral on (a, b) is obtained by repeatedly applying (4.1.2). Since a single application of the Simpson Rule needs two subintervals, it is clear

---

[1] Thomas Simpson, a British scientist, 1710–1761.

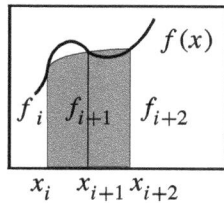

Figure 4.1.2. Simpson Rule.

that the total number of subintervals must be even. Lastly, the following formula is obtained for the integral of $f(x)$ on (a, b)

$$\int_a^b f(x)\ dx \approx (h/3)(f_0 + 4f_1 + 2f_2 + \cdots + 2f_{n-2} + 4f_{n-1} + f_n). \quad (4.1.3)$$

Let us present two Matlab functions able to calculate defined integrals: integral and int. The second function uses symbolic computation. The essential syntax for both is outlined below.

integral(f, a, b)	int(f(x), x, a, b)
f integrand function	f(x) integrand function
a, b integration limits	a, b integration limits
	x integration variable

Example 4.1.1 Consider the integrals

$$\int_a^b (px + q)\ dx, \quad \int_a^b \frac{x + 1}{\sqrt{x}}\ dx. \quad (4.1.4)$$

The two following listings apply the integral and int functions to calculate Integrals (4.1.4). In the first, anonymous functions are used to define integrand functions. Local functions are used in the second.

```
function integral_1
% This is the function file integral_1.m.
% Integral and int Matlab functions are applied to calculate two integrals.
% Anonymous functions are used to define integrand functions.
a = 1; b = 4; p =1 ; q = 2;
f1 = @(x)        p.*x+q;
f2 = @(x)        (1+x)./sqrt(x);
Q1 = integral(f1,a,b);
```

```
Q2 = integral(f2,a,b);
fprintf('integral_1 = %g; integral_2 = %g\n', Q1, Q2)
x = sym('x');% The int function needs symbolic variables.
E1 = int(f1(x),x,a,b);
E2 = int(f2(x),x,a,b);
fprintf('int_1 = %g; int_2 = %g\n', double(E1), double(E2))
end

function integral_2
% This is the function file integral_2.m.
% Integral and int Matlab functions are applied to calculate two integrals.
% Local functions are used to define integrand functions.
a = 1; b = 4; p =1 ; q = 2;
Q1 = integral(@(x)f1(x, p, q), a, b);
Q2 = integral(@(x)f2(x),a,b);
x = sym('x')
    E1 = int(f1(x, p, q), x, a, b);
    E2 = int(f2(x),x,a,b);
fprintf('integral_1 = %g; integral_2 = %g\n', Q1, Q2)
fprintf('int_1 = %g; int_2 = %g\n', double(E1), double(E2))
end

% ———— Local functions ————
function f = f1(x, p, q)
f = p.*x + q;
end
function f = f2(x)
f = (1+x)./sqrt(x);
end
```

Other applications are suggested in Exercise 4.4.2.

Example 4.1.2   Let us present a function that applies Simpson Rule to calculate integrals.

```
function integral = simpson(f, a, b, n, varargin)
% This is the function file simpson.m.
% Simpson Rule is applied to calculate integrals.
% integral = simpson(f, a, b, n, p1, p2,...)
% f is the integrand function,
% a, b are the integration limits,
```

```
% n is an even number,
% p1, p2,... are parameters different from the previous ones.
nn = n;% The input value of n is saved in nn.
w = 0;% The test variable w is initialized.
n = double(uint16(abs(n)));% n is modified to a positive integer, if it is not.
if n ~= nn
    w = 1;% if n ≠ nn, the value of w changes to 1.
end
if rem(n,2) ~= 0
    n = n+1;% n is modified to an even number, if it is not.
    w = 1;
end
if n == 0
    n = 20; % if n = 0, the value of n changes to 20.
    w = 1;
end
x = linspace(a,b,n+1); h = (b - a)/n; integral = 0;
for i = 1:2:n-1 % Simpson Rule.
    fi = feval(f, x(i), varargin{:});
    fip1 = feval(f, x(i+1), varargin{:});
    fip2 = feval(f, x(i+2), varargin{:});
    integral = integral + (fi + 4*fip1 + fip2)*h/3;
end
if w > 0 % This means that n is changed. The change must be sent to user.
    msg = 'n must be a positive even number; input n changed to:';
    msg = strcat(msg, num2str(n));
    warning(msg); % The message is sent to the User.
end
end
```

Example 4.1.3 The following listing applies the simpson function to calculate Integrals (4.1.4).

```
function integral_3
% This is the function file integral_3.m.
% Simpson function is called to calculate two integrals.
a = 1; b = 4; p = 1; q = 2; n = 18;
f1 = @(x)        p.*x+q;
f2 = @(x)        (1+x)./sqrt(x);
```

```
S1 = simpson(f1,a,b,n);
S2 = simpson(f2,a,b,n);
fprintf('simpson_1 = %g; simpson_2 = %g\n', S1, S2)
end
```

The next example considers the Bessel[2] function J_n with $n = 1$. The Bessel functions J_n are characterized by a sinusoidal behavior. Some of them, J_0, J_1 and J_2, are plotted in Fig. 4.1.3. The Matlab code that produces Fig. 4.1.3 is suggested in Exercise 4.4.3.

Example 4.1.4   Consider the integral function

$$g(x) = \int_0^{1-x} (x/z)\sin(t)J_1(z)\ dt, \quad z = \sqrt{(1-t)^2 - x^2}, \quad 0 \le x \le 1.$$

The following listing applies the integral and simpson functions to calculate $g(x)$. The graphs of $g(x)$ generated by the functions are plotted in Fig. 4.1.4. See Exercise 4.4.4 devoted to a limit that occurs in the listing.

```
function Q = integral_4
% This is the function file integrale4.m.
% Integral and simpson functions are called to calculate the integral function.
```

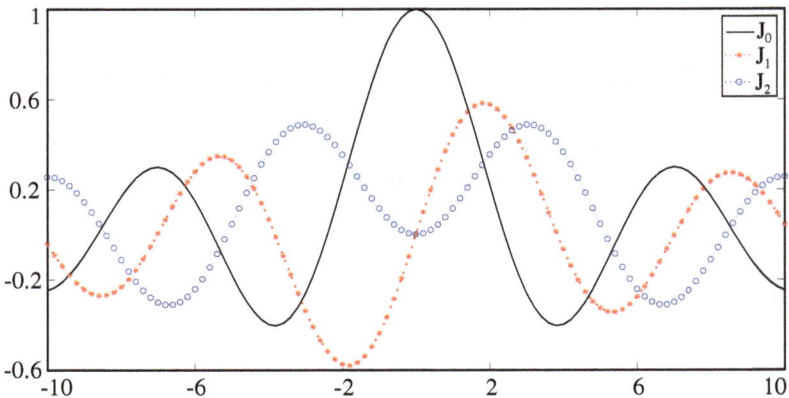

Figure 4.1.3.   Bessel functions.

---

[2] Friedrich Wilhelm Bessel, a German scientist, 1784–1846. He introduced the functions that bear his name, together with Daniel Bernoulli.

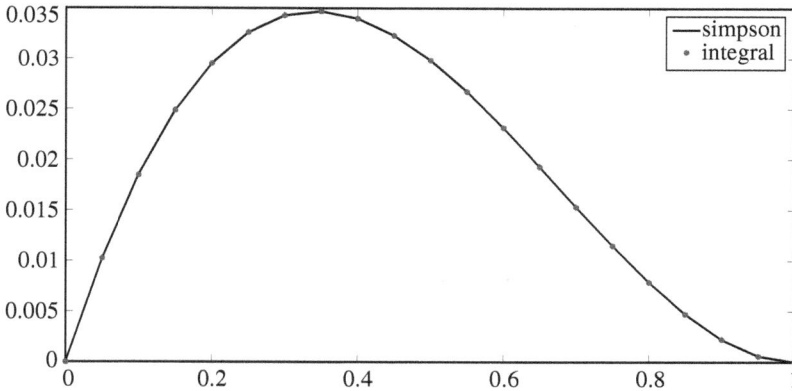

Figure 4.1.4. Graph of the integral function.

```
L = 1; n = 10; nx = 20;
x = linspace(0,L,nx+1); a = x(1);
S = zeros(nx+1,1); Q = zeros(nx+1,1);
for i=1:nx+1
    b =1-x(i);
    S(i) = simpson(@f, a, b, n, x(i));
    Q(i) = integral(@(t)f(t, x(i)), a, b);
end
plot(x,S,'k',x,Q,'r*','LineWidth',2);
legend('simpson','integral','Location','NorthEast');
end
% ———— Local function ————
function y = f(t, x)
z = sqrt((1-t).^2 - x.^2);
if z > 0
    y = sin(t).*besselj(1,z).*x./z;
else
    y = sin(t).*x/2; % besselj(1,z)/z -> .5 when z-> 0. See Exercise 4.4.4.
end
end
```

Consider the double integral

$$\int_a^b dx \int_{\alpha(x)}^{\beta(x)} f(x,y) \, dy. \tag{4.1.5}$$

Let us present two Matlab functions able to calculate Integral (4.1.5): integral2 and int. The syntax is outlined below.

---

integral2(f, a, b, α, β)　　　　int(int(f(x,y), y, α, β), x, a, b)

f integrand function　　　　f(x,y) integrand function

a, b integration limits on x

α, β integration limits on y

x, y integration variables

Example 4.1.5 Consider the integral

$$\int_a^b dx \int_{x^2/4}^{2\sqrt{x}} xy\ dy. \qquad (4.1.6)$$

The following listing applies the integral2 and int functions to calculate Integral (4.1.6).

```
function integral_2D_1
% This is the function file integral_2D_1.m.
% Integral2 and int functions are called to calculate the integral.
% Anonymous functions are used to define f(x, y), alfa(x), beta(x).
a = 0; b = 4;
alfa = @(x)      x.²/4;;
beta = @(x)      2*sqrt(x);
f = @(x, y)      x.*y;
Q = integral2(f, a, b, alfa, beta);
x = sym('x'); y = sym('y');
E = int(int(f(x,y), y, alfa, beta), x, a, b);
fprintf('integral2 = %g; int = %g\', Q, double(E))
end
```

Another application is suggested in Exercise 4.4.5.

Consider Integral (4.1.5). By setting

$$g(x) = \int_{\alpha(x)}^{\beta(x)} f(x,y)\ dy, \qquad (4.1.7)$$

Integral (4.1.5) simplifies to

$$\int_a^b dx \int_{\alpha(x)}^{\beta(x)} f(x,y)\ dy = \int_a^b g(x)\ dx. \qquad (4.1.8)$$

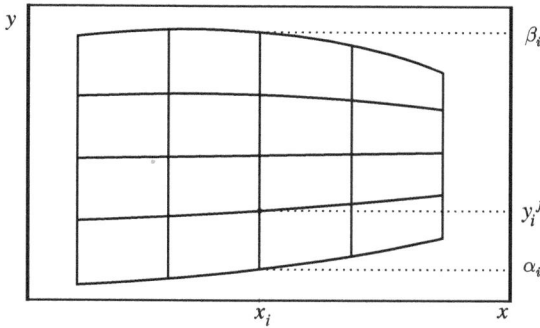

Figure 4.1.5. Mesh for 2D integrals.

Formula (4.1.8) is used to generalize the Simpson Rule to a 2D Integral (4.1.5). The interval (a, b) is divided into n subintervals (x_i, x_{i+1}), $x_0 = a$, $x_n = b$, with equal $h = x_{i+1} - x_i$ spacing. See Fig. 4.1.5. In addition, for any $i = 0, \ldots, n$, the interval $(\alpha_i, \beta_i) = (\alpha(x_i), \beta(x_i))$ is divided into m subintervals (y_i^j, y_i^{j+1}), $y_i^0 = \alpha_i$, $y_i^m = \beta_i$, with equal h_i spacing. Let us apply the Simpson Rule and recall that n and m must be even. By considering (4.1.8), Integral (4.1.5) is approximated by

$$\int_a^b dx \int_{\alpha(x)}^{\beta(x)} f(x, y) \, dy \approx \frac{h}{3}(g_0 + 4g_1 + 2g_2 + \cdots + 2g_{n-2} + 4g_{n-1} + g_n) \quad (4.1.9)$$

with g_i given by

$$g_i \approx (h_i/3)(f_i^0 + 4f_i^1 + 2f_i^2 + \cdots + 2f_i^{m-2} + 4f_i^{m-1} + f_i^m). \quad (4.1.10)$$

A function that generalizes the simpson function to a 2D case is illustrated in the following example.

Example 4.1.6 The following function applies Formulas (4.1.9)–(4.1.10) to solve Integral (4.1.5).

```
function integral = simpson2d(f, a, b, alfa, beta, n, m, varargin)
% This is the function file simpson2d.m.
% Simpson Rule is applied to calculate double integrals.
% integral = simpson2d(f, a, b, alfa, beta, n, m, p1, p2,...)
% f is the integrand function,
% a, b are the integration limits on x,
% alfa, beta are the integration limits on y,
```

```
% n, m are even numbers,
% p1, p2,... are parameters different from the previous ones.
nn = n; mm = m;
w = 0;
n = double(uint16(abs(n))); m = double(uint16(abs(m)));
if n ~= nn || m ~= mm
    w = 1;
end
if rem(n,2)~ = 0
    n = n + 1; w = 1;
end
if rem(m,2)~ = 0
    m = m + 1; w = 1;
end
if n == 0
    n = 10; w = 1;
end
if m == 0
    m = 10; w = 1;
end
x = linspace(a, b, n+1); h = (b-a)/n;
g = zeros(n+1,1);
for i = 1:n+1
    xx = x(i); c = feval(alfa,xx); d = beta(xx);
    yi = linspace(c, d, m+1); hi = (d-c)/m;
    for j = 1:2:m-1
        fi = feval(f, xx, yi(j), varargin{:});
        fip1 = feval(f, xx, yi(j+1), varargin{:});
        fip2 = feval(f, xx, yi(j+2), varargin{:});
        g(i) = g(i) + (fi + 4*fip1 + fip2)*hi/3;
    end
end
integral = 0;
for i = 1:2:n-1
    integral = integral + (g(i) + 4*g(i+1) + g(i+2))*h/3;
end
if w>0
    msg = 'n, m must be positive even numbers; n, m changed to:';
```

```
    msg = strcat(msg, num2str(n), ';', num2str(m));
    warning(msg);
  end
end
```

Example 4.1.7   The following listing calls the simpson2d function to calculate Integral (4.1.6).

```
function integral_2D_3
% This is the function file integral_2D_3.m.
% Simpson2d function is called to calculate the integral.
% Anonymous functions are used to define f(x, y), alfa(x), beta(x).
a = 0; b = 4; m = 10; n = 30;
alfa = @(x)        x.^2/4;
beta = @(x)        2*sqrt(x);
f = @(x,y)         x.*y;
S = simpson2d(f, a, b, alfa, beta, n, m);
fprintf(' simpson2d = %g\n', S)
end
```

Example 4.1.8   Consider the integral function

$$g(x) = \int_x^L d\xi \int_0^{x-\xi+L} f(\xi, t) J_0(z) \, dt, \quad 0 < x < L,$$

where

$$z = \sqrt{(L-t)^2 - (x-\xi)^2}, \quad f(\xi, \tau) = \begin{cases} \sin(t - \xi), & \text{if } \tau > \xi, \\ 0, & \text{if } \tau \le \xi. \end{cases}$$

The following listing calls the integral2 and simpson2d functions to calculate $g(x)$. The graphs of $g(x)$ returned by the functions are shown in Fig. 4.1.6.

```
function integral_2D_4
% This is the function file integral_2D_4.m.
% Simpson2d and integral2 functions are called to calculate
% the integral function.
L = 5; m = 20; n = 44;
nx = 20; x = linspace(0,L,nx+1);
alfa = @(xi)    0*xi;
S = zeros(nx+1,1); Q = zeros(nx+1,1);
```

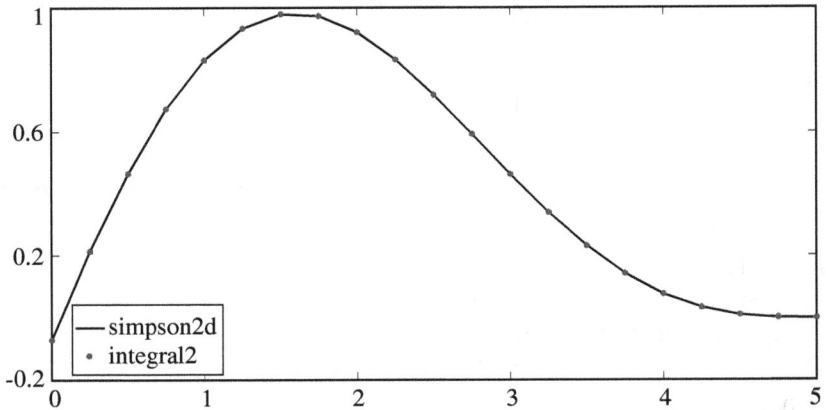

Figure 4.1.6.　Integral function.

```
for i=1:nx+1
    a = x(i);
    beta = @(xi)    a + L - xi;
    f = @(xi, t)    besselj(0,sqrt((L - t).^2 - (a - xi).^2)).*sin(t - xi).*(t - xi> 0);
    Q(i) = integral2(f, a, L, alfa, beta);
    S(i) = simpson2d(f, a, L, alfa, beta, n, m);
end
plot(x,S,'k',x,Q,'r*','LineWidth',2);
legend('simpson2d','integral2','Location','SouthWest');
end
```

4.2　Finite Element Method

4.2.1　*Axial Motion of a Bar*

Consider the axial motion of a prismatic bar having a straight longitudinal axis and constant cross-sectional area A, subjected to an axial distributed force per unit length $F(x,t)$. In Fig. 4.2.1, $U(x,t)$ is the displacement and $\sigma(x,t)$ is the stress. The equation for the axial motion of a bar is derived from Newton's Second Law

$$A \int_{x_1}^{x_2} \rho \frac{\partial^2 U}{\partial t^2} dx = A\sigma(x_2, t) - A\sigma(x_1, t) + \int_{x_1}^{x_2} F dx, \qquad (4.2.1)$$

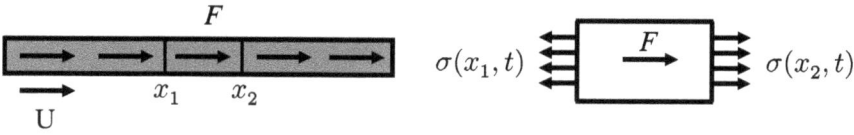

Figure 4.2.1. Axial motion.

where ρ is the density of the bar. From the previous equation, it follows that

$$A \int_{x_1}^{x_2} \rho \frac{\partial^2 U}{\partial t^2} dx = A \int_{x_1}^{x_2} \frac{\partial \sigma}{\partial x} + \int_{x_1}^{x_2} F dx.$$

Note that the previous equation holds for any control volume $V = A(x_2 - x_1)$. Therefore, it implies

$$A\rho \frac{\partial^2 U}{\partial t^2} = A \frac{\partial \sigma}{\partial x} + F. \tag{4.2.2}$$

If the bar is a linear-elastic material, then the relationship between stress σ and strain $\varepsilon = \partial U / \partial x$ is given by *Hooke's*[3] *law*

$$\sigma = E\varepsilon = E \frac{\partial U}{\partial x}, \tag{4.2.3}$$

where E is Young's[4] modulus. Substituting (4.2.3) into (4.2.2), one arrives at the equation for the axial motion of a bar

$$A\rho \frac{\partial^2 U}{\partial t^2} = A \frac{\partial}{\partial x} \left(E \frac{\partial U}{\partial x} \right) + F. \tag{4.2.4}$$

Equation (4.2.4) is named the *wave equation*. It is a hyperbolic partial differential equation. If E is constant, Eq. (4.2.4) simplifies to

$$A\rho \frac{\partial^2 U}{\partial t^2} - AE \frac{\partial^2 U}{\partial x^2} = F, \tag{4.2.5}$$

$$\frac{\partial^2 U}{\partial t^2} - c^2 \frac{\partial^2 U}{\partial x^2} = F/A/\rho,$$

where $c = \sqrt{E/\rho}$ is the velocity of wave propagation.

---

[3]Robert Hooke, a British scientist, 1635–1761. He introduced the law of elasticity that bears his name.
[4]Thomas Young, a British scientist, 1773–1829. He introduced the constant of elasticity that bears his name.

Suppose that the bar is subjected to axial concentrated forces too

$$\{(P_i, F_i), \quad i = 1, \ldots, N\}.$$

In this case Eq. (4.2.1) is modified to

$$A \int_{x_1}^{x_2} \rho \frac{\partial^2 U}{\partial t^2} dx = A\sigma(x_2, t) - A\sigma(x_1, t) + \int_{x_1}^{x_2} F dx + \sum_{i=n_1}^{n_2} F_i, \quad (4.2.6)$$

where F_i, $i = n_1, \ldots, n_2$, are the axial concentrated forces between x_1 and x_2. To understand what happens in this situation, let us consider an interval where there is only one concentrated force, say (x_h, F_h). Assuming, without loss of generality, that $x_1 = x_h - \Delta x$ and $x_2 = x_h + \Delta x$, Eq. (4.2.6) is written as

$$A \int_{x_h - \Delta x}^{x_h + \Delta x} \rho \frac{\partial^2 U}{\partial t^2} dx = A\sigma(x_h + \Delta x, t) - A\sigma(x_h - \Delta x, t) + \int_{x_h - \Delta x}^{x_h + \Delta x} F dx + F_h.$$

For $\Delta x \to 0$, the previous equation gives

$$A\sigma(x_h^+, t) - A\sigma(x_h^-, t) = -F_h, \quad (4.2.7)$$

i.e., the stress σ is discontinuous for $x = x_h$. Therefore, σ cannot be differentiated for $x = x_h$ and U cannot be a (classical) solution of Eq. (4.2.5).

In Statics Eq. (4.2.5) simplifies to

$$-AEU''(x) = F(x). \quad (4.2.8)$$

Appropriate boundary conditions are associated to Eq. (4.2.8). For example, for a bar with both ends fixed, they are (Fig. 4.2.2)

$$U(x_A) = 0, \quad U(x_B) = 0. \quad (4.2.9)$$

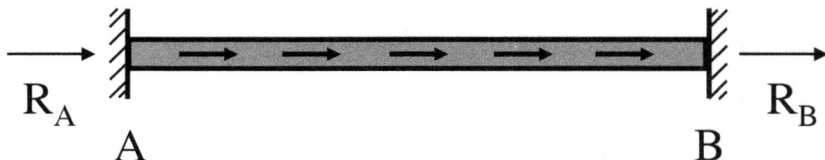

Figure 4.2.2.   Fixed end-fixed end bar.

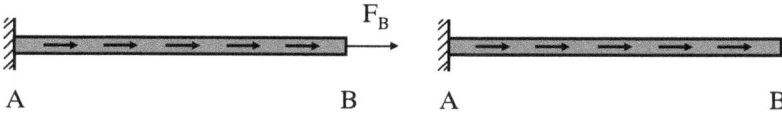

Figure 4.2.3. Fixed end-free end bar.

Moreover, the constraint reactions R_A and R_B are expressed as

$$R_A = -A\sigma(x_A) = -AEU'(x_A), \quad R_B = A\sigma(x_B) = AEU'(x_B). \quad (4.2.10)$$

The first relationship in (4.2.10) is derived by considering (4.2.1) on the control volume $V = A(x_2 - x_1)$ with $x_1 = x_A$ and $x_2 = x_A + \Delta x$

$$A\sigma(x_A + \Delta x) + R_A + \int_{x_A}^{x_A + \Delta x} F dx = 0.$$

When $\Delta x \to \quad 0$, the previous equation gives $(4.2.10)_1$. The second relationship in (4.2.10) is similarly derived.

Consider a bar with one end fixed and the other end free. If the free end is subjected to a known external force F_B, the boundary conditions are (Fig. 4.2.3, left)

$$U(x_A) = 0, \quad AEU'(x_B) = F_B. \quad (4.2.11)$$

The second boundary Condition follows from $(4.2.10)_2$. If $F_B = 0$, boundary Conditions (4.2.11) simplify to (Fig. 4.2.3, right)

$$U(x_A) = 0, \quad U'(x_B) = 0.$$

See Exercises 4.4.6–4.4.10.

4.2.2 Weak Solution

The equation governing the axial displacements of a bar will be assumed as an equation model for the introduction to the FEM. Therefore, consider the equation

$$-U''(x) = F(x), \quad 0 < x < L, \quad (4.2.12)$$

where the constant AE was included in F. Multiply Eq. (4.2.12) by a smooth function $v(x)$, named *test function*, and integrate over $[0, L]$

$$-\int_0^L U''(x)v(x) \, dx = \int_0^L F(x)v(x) \, dx. \quad (4.2.13)$$

Integrate by parts, the first integral and obtain

$$\int_0^L U'(x)v'(x)\ dx = \int_0^L F(x)v(x)\ dx + [U'v]_0^L. \tag{4.2.14}$$

Equation (4.2.14) is the *weak form* of differential Eq. (4.2.12). By contrast, the original form is named the *strong form*. It is important to note that the integration by parts has lowered the greatest order of differentiation, with the consequence that solutions to the weak form can now exist that do not have the necessary regularity to be solutions to the strong form. A solution to the weak form is named a *weak solution*. A solution to the strong form is named a *classical solution* or *strong solution*.

The FEM considers the weak form. The approximated solution u is expressed as a finite series

$$u(x) = \sum_{j=0}^n u_j \Phi_j(x), \tag{4.2.15}$$

where u_j are unknown coefficients and Φ_j are known functions. Imposing that u satisfies Eq. (4.2.14) with $v = \Phi_i$, $\forall \Phi_i$ yields

$$\sum_{j=0}^n u_j \int_0^L \Phi_j'(x)\Phi_i'(x)\ dx = \int_0^L F(x)\Phi_i(x)\ dx + [u'\Phi_i]_0^L, \quad \forall \Phi_i. \tag{4.2.16}$$

By introducing the definitions

$$K_{ij} = \int_0^L \Phi_i'(x)\Phi_j'(x)\ dx, \quad f_i = \int_0^L F(x)\Phi_i(x)\ dx + [u'\Phi_i]_0^L. \tag{4.2.17}$$

System (4.2.16) can be written in matrix notation

$$\sum_{j=0}^n K_{ij}u_j = f_i, \quad i = 0,\dots,n, \tag{4.2.18}$$

and, in compact form, as follows

$$K\mathbf{u} = \mathbf{f}, \tag{4.2.19}$$

where $\mathbf{u}$ and $\mathbf{f}$ are column vectors. The matrix K is named the *stiffness matrix* and the vector $\mathbf{f}$ *load vector*. The names come from Mechanics of Structures where the FEM was initially introduced.

Consider Eq. (4.2.13). If F is a force and v is a displacement, Eq. (4.2.13) is the *Principle of Virtual Work*. The approximating solution

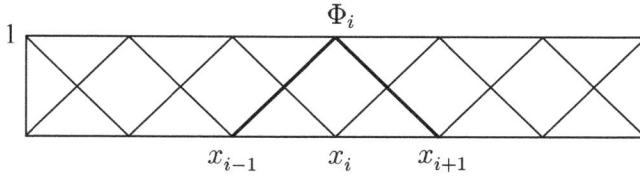

Figure 4.2.4.   Shape functions.

provided by the FEM is in agreement with the weak form of such a principle. The Principle of Virtual Work was introduced by Lagrange[5].

4.2.3   *Shape Functions*

The *shape functions* Φ_i are the linear functions defined by (see Fig. 4.2.4)

$$\Phi_i(x) = \begin{cases} (x - x_{i-1})/(x_i - x_{i-1}), & x_{i-1} \le x \le x_i, \\ (x_{i+1} - x)/(x_{i+1} - x_i), & x_i \le x \le x_{i+1}, \\ 0, & x \notin [x_{i-1}, x_{i+1}]. \end{cases} \tag{4.2.20}$$

For a regular grid with step $h = x_{i+1} - x_i$, Formula (4.2.20) simplifies to

$$\Phi_i(x) = \begin{cases} (x - x_{i-1})/h, & x_{i-1} \le x \le x_i, \\ (x_{i+1} - x)/h, & x_i \le x \le x_{i+1}, \\ 0, & x \notin [x_{i-1}, x_{i+1}]. \end{cases} \tag{4.2.21}$$

In addition, the derivative is

$$\Phi_i'(x) = \begin{cases} 1/h, & x_{i-1} < x < x_i, \\ -1/h, & x_i < x < x_{i+1}, \\ 0, & x \notin]x_{i-1}, x_{i+1}[. \end{cases} \tag{4.2.22}$$

From Definition (4.2.20), it follows that

$$\Phi_i(x_j) = \delta_{ij} = \begin{cases} 1, & i = j, \\ 0, & i \ne j, \end{cases} \tag{4.2.23}$$

where δ_{ij} is a *Kronecker*[6] δ *function*. Property (4.2.23) characterizes the shape functions. Indeed, if it is assumed that Φ_i is linear

---

[5]Giuseppe Luigi Lagrange, an Italian scientist, 1736–1813. He studied at the University of Turin. He was Professor at the University of Paris and the University of Berlin. He made significant contributions in Mechanics and the Calculus of Variations. He published *Mécanique Analitique*, the most comprehensive treatise since Newton's on Mechanics.
[6]Leopold Kronecker, a German scientist, 1823–1891. He made scientific contributions in Algebra and Logic.

$$\Phi_i(x) = a_i x + b_i, \tag{4.2.24}$$

and has Property (4.2.23), then Φ_i is given by (4.2.20). To prove the statement above, use Property (4.2.23) in (4.2.24) and obtain

$$\begin{aligned} \Phi_i(x_i) = 1, \quad a_i x_i + b_i = 1, \\ \Phi_i(x_{i+1}) = 0, \ a_i x_{i+1} + b_i = 0, \end{aligned} \quad x_i \le x \le x_{i+1}.$$

Solving with respect to a_i, b_i and the result substituted into (4.2.24) yields

$$\Phi_i(x) = -\frac{x}{x_{i+1} - x_i} + \frac{x_{i+1}}{x_{i+1} - x_i} = \frac{x_{i+1} - x}{x_{i+1} - x_i}, \quad x_i \le x \le x_{i+1},$$

that is the first part of the desired result. Similarly, from

$$\begin{aligned} \Phi_i(x_i) = 1, \quad a_i x_i + b_i = 1, \\ \Phi_i(x_{i-1}) = 0, \ a_i x_{i-1} + b_i = 0, \end{aligned} \quad x_{i-1} \le x \le x_i,$$

it follows that

$$\Phi_i(x) = \frac{x - x_{i-1}}{x_i - x_{i-1}}, \quad x_{i-1} \le x \le x_i.$$

The interval $[x_i, x_{i+1}]$ is named *element* and the point x_i is named *node* in the FEM context. The shape functions are a basis for the linear functions defined on the element (Fig. 4.2.5). Indeed, consider the linear function

$$u(x) = ax + b, \quad x \in e_i = [x_i, x_{i+1}], \tag{4.2.25}$$

and let $u_i = u(x_i)$ be the values of u in the nodes x_i. The coefficients a and b can be expressed in terms of u_i and u_{i+1} by considering the system

$$\begin{cases} ax_i + b = u_i, \\ ax_{i+1} + b = u_{i+1}. \end{cases}$$

Solving with respect to a and b and the result substituted into (4.2.25) yields

$$u(x) = \frac{u_{i+1} - u_i}{x_{i+1} - x_i} x + \frac{u_i x_{i+1} - u_{i+1} x_i}{x_{i+1} - x_i} = \frac{x - x_i}{x_{i+1} - x_i} u_{i+1} + \frac{x_{i+1} - x}{x_{i+1} - x_i} u_i,$$

$$u(x) = u_i \Phi_i(x) + u_{i+1} \Phi_{i+1}(x), \quad x \in e_i = [x_i, x_{i+1}], \tag{4.2.26}$$

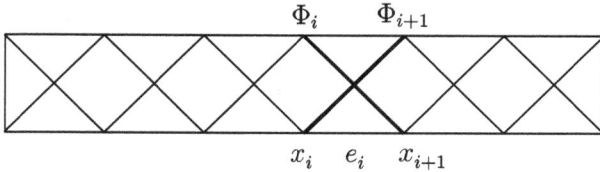

Figure 4.2.5. Shape functions on an element.

proving that u is a linear combination of Φ_i. It remains to show that

$$u(x) = 0, \ x \in e_i = [x_i, x_{i+1}], \ \Leftrightarrow \ u_i = u_{i+1} = 0. \qquad (4.2.27)$$

The condition $\Leftarrow$ is apparent. Prove $\Rightarrow$. Evaluate $u(x_i)$ from (4.2.26) and obtain $u(x_i) = u_i$. The hypothesis $u(x_i) = 0$ implies $u_i = 0$. Similarly, $u_{i+1} = 0$ is derived from $u(x_{i+1}) = 0$.

Let $u(x)$ be a continuous function defined on $[0, L]$. If the function is linear on each element $e_i = [x_i, x_{i+1}]$, the function is named *piecewise linear*. Consider the approximating solution in the FEM

$$u(x) = \sum_{j=0}^{n} u_j \Phi_j(x), \quad x \in [0, L], \qquad (4.2.28)$$

where Φ_j is the shape function. From the reasoning above, we immediately realize that the shape functions are a basis for the piecewise linear function on $[0, L]$. Therefore, the FEM approximates the exact solution with a piecewise linear function.

4.2.4 *Boundary Value Problems*

Let us discuss the Dirichlet problem

$$-U'' = F(x), \quad 0 < x < L, \quad U(0) = U_0, \quad U(L) = U_L. \qquad (4.2.29)$$

Consider the weak form written in terms of the shape functions defined in the previous section

$$\int_0^L u'(x)\Phi_i'(x) \ dx = \int_0^L F(x)\Phi_i(x) \ dx + [u'\Phi_i]_0^L, \ i = 0, \dots, n. \qquad (4.2.30)$$

The approximating solution u is expressed as

$$u(x) = \sum_{j=0}^{n} u_j \Phi_j(x), \qquad (4.2.31)$$

with u_j unknown coefficients. Note that

$$u(0) = u_0, \quad u(L) = u_n,$$

because of Property (4.2.23) of the shape functions. Therefore, if it is assumed that

$$u_0 = U_0, \quad u_n = U_L,$$

the approximating solution satisfies the boundary conditions in (4.2.29) and can be written as

$$u(x) = \sum_{j=1}^{n-1} u_j \Phi_j(x) + U_0 \Phi_0(x) + U_L \Phi_n(x), \qquad (4.2.32)$$

where the last two terms are known. Substituting (4.2.32) into (4.2.30) yields

$$\sum_{j=1}^{n-1} u_j \int_0^L \Phi_i'(x)\Phi_j'(x) \, dx + U_0 \int_0^L \Phi_i'(x)\Phi_0'(x) \, dx + U_L \int_0^L \Phi_i'(x)\Phi_n'(x) \, dx$$

$$= \int_0^L F(x)\Phi_i(x) \, dx, \forall \Phi_i, \quad i = 1, \ldots, n-1. \qquad (4.2.33)$$

as

$$\Phi_i(0) = 0, \quad \Phi_i(L) = 0, \quad i = 1, \ldots, n-1.$$

Setting

$$K_{ij} = \int_0^L \Phi_i'(x)\Phi_j'(x) \, dx, \quad f_i = \int_0^L F(x)\Phi_i(x) \, dx, \quad i, j = 1, \ldots, n-1,$$

$$g_i = U_0 \int_0^L \Phi_i'(x)\Phi_0'(x) \, dx + U_L \int_0^L \Phi_i'(x)\Phi_n'(x) \, dx, \quad i = 1, \ldots, n-1,$$

System (4.2.33) is written as

$$\sum_{j=1}^{n-1} K_{ij} u_j = f_i - g_i, \quad i = 1, \ldots, n-1.$$

It is a linear algebraic system of $n - 1$ equations in the $n - 1$ unknowns $u_1, \ldots, u_{n-1}$. Let us evaluate the elements of the symmetric matrix $K_{i,j}$. Consider the main diagonal

$$K_{i,i} = \int_{x_{i-1}}^{x_{i+1}} (\Phi'_i)^2(x) \, dx = \int_{x_{i-1}}^{x_{i+1}} (1/h^2) \, dx = \frac{2}{h}, \quad i = 1, \ldots, n-1, \quad (4.2.34)$$

as the support of Φ_i is the interval $[x_{i-1}, x_{i+1}]$. Consider $K_{i,i+1}$

$$K_{i,i+1} = \int_{x_i}^{x_{i+1}} \Phi'_i(x) \Phi'_{i+1}(x) dx = -\int_{x_i}^{x_{i+1}} \frac{dx}{h^2} = -\frac{1}{h}, \quad i = 1, \ldots, n-2,$$

$$(4.2.35)$$

as the support of $\Phi_i \Phi_{i+1}$ is the interval $[x_i, x_{i+1}]$. The remaining elements are zero. For example, consider

$$K_{1,3} = \int_0^L \Phi'_1(x) \Phi'_3(x) \, dx.$$

The support of Φ_1 is the interval $[x_0, x_2]$, the support of Φ_3 is the interval $[x_2, x_4]$, their intersection is void and $K_{1,3} = 0$. Let us evaluate the elements of the vector g_i. Consider

$$g_1 = U_0 \int_0^L \Phi'_1(x) \Phi'_0(x) \, dx + U_L \int_0^L \Phi'_1(x) \Phi'_n(x) \, dx. \quad (4.2.36)$$

The supports of Φ_0, Φ_1 and Φ_n are the intervals $[x_0, x_1]$, $[x_0, x_2]$ and $[x_{n-1}, x_n]$, respectively. Hence, Formula (4.2.36) simplifies to

$$g_1 = U_0 \int_{x_0}^{x_1} \Phi'_0(x) \Phi'_1(x) \, dx = -U_0/h. \quad (4.2.37)$$

Consider

$$g_{n-1} = U_0 \int_0^L \Phi'_{n-1}(x) \Phi'_0(x) \, dx + U_L \int_0^L \Phi'_{n-1}(x) \Phi'_n(x) \, dx. \quad (4.2.38)$$

The supports of Φ_0, Φ_{n-1} and Φ_n are the intervals $[x_0, x_1]$, $[x_{n-2}, x_n]$ and $[x_{n-1}, x_n]$, respectively. Hence, Formula (4.2.38) simplifies to

$$g_{n-1} = U_L \int_{x_{n-1}}^{x_n} \Phi'_{n-1}(x) \Phi'_n(x) \, dx = -U_L/h. \quad (4.2.39)$$

In addition, $g_i = 0$, $i = 2, \ldots, n - 2$. For example, consider

$$g_2 = U_0 \int_0^L \Phi'_2(x) \Phi'_0(x) \, dx + U_L \int_0^L \Phi'_2(x) \Phi'_n(x) \, dx.$$

The supports of Φ_0, Φ_2 and Φ_n are the intervals $[x_0, x_1]$, $[x_1, x_3]$ and $[x_{n-1}, x_n]$, respectively. The integrals are zero and $g_2 = 0$. In conclusion, by considering (4.2.34)–(4.2.39), one arrives at the following algebraic system for the unknown coefficients u_j

$$\frac{1}{h}\begin{bmatrix} 2 & -1 & & \\ -1 & 2 & -1 & \\ & . & . & . \\ & & -1 & 2 \end{bmatrix}\begin{bmatrix} u_1 \\ u_2 \\ . \\ u_{n-1} \end{bmatrix} = \begin{bmatrix} f_1 \\ f_2 \\ . \\ f_{n-1} \end{bmatrix} + \frac{1}{h}\begin{bmatrix} U_0 \\ 0 \\ . \\ U_L \end{bmatrix}. \qquad (4.2.40)$$

Example 4.2.1   The following listing provides a function that applies the FEM to the Dirichlet Problem (4.2.29).

```
function u = fem_dd(L, u0, uL, F, n)
% This is the function file fem_dd.m.
% FEM is applied to solve the Dirichlet problem:
% -U"(x) = F(x), U(0) = U0, U(L) = UL.
% The input arguments are: length, boundary conditions, forcing term,
% number of elements. The function returns a vector with the solution.
% Initialization
h = L/n; x = linspace(0,L,n+1);
u = zeros(n+1,1); f = zeros(n-1,1);
Phil = @(xi, x1, x2)      (xi-x1)/(x2-x1);
Phir = @(xi, x2, x3)      -(xi-x3)/(x3-x2);
% FEM
KK = [-ones(n-1,1)  2*ones(n-1,1)  -ones(n-1,1)];
K = spdiags(KK, -1:1, n-1, n-1)/h;
for i = 2:n
    f(i-1) = integral(@(xi)F(xi).*Phil(xi,x(i-1),x(i)), x(i-1), x(i));
    f(i-1) = f(i-1) + integral(@(xi)F(xi).*Phir(xi,x(i),x(i+1)), x(i), x(i+1));
end
f(1) = f(1) + u0/h; f(n-1) = f(n-1) + uL/h;
u(2:end-1) = K\f;
u(1) = u0; u(n+1) = uL;
end
```

Example 4.2.2   Consider the Dirichlet problem

$$-U'' = c\sin(\omega x), \quad 0 < x < L, \quad U(0) = U_0, \quad U(L) = U_L. \qquad (4.2.41)$$

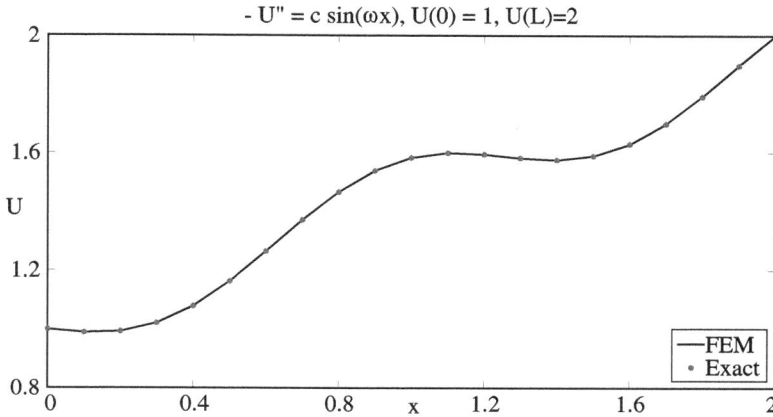

Figure 4.2.6. Graph of the approximating solution of Problem (4.2.41).

The following listing applies the fem_dd function to Problem (4.2.41). The graph of the approximating solution is shown in Fig. 4.2.6.

```
function u = fem_dd_ex1
% This is the function file fem_dd_ex1.m.
% fem_dd function is called to solve the special Dirichlet problem:
% -U''(x) = c * sin(om*x), U(0) = U0, U(L) = UL.
% Analytical solution:
% U = c*sin(om*x)/om^2 + x*(UL- U0 -c*sin(om*L)/om^2)/L +U0;
L = 2; n = 20; U0 = 1; UL = 2; om = 5; c = -3;
F = @(xi)        c*sin(xi*om);
u = fem_dd(L, U0, UL, F, n);
x = linspace(0,L,n+1);
U = c*sin(om*x')/om^2 + x'*(UL - U0 - c*sin(om*L)/om^2)/L + U0;
fprintf( 'Maximum error = %g\n',max(abs(U - u)))
plot(x,u,'k',x,U,'r*','LineWidth',2);
legend('FEM','Exact','Location','SouthEast');
xlabel('x'); ylabel('U');
title(['-U''= c sin(\omega x), U(0)=',num2str(U0),', U(L)=',num2str(UL)]);
end
```

Another application is suggested in Exercise 4.4.11.

Let us discuss the Dirichlet–Neumann problem

$$-U'' = F(x), \quad 0 < x < L, \quad U(0) = U_0, \quad U'(L) = U'_L. \tag{4.2.42}$$

Consider the weak form

$$\int_0^L u'(x)\Phi'_i(x)\,dx = \int_0^L F(x)\Phi_i(x)\,dx + U'_L\Phi_i(L) - U'(0)\Phi_i(0), \ i = 0, \dots, n. \tag{4.2.43}$$

The approximating solution u that satisfies the boundary condition $u(0) = U_0$ is given by

$$u(x) = \sum_{j=1}^n u_j\Phi_j(x) + U_0\Phi_0(x). \tag{4.2.44}$$

Substituting (4.2.44) into (4.2.43) yields

$$\sum_{j=1}^n u_j \int_0^L \Phi'_i(x)\Phi'_j(x)\,dx + U_0 \int_0^L \Phi'_i(x)\Phi'_0(x)\,dx = U'_L\Phi_i(L)$$

$$+ \int_0^L F(x)\Phi_i(x)\,dx, \quad \forall \Phi_i, \quad i = 1, \dots, n. \tag{4.2.45}$$

as

$$\Phi_i(0) = 0, \quad i = 1, \dots, n.$$

By introducing the definitions

$$K_{ij} = \int_0^L \Phi'_i(x)\Phi'_j(x)\,dx, \quad f_i = \int_0^L F(x)\Phi_i(x)\,dx, \quad i, j = 1, \dots, n,$$

$$g_i = U'_L\Phi_i(L) - U_0 \int_0^L \Phi'_i(x)\Phi'_0(x)\,dx, \quad i = 1, \dots, n,$$

System (4.2.45) is written as

$$\sum_{j=1}^n K_{ij}u_j = f_i + g_i, \quad i = 1, \dots, n, \ \Leftrightarrow \ K\mathbf{u} = \mathbf{f} + \mathbf{g}. \tag{4.2.46}$$

It is a linear algebraic system of n equations with n unknowns $u_1, \dots, u_n$. The elements of K different from $K_{n,n}$ were calculated in Formulas (4.2.34)–(4.2.35). Let us evaluate $K_{n,n}$

$$K_{n,n} = \int_{x_{n-1}}^{x_n} (\Phi'_n)^2(x)\,dx = \int_{x_{n-1}}^{x_n} (1/h^2)\,dx = 1/h, \tag{4.2.47}$$

as the support of Φ_n is the interval $[x_{n-1}, x_n]$. The elements of the vector **f** are given by

$$f_i = \int_{x_{i-1}}^{x_{i+1}} F(x)\Phi_i(x)\,dx, \quad i = 1, \ldots, n-1, \quad f_n = \int_{x_{n-1}}^{x_n} F(x)\Phi_n(x)\,dx,$$

(4.2.48)

as the supports of Φ_i, $i = 1, \ldots, n-1$, are the intervals $[x_{i-1}, x_{i+1}]$ and the support of Φ_n is the interval $[x_{n-1}, x_n = L]$. Let us evaluate the elements of the vector **g**. Consider g_1 and note that $\Phi_1(L) = 0$. Therefore,

$$g_1 = -U_0 \int_{x_0}^{x_1} \Phi_1'(x)\Phi_0'(x)\,dx = U_0/h.$$

Consider g_n. Note that $\Phi_n(L) = 1$ and the support of $\Phi_n'(x)\Phi_0'(x)$ is void. Therefore,

$$g_n = U_L'.$$

In addition,

$$g_i = 0, \quad i = 2, \ldots, n-1$$

as $\Phi_i(L) = 0$, $i = 2, \ldots, n-1$, and the supports of $\Phi_i'(x)\Phi_0'(x)$, $i = 2, \ldots, n-1$, are void. By considering the previous results, linear System (4.2.46) is specified in

$$\frac{1}{h}\begin{bmatrix} 2 & -1 & & & \\ -1 & 2 & -1 & & \\ & \cdot & \cdot & \cdot & \\ & & -1 & 2 & -1 \\ & & & -1 & 1 \end{bmatrix}\begin{bmatrix} u_1 \\ u_2 \\ \cdot \\ u_{n-1} \\ u_n \end{bmatrix} = \begin{bmatrix} f_1 \\ f_2 \\ \cdot \\ f_{n-1} \\ f_n \end{bmatrix} + \begin{bmatrix} U_0/h \\ 0 \\ \cdot \\ 0 \\ U_L' \end{bmatrix}. \quad (4.2.49)$$

Example 4.2.3 The following listing presents a function that applies the FEM to the Dirichlet–Neumann Problem (4.2.42).

```
function u = fem_dn(L, u0, uxL, F, n)
% This is the function file fem_dn.m.
% The FEM is applied to solve the Dirichlet–Neumann problem:
% -U"(x) = F(x), U(0) = U0, Ux(L) = UxL.
% The input arguments are: length, boundary conditions, forcing term,
% number of elements. The function returns a vector with the solution.
% Initialization
h = L/n; x = linspace(0,L,n+1);
```

```
u = zeros(n+1,1); f = zeros(n,1);
Phil = @(xi, x1, x2)          (xi-x1)/(x2-x1);
Phir = @(xi, x2, x3)          -(xi-x3)/(x3-x2);
% FEM
KK = [-ones(n,1)  [2*ones(n-1,1);1]  -ones(n,1)];
K = spdiags(KK, -1:1, n, n)/h;
for i = 2:n
    f(i-1) =integral(@(xi)F(xi).*Phil(xi,x(i-1),x(i)), x(i-1), x(i));
    f(i-1) = f(i-1) + integral(@(xi)F(xi).*Phir(xi,x(i),x(i+1)), x(i), x(i+1));
end
f(n) = integral(@(xi)F(xi).*Phil(xi,x(n),x(n+1)), x(n), x(n+1)) + uxL;
f(1) = f(1) + u0/h;
u(2:end) = K\f;
u(1) = u0;
end
```

Example 4.2.4   Consider the Dirichlet–Neumann problem

$$-U'' = cx^2, \quad 0 < x < L, \quad U(0) = U_0, \quad U'(L) = U'_L. \qquad (4.2.50)$$

The following listing applies the fem_dn function to Problem (4.2.50). The graph of the approximating solution is shown in Fig. 4.2.7.

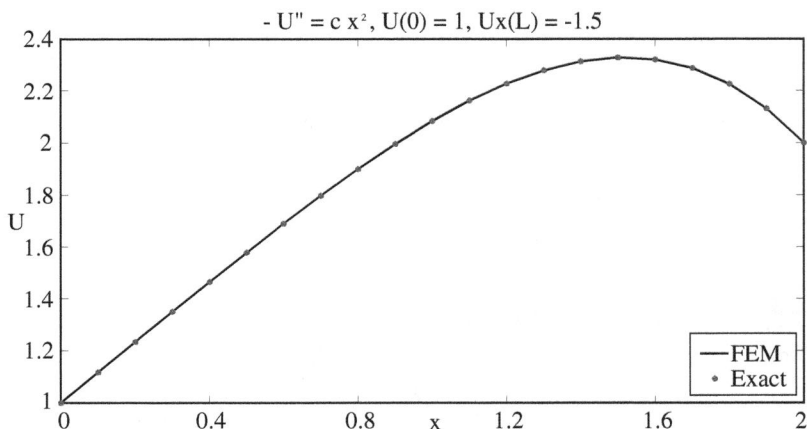

Figure 4.2.7.   Graph of the approximating solution of the Problem (4.2.50).

```
function u = fem_dn_ex1
% This is the function file fem_dn_ex1.m.
% fem_dn function is called to solve the special Dirichlet-Neumann problem:
% -U"(x) = c * x², U(0) = U0, Ux(L) = UxL.
% Analytical solution: U = -c*x.⁴/12 + (c*L³/3 + UxL)*x + U0;
L = 2; n = 20; U0 = 1; UxL = -1.5; c = 1;
F = @(xi)         c*xi.²;
u = fem_dn(L, U0, UxL, F, n);
x = linspace(0,L,n+1);
U = -c*x'.⁴/12 + (c*L³/3 + UxL)*x' + U0;
fprintf( 'Maximum error = %g\n',max(abs(U - u)))
plot(x,u,'k',x,U,'r*','LineWidth',2);
legend('FEM','Exact','Location','SouthEast');
xlabel('x'); ylabel('U');
title(['- U" = c x^2, U(0) = ',num2str(U0),', Ux(L) = ', num2str(UxL)]);
end
```

Another application is suggested in Exercise 4.4.12.

Consider the *pure Neumann problem*

$$-U'' = F(x), \quad 0 < x < L, \quad U'(0) = U_0', \quad U'(L) = U_L'. \qquad (4.2.51)$$

Integrate the differential equation over $(0, L)$ and use the boundary conditions

$$U_0' - U_L' = \int_0^L F(x)dx. \qquad (4.2.52)$$

The two constants U_0' and U_L', and the function $F(x)$ are known assigned data. Therefore, Relationship (4.2.52) is a compatibility condition for Problem (4.2.51). If such a condition is not satisfied from the data, the pure Neumann problem has no solution. In addition, suppose that Condition (4.2.52) is satisfied and let U_1 be a solution to Problem (4.2.51). Consider the function

$$U_2 = U_1 + c, \qquad (4.2.53)$$

where c is an constant. As the derivatives of U_2 are the same as those of U_1, function U_2 is the solution to Problem (4.2.51) as well. Let us summarize. Problem (4.2.51) has no solutions unless Condition (4.2.52) is satisfied. When the compatibility condition on data is satisfied, Problem (4.2.51) has infinite solutions in agreement with Formula (4.2.53). Both Formulas

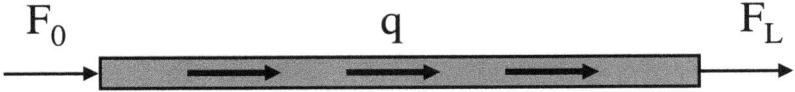

Figure 4.2.8.   Pure Neumann problem in Statics.

(4.2.52) and (4.2.53) have an interesting meaning in Statics, illustrated below. Rewrite (4.2.51) using the equation for the axial displacement of a bar in Statics

$$-AEU'' = q(x),\ \ 0 < x < L,\ \ -AEU'(0) = F_0,\ \ AEU'(L) = F_L, \quad (4.2.54)$$

where $q(x)$ is the axial load and F_0 and F_L are known external forces (Fig. 4.2.8). Integrate the differential equation over $(0, L)$ and use the boundary conditions

$$F_0 + F_L + \int_0^L q(x)dx = 0. \quad (4.2.55)$$

Relationship (4.2.55) is the version in Statics of compatibility Condition (4.2.52). The meaning is clear. An equilibrium implies that the total sum of the forces must be zero. The meaning of (4.2.53) in Statics is clear as well. The displacement is determined unless a rigid translation does not influence the stress.

4.2.5   *Axial Displacement and Stress in a Bar*

Consider the equation for the axial displacement of a bar in Statics (Sec. 4.2.1)

$$-AEU'' = q(x), \quad 0 < x < L, \quad (4.2.56)$$

where $q(x)$ is the axial load and L is the length of the bar. Appropriate boundary conditions are associated to Eq. (4.2.56). For example, for the fixed end-fixed end bar in Fig. 4.2.9, the boundary conditions are

$$U(0) = 0, \quad U(L) = 0. \quad (4.2.57)$$

Boundary value Problem (4.2.56)–(4.2.57) is similar to those already discussed in the previous sections. Therefore, the FEM is able to provide the axial displacement of a bar. Now, consider the axial stress $\sigma = EU'$ that may be even more important in engineering applications. It would be desirable that the FEM could calculate σ at least with the same accuracy as

U. Let us show that σ can be calculated by using the coefficients u_i again, i.e., with the same accuracy. Consider the element $e_i = [x_i, x_{i+1}]$ and the related approximated solution

$$u = u_i \Phi_i + u_{i+1} \Phi_{i+1}.$$

Inserting the previous function in the weak formula restricted to e_i yields (see Sec. 4.2.2)

$$u_i \int_{x_i}^{x_{i+1}} \Phi_i' \Phi_i' \, dx + u_{i+1} \int_{x_i}^{x_{i+1}} \Phi_{i+1}' \Phi_i' \, dx = \int_{x_i}^{x_{i+1}} F\Phi_i \, dx + [U'\Phi_i]_{x_i}^{x_{i+1}},$$

$$u_i \int_{x_i}^{x_{i+1}} \Phi_i' \Phi_{i+1}' dx + u_{i+1} \int_{x_i}^{x_{i+1}} \Phi_{i+1}' \Phi_{i+1}' dx = \int_{x_i}^{x_{i+1}} F\Phi_{i+1} dx + [U'\Phi_{i+1}]_{x_i}^{x_{i+1}}.$$

Hence,

$$
\begin{aligned}
(u_i - u_{i+1})/h &= \int_{x_i}^{x_{i+1}} F\Phi_i \, dx - U'(x_i), \\
(-u_i + u_{i+1})/h &= \int_{x_i}^{x_{i+1}} F\Phi_{i+1} \, dx + U'(x_{i+1}).
\end{aligned}
\tag{4.2.58}
$$

After finding u_i, System (4.2.58) gives the stress on the ends of any element e_i. Lastly, the reactive forces are derived from Formulas (4.2.10).

Example 4.2.5   Consider the bar in Fig. 4.2.9 and the related boundary value Problem (4.2.56)–(4.2.57). Suppose the bar is subjected to the trapezoidal load

$$q(x) = q_A + (q_B - q_A)x/L. \tag{4.2.59}$$

The following listing applies the FEM to calculate displacement, stress with Formula (4.2.58), and reactive forces. The graphs of the solutions are shown in Fig. 4.2.10. The analytical solution is provided in Exercise 4.4.6.

```
function [u, sigma] = stress_1
% This is the function file stress_1.m.
% FEM is applied to solve the following problem:
% -AEU"(x) = q(x), q(x) = qA + (qB - qA)*x/L, U0 = U(L) = 0,
```

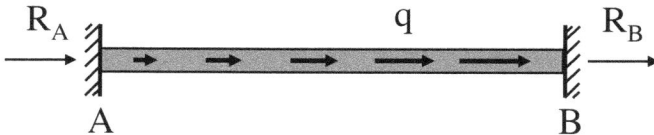

Figure 4.2.9.   Fixed end-fixed end bar subjected to axial load.

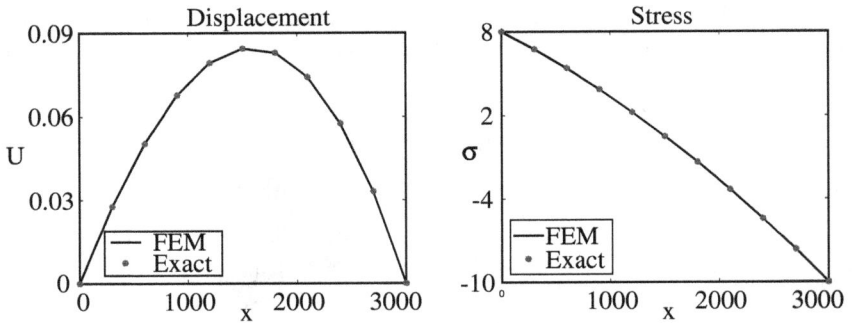

Figure 4.2.10. Displacement and stress of the bar subjected to axial load (4.2.59).

```
% Analytical solution:
% U = qA/2/A/E*(Lx - x²) + (qB - qA)/6/A/E/L(L²x - x³);
% Initialization
n = 10; u0 = 0; uL = 0;
L = 3000; % mm
E = 80000; % N/mm²
A = 1000; % mm²
qA = 4; % N/mm
qB = 8; % N/mm
Phil = @(xi, x1, x2)      (xi-x1)/(x2-x1);
Phir = @(xi, x2, x3)      -(xi-x3)/(x3-x2);
F = @(xi)      qA/A/E + (qB - qA)*xi/L/A/E;
h = L/n; x = linspace(0,L,n+1);
u = zeros(n+1,1); sigma = zeros(n+1,1); f = zeros(n-1,1);
U = qA/2/A/E*(L*x'-x'.²) + (qB - qA)/6/A/E/L*(L²*x' - x'.³);
S = qA/A*(L/2-x') + (qB - qA)/A*(L/6-x'.²/2/L);% Exact stress
% FEM
KK = [-ones(n-1,1)  2*ones(n-1,1)  -ones(n-1,1)];
K = spdiags(KK, -1:1, n-1, n-1)/h;
for i = 2:n
    f(i-1) =integral(@(xi)F(xi).*Phil(xi,x(i-1),x(i)), x(i-1), x(i));
    f(i-1) = f(i-1) + integral(@(xi)F(xi).*Phir(xi,x(i),x(i+1)), x(i), x(i+1));
end
f(1) = f(1) + u0/h; f(n-1) = f(n-1) + uL/h;
u(2:end-1) = K\f;
```

```
u(1) = u0; u(n+1) = uL;
for i=1:n
    sigma(i,1) = E*integral(@(xi)F(xi).*Phir(xi,x(i),x(i+1)), x(i), x(i+1))...
    - u(i)*E/h + u(i+1)*E/h;
end
sigma(i+1) = -E*integral(@(xi)F(xi).*Phil(xi,x(n),x(n+1)), x(n), x(n+1))...
    - u(n)*E/h;
rA = -A*sigma(1); rB = A*sigma(n+1);% Reactive forces
plot(x,u,'k',x,U,'r*','LineWidth',2);
legend('Fem','Exact','Location','SouthWest');
xlabel('x'); ylabel('U'); title('Displacement');
figure(2); plot(x,sigma,'k',x,S,'r*','LineWidth',2);
legend('Fem','Exact','Location','SouthWest'); xlabel('x');
ylabel('\sigma'); title('Stress');
fprintf('RA=%f\n',rA)
fprintf('RB=%f\n',rB)
fprintf( 'Maximum error for displacement = %g\n',max(abs(U - u)))
fprintf( 'Maximum error for stress= %g\n',max(abs(S - sigma)))
end
```

Example 4.2.6 Consider the bar in Fig. 4.2.11 and the related boundary value problem

$$-AEU'' = q(x), \quad 0 < x < L, \quad U(0) = 0, \quad U'(L) = 0, \qquad (4.2.60)$$

where $q(x)$ is the triangular load $q(x) = q_A(L - x)/L$. The analytical solution of Problem (4.2.60) is given in Exercise 4.4.7. A function that applies the FEM to calculate displacement and stress is illustrated in the worked Exercise 4.4.13. The graphs of displacement and stress are shown in Fig. 4.2.12.

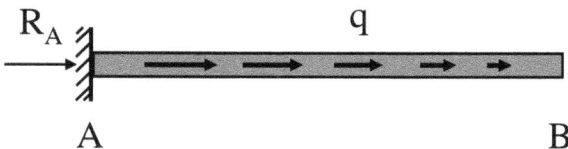

Figure 4.2.11. Fixed end-free end bar subjected to an axial distributed load.

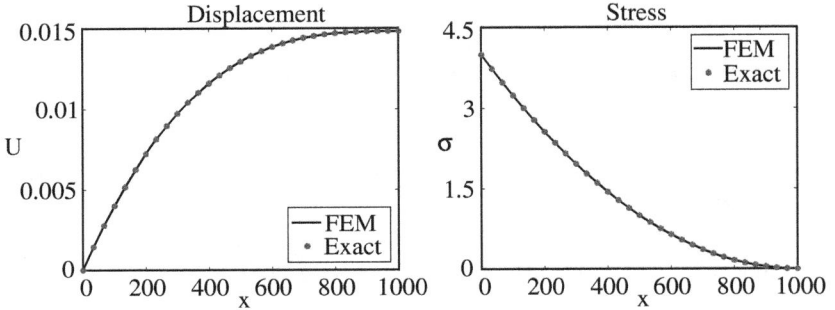

Figure 4.2.12.   Displacement and stress of the bar subjected to an triangular load.

4.2.6   *Concentrated Force and Dirac Function*

Consider a function $v(x)$ with compact support K. The *Dirac function* δ is defined as

$$\int_K \delta(x - \bar{x})v(x)dx = v(\bar{x}). \qquad (4.2.61)$$

For $\bar{x} = 0$, Definition (4.2.61) gives

$$\int_K \delta(x)v(x)dx = v(0). \qquad (4.2.62)$$

The δ function was introduced by Dirac[7] in Quantum Mechanics. Some years later, it was outlined that the δ function is a *generalized function*, or *distribution* Schwartz (1950).

Let us show that δ can model the concentrated forces on a bar. Consider a fixed end-fixed end bar in Statics (Fig. 4.2.13). The bar has length L and is subjected to the axial concentrated force (x_h, F_h). The formula for the axial displacement was derived in Exercise 4.4.9 and is expressed as

$$U(x) = \begin{cases} (L - x_h)F_h x/(AEL), & \text{if } 0 < x < x_h, \\ (L - x)F_h x_h/(AEL), & \text{if } x_h < x < L. \end{cases} \qquad (4.2.63)$$

---

[7]Paul Adrien Maurice Dirac, a British scientist, 1902–1984. He was Professor at the University of Cambridge. His most important book is *The Principles of Quantum Mechanics*. He was awarded the Nobel Prize in 1933.

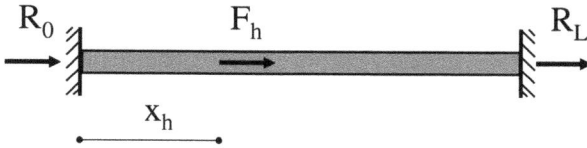

Figure 4.2.13. Fixed end-fixed end bar subjected to an axial concentrated force.

Hence, stress and reactive forces

$$\sigma(x) = EU'(x) \begin{cases} (L - x_h)F_h/(AL), & \text{if } 0 < x < x_h, \\ -F_h x_h/(AL), & \text{if } x_h < x < L, \end{cases} \quad (4.2.64)$$

$$R_0 = -(L - x_h)F_h/L, \quad R_L = -F_h x_h/L. \quad (4.2.65)$$

Note that U' is discontinuous in x_h:

$$AEU'(x_h^+) - AEU'(x_h^-) = -F_h. \quad (4.2.66)$$

Formula (4.2.66) follows from the general Result (4.2.7). Of course, it follows from the particular Result (4.2.64) as well. As a consequence, U cannot be a classical solution of the bar equation on the whole interval $(0, L)$. Let us show that U is a weak solution. Indeed, consider a smooth test function v with support on $K = [0, L]$ and note that

$$\int_0^L U'v' \, dx = \int_0^{x_h} U'v' \, dx + \int_{x_h}^L U'v' \, dx$$

$$= \frac{(L - x_h)F_h}{AEL} v(x_h) + \frac{F_h x_h}{AEL} v(x_h) = \frac{F_h v(x_h)}{AE}.$$

Hence,

$$\int_0^L U'v' \, dx = \frac{F_h}{AE} \int_0^L \delta(x - x_h)v(x) \, dx, \quad (4.2.67)$$

because of (4.2.61). Formula (4.2.67) shows that U is a weak solution of

$$-AEU'' = F_h \delta(x - x_h). \quad (4.2.68)$$

As the FEM considers weak solutions, it can be applied to find the axial displacements of a bar subjected to axial concentrated forces.

Example 4.2.7   As first application, consider the bar in Fig. 4.2.13. The axial displacement is a weak solution of Eq. (4.2.68) with homogeneous boundary conditions

$$U(0) = 0, \quad U(L) = 0. \tag{4.2.69}$$

Let us apply the FEM. The approximating solution is expressed as

$$u(x) = \sum_{j=1}^{n-1} u_j \Phi_j(x).$$

The unknown coefficients u_j are found by solving the system

$$\frac{1}{h} \begin{bmatrix} 2 & -1 & & \\ -1 & 2 & -1 & \\ & \cdot & \cdot & \cdot \\ & & -1 & 2 \end{bmatrix} \begin{bmatrix} u_1 \\ u_2 \\ \cdot \\ u_{n-1} \end{bmatrix} = \begin{bmatrix} f_1 \\ f_2 \\ \cdot \\ f_{n-1} \end{bmatrix}, \tag{4.2.70}$$

where

$$f_i = \frac{F_h}{AE} \int_0^L \delta(x - x_h)\Phi_i(x) \, dx = \frac{F_h}{AE}\Phi_i(x_h).$$

Since $\Phi_i(x_h) \neq 0$ only when $x_h \in (x_{i-1}, x_{i+1})$, it is convenient to consider a grid where x_h is the same as the node. Therefore, we get

$$f_i = \Phi_i(x_h)F_h/(AE) = \begin{cases} F_h/(AE), & \text{if } i = h, \\ 0, & \text{if } i \neq h. \end{cases}$$

Moreover, as linear displacements are expected, we can consider the simple grid illustrated in Fig. 4.2.14, where $x_h = h = L/3$. In such a case, System

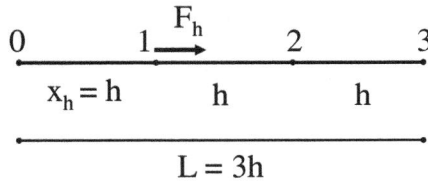

Figure 4.2.14.   Simple grid with $x_h = x_1$.

(4.2.70) simplifies to

$$\begin{cases} 2u_1/h - u_2/h = F_h/(AE), \\ -u_1/h + 2u_2/h = 0. \end{cases}$$

Solving the previous system yields

$$u_1 = 2hF_h/(3AE), \quad u_2 = hF_h/(3AE).$$

The values of u_i are the same as U_i obtained from (4.2.63). To understand why, let us recall that the FEM approximates the exact solution with a piecewise linear function. If the analytical solution is also piecewise linear, then $U_i = u_i$. Lastly, a rapid calculation shows that stress and reactive forces are exact too, as expected.

Example 4.2.8 Another application is given by the fixed end-free end bar in Statics (Fig. 4.2.15). The bar has length L and is subjected to the axial concentrated force (x_h, F_h). The axial displacements are weak solutions of Eq. (4.2.68) with boundary conditions

$$U(0) = 0, \quad U'(L) = 0. \tag{4.2.71}$$

The analytical solution was derived in Exercise 4.4.10 and is expressed as

$$U(x) = \begin{cases} F_h x/(AE), & \text{if } 0 < x < x_h, \\ F_h x_h/(AE), & \text{if } x_h < x < L. \end{cases} \tag{4.2.72}$$

Hence, stress and reactive force

$$\sigma(x) = \begin{cases} F_h/A, & \text{if } 0 < x < x_h, \\ 0, & \text{if } x_h < x < L, \end{cases} \quad R_0 = -F_h. \tag{4.2.73}$$

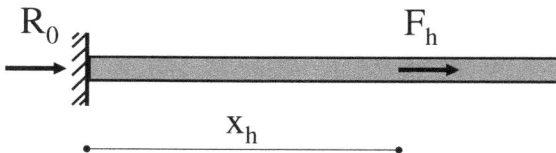

Figure 4.2.15. Fixed end-free end bar subjected to axial concentrated force.

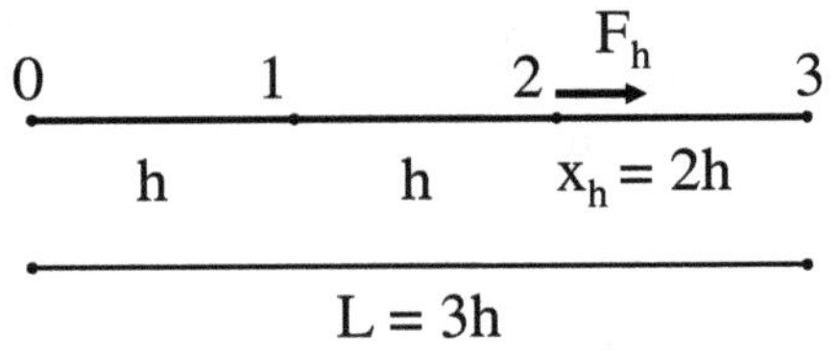

Figure 4.2.16.   Simple grid with $x_h = x_2$.

Consider the FEM. The approximating solution is expressed as (Sec. 4.2.4)

$$u(x) = \sum_{j=1}^{n} u_j \Phi_j(x).$$

The unknown coefficients satisfy the system

$$\frac{1}{h}
\begin{bmatrix}
2 & -1 & & & & \\
-1 & 2 & -1 & & & \\
 & \cdot & \cdot & \cdot & & \\
 & & -1 & 2 & -1 \\
 & & & -1 & 1
\end{bmatrix}
\begin{bmatrix}
u_1 \\ u_2 \\ \cdot \\ u_{n-1} \\ u_n
\end{bmatrix}
=
\begin{bmatrix}
f_1 \\ f_2 \\ \cdot \\ f_{n-1} \\ f_n
\end{bmatrix},
\qquad (4.2.74)$$

where

$$f_i = \frac{F_h}{AE} \int_0^L \delta(x - x_h)\Phi_i(x)\, dx = \frac{F_h}{AE}\Phi_i(x_h).$$

Consider the simple grid in Fig. 4.2.16. It is

$$f_i = \begin{cases} F_h/(AE), & \text{if } i = h, \\ 0, & \text{if } i \neq h, \end{cases}$$

and

$$\begin{cases} 2u_1/h - u_2/h = 0, \\ -u_1/h + 2u_2/h - u_3/h = F_h/(AE), \\ -u_2/h + u_3/h = 0. \end{cases}$$

Solving the previous system yields

$$u_1 = hF_h/(AE), \quad u_2 = u_3 = 2hF_h/(AE).$$

These values are the same as the analytical values U_i in (4.2.72), as expected. Another application is suggested in Exercise 4.4.14.

4.3 Partial Differential Equations

This section introduces the FEM for some parabolic and hyperbolic partial differential equations.

4.3.1 *Diffusion Equation*

Consider the Dirichlet problem for the diffusion equation

$$U_t - \alpha U_{xx} = F(x,t), \quad 0 < x < L, \quad 0 < t \leq T, \tag{4.3.1}$$
$$U(x,0) = \varphi(x), \quad 0 \leq x \leq L, \tag{4.3.2}$$
$$U(0,t) = G_1(t), \quad U(L,t) = G_2(t), \quad 0 < t \leq T. \tag{4.3.3}$$

Multiply Eq. (4.3.1) by the smooth test function $v(x)$

$$U_t v - \alpha U_{xx} v - F v = 0, \quad \Leftrightarrow \quad U_t v - \alpha (U_x v)_x + \alpha U_x v' - F v = 0,$$

and integrate over $[0, L]$

$$\int_0^L U_t, v dx + \int_0^L \alpha U_x v' dx = \int_0^L F v dx + \alpha \left[U_x v\right]_{x=0}^{x=L}. \tag{4.3.4}$$

Equation (4.3.4) is the *weak form* of Eq. (4.3.1) that is named *strong form* by contrast. The FEM considers Eq. (4.3.4). The approximating solution to the Dirichlet problem is expressed as a finite series

$$u(x,t) = \sum_{j=0}^n u_j(t) \Phi_j(x), \tag{4.3.5}$$

where $u_j(t)$, $j = 0, ..., n$, are unknown functions and $\Phi_j(x)$, $j = 0, ..., n$, are the shape functions introduced in Sec. 4.2.3. Note that $u(0,t) = u_0(t)$ and $u(L,t) = u_n(t)$ because of Property (4.2.23) of the shape functions. Therefore, if it is assumed $u_0(t) = G_1(t)$ and $u_n(t) = G_2(t)$, the function u satisfies boundary Conditions (4.3.3). In conclusion,

$$u(x,t) = \sum_{j=1}^{n-1} u_j(t) \Phi_j(x) + G_1(t) \Phi_0(x) + G_2(t) \Phi_n(x), \tag{4.3.6}$$

where the last two terms are known. Substitute $u(x,t)$ into Eq. (4.3.4) and assume $v = \Phi_i$, $i = 1, \ldots, n-1$,

$$\sum_{j=1}^{n-1} \dot{u}_j(t) \int_0^L \Phi_j \Phi_i dx + \sum_{j=1}^{n-1} u_j(t) \int_0^L \alpha \Phi_j' \Phi_i' dx = \int_0^L F \Phi_i dx - \dot{G}_1(t) \int_0^L \Phi_0 \Phi_i dx$$

$$-\dot{G}_2(t)\int_0^L \Phi_n\Phi_i dx - \alpha G_1(t)\int_0^L \Phi_0'\Phi_i' dx - \alpha G_2(t)\int_0^L \Phi_n'\Phi_i' dx,$$

$$i = 1,\ldots,n-1,$$

as $\Phi_i(0) = 0$, $\Phi_i(L) = 0$, $i = 1,\ldots,n-1$. Hence, in matrix notation

$$\sum_{j=1}^{n-1} M_{ij}\dot{u}_j(t) + \sum_{j=1}^{n-1} K_{ij}u_j(t) = f_i(t) - g_i(t), \quad i = 1,\ldots,n-1, \quad (4.3.7)$$

where

$$M_{ij} = \int_0^L \Phi_j\Phi_i dx, \quad K_{ij} = \int_0^L \alpha\Phi_j'\Phi_i' dx, \quad i,j = 1,\ldots,n-1, \quad (4.3.8)$$

$$f_i(t) = \int_0^L F,\Phi_i dx, \quad g_i(t) = \dot{G}_1(t)\int_0^L \Phi_0\Phi_i dx + \dot{G}_2(t)\int_0^L \Phi_n\Phi_i dx$$

$$+ \alpha G_1(t)\int_0^L \Phi_0'\Phi_i' dx + \alpha G_2(t)\int_0^L \Phi_n'\Phi_i' dx, \quad i = 1,\ldots,n-1. \quad (4.3.9)$$

In compact form, System (4.3.7) is written as

$$M\dot{\mathbf{u}}(t) + K\mathbf{u}(t) = \mathbf{f}(t) - \mathbf{g}(t), \quad (4.3.10)$$

where $\mathbf{u}$, $\mathbf{f}$ and $\mathbf{g}$ are column vectors. The matrix M is named a *mass matrix* with reference to Mechanics. The initial conditions $u_i(0)$ for the system of ordinary differential Eqs. (4.3.10) are derived from (4.3.2). Indeed, note that

$$u(x_i,0) = \sum_{j=1}^n u_j(0)\Phi_j(x_i) = u_i(0), \quad i = 1,\ldots,n.$$

Therefore, from (4.3.2), it follows

$$u_i(0) = \varphi(x_i) = \varphi_i, \quad i = 1,\ldots,n. \quad (4.3.11)$$

The elements $f_i(t)$ of the vector $\mathbf{f}$ are given by

$$f_i(t) = \int_{x_{i-1}}^{x_{i+1}} F(x,t)\Phi_i(x)\,dx, \quad i = 1,\ldots,n-1, \quad (4.3.12)$$

as the support of Φ_i is the interval $[x_{i-1},x_{i+1}]$. Let us evaluate the vector $\mathbf{g}$. Consider $g_1(t)$ and note that the supports of Φ_0, Φ_1 and Φ_n

are the intervals $[x_0, x_1]$, $[x_0, x_2]$ and $[x_{n-1}, x_n]$, respectively. Therefore,

$$g_1(t) = \dot{G}_1(t) \int_{x_0}^{x_1} \Phi_0(x)\Phi_1(x) \, dx + \alpha G_1(t) \int_{x_0}^{x_1} \Phi_0'(x)\Phi_1'(x) \, dx,$$

$$g_1(t) = \dot{G}_1(t)h/6 - \alpha G_1(t)/h. \tag{4.3.13}$$

Consider $g_{n-1}(t)$ and note that the supports of Φ_0, Φ_{n-1} and Φ_n are the intervals $[x_0, x_1]$, $[x_{n-2}, x_n]$ and $[x_{n-1}, x_n]$, respectively. Therefore,

$$g_{n-1}(t) = \dot{G}_2(t) \int_{x_{n-1}}^{x_n} \Phi_n(x)\Phi_{n-1}(x) \, dx + \alpha G_2(t) \int_{x_{n-1}}^{x_n} \Phi_n'(x)\Phi_{n-1}'(x) \, dx,$$

$$g_{n-1}(t) = \dot{G}_2(t)h/6 - \alpha G_2(t)/h. \tag{4.3.14}$$

The remaining elements of $\mathbf{g}$ are zero

$$g_i(t) = 0, \quad i = 2, \ldots, n-2. \tag{4.3.15}$$

The matrices M and K are given by

$$M = \frac{h}{3} \begin{bmatrix} 2 & 1/2 & & \\ 1/2 & 2 & 1/2 & \\ & . & . & \\ & & 1/2 & 2 \end{bmatrix}, \quad K = \frac{\alpha}{h} \begin{bmatrix} 2 & -1 & & \\ -1 & 2 & -1 & \\ & . & . & \\ & & -1 & 2 \end{bmatrix}. \tag{4.3.16}$$

K was evaluated in Sec. 4.2.4. The elements of M are calculated in worked Exercise 4.4.15. The mass matrix is nonsingular. System (4.3.10) can be written in normal form

$$\dot{\mathbf{u}}(t) = -M^{-1}K\mathbf{u}(t) + M^{-1}[\mathbf{f}(t) - \mathbf{g}(t)]. \tag{4.3.17}$$

Example 4.3.1 The following listing presents a function that applies the FEM to Problem (4.3.1)–(4.3.3). The listing considers the initial-boundary conditions

$$U(x, 0) = \sin(\pi x/L), \quad 0 \le x \le L, \tag{4.3.18}$$

$$U(0, t) = 0, \quad U(L, t) = 0, \quad 0 < t \le T. \tag{4.3.19}$$

However, the code can be easily modified for other applications. The graph of the numerical solution is shown in Fig. 4.3.1.

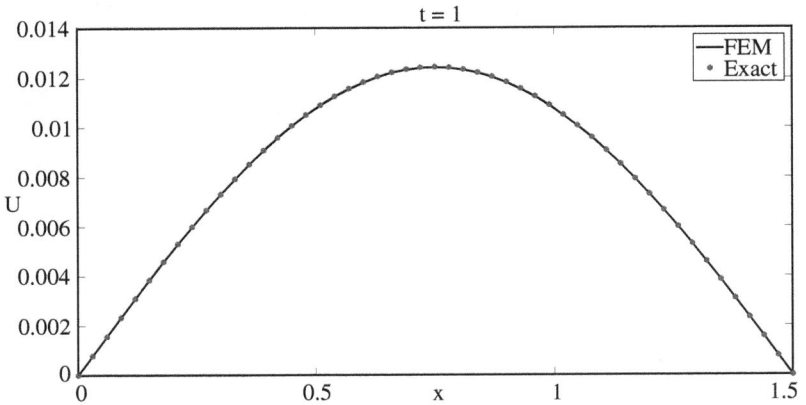

Figure 4.3.1. Graph of the numerical solution to Problem (4.3.18) and (4.3.19).

```
function u = diffusion_d
% This is the function file diffusion_d.m.
% FEM is applied to the following Dirichlet problem:
% Ut - alpha Uxx = 0, U(x,0) = phi(x), U(0,t) = G1(t), U(L,t) = G2(t),
% where phi(x) = sin(pi*x), G1(t) = 0, G2(t)=0.
% Analytical solution: U = sin(pi*x)*exp(-pi^2*t).
% The code can be easily modified for other applications. It is
% enough to modify the lines related to the initial-boundary conditions
% and to eliminate the line related to the analytical solution.
% Initialization
alpha = 1; L = 1.5; T = 1; nx = 50;
phi = @(x)          sin(pi*x/L);
G1 = @(t)           0*t;
G2 = @(t)           0*t;
G1t = @(t)          0*t;
G2t = @(t)          0*t;
dx = L/nx; x = linspace(0,L,nx+1);
KK = [-ones(nx-1,1)  2*ones(nx-1,1)  -ones(nx-1,1)];
K = spdiags(KK, -1:1, nx-1, nx-1)*alpha/dx;
MM = [ones(nx-1,1)/6  2/3*ones(nx-1,1)  ones(nx-1,1)/6];
M = spdiags(MM, -1:1, nx-1, nx-1)*dx;
B = M^(-1); A = B*K;
g = zeros(nx-1,1); u = feval(phi,x');
% FEM
```

```
[~,y] = ode45(@system,[0 T],u(2:nx),[ ],A,B,g,dx,alpha,G1,G2,G1t,G2t);
u(2:nx) = y(end,:);
u(1) = G1(T); u(nx+1) = G2(T);
U = sin(pi*x'/L)*exp(-(pi/L)^2*T);
plot(x,u,'k',x,U,'r*','LineWidth',2);
legend('FEM','Exact'); xlabel('x'); ylabel('U');
title(['t = ',num2str(T)]);
fprintf( 'Maximum error= %g\n',max(abs(U - u)))
end
% ———— Local function ————
function Du = system(t, u, A, B, g, dx, alpha, G1, G2, G1t, G2t)
g(1) = G1t(t)*dx/6 - alpha*G1(t)/dx;
g(end) = G2t(t)*dx/6 - alpha*G2(t)/dx;
Du = -A*u - B*g;
end
```

Another application is suggested in Exercise 4.4.16.

Consider the Neumann problem

$$U_t - \alpha U_{xx} = F(x,t), \quad 0 < x < L, \quad 0 < t \le T, \tag{4.3.20}$$

$$U(x,0) = \varphi(x), \quad 0 \le x \le L, \tag{4.3.21}$$

$$-U_x(0,t) = G_1(t), \quad U_x(L,t) = G_2(t), \quad 0 < t \le T. \tag{4.3.22}$$

The weak form is given by

$$\int_0^L U_t v\,dx + \int_0^L \alpha U_x v'\,dx = \int_0^L Fv\,dx + \alpha G_1(t)v(0) + \alpha G_2(t)v(L). \tag{4.3.23}$$

The approximating solution to the Neumann problem is expressed as a finite series

$$u(x,t) = \sum_{j=0}^n u_j(t)\Phi_j(x), \tag{4.3.24}$$

where $u_j(t)$, $j = 0, ..., n$, are unknown functions and $\Phi_j(x)$, $j = 0, ..., n$, are the shape functions introduced in Sec. 4.2.3. Substitute $u(x,t)$ into Eq. (4.3.23) and assume $v = \Phi_i$, $i = 0, \ldots, n$,

$$\sum_{j=0}^n \dot{u}_j(t) \int_0^L \Phi_j \Phi_i\,dx + \sum_{j=0}^n u_j(t) \int_0^L \alpha \Phi_j' \Phi_i'\,dx = \int_0^L F\Phi_i\,dx +$$

$$+\alpha G_1(t)\Phi_i(0) + \alpha G_2(t)\Phi_i(L), \quad \forall \Phi_i, \quad i = 0, \ldots, n.$$

Hence, the matrix notation

$$\sum_{j=0}^{n} M_{ij}\dot{u}_j(t) + \sum_{j=0}^{n} K_{ij}u_j(t) = f_i(t) + g_i(t), \quad i = 0, \ldots, n, \qquad (4.3.25)$$

where

$$M_{ij} = \int_0^L \Phi_j \Phi_i dx, \;\; K_{ij} = \int_0^L \alpha \Phi'_j \Phi'_i dx, \;\; f_i(t) = \int_0^L F\Phi_i)dx, \;\; i, j = 0, \ldots, n,$$

$$g_i(t) = \alpha G_1(t)\Phi_i(0) + \alpha G_2(t)\Phi_i(L), \quad i = 0, \ldots, n.$$

In compact form, System (4.3.25) is written as

$$M\dot{u}(t) + K u(t) = f(t) + g(t), \qquad (4.3.26)$$

where **u**, **f** and **g** are column vectors. The elements of **f** are given by

$$f_0(t) = \int_{x_0}^{x_1} \Phi_i dx, \quad f_n(t) = \int_{x_{n-1}}^{x_n} F\Phi_i dx,$$

$$f_i(t) = \int_{x_{i-1}}^{x_{i+1}} F\Phi_i dx, \; i = 1, \ldots, n-1,$$

as the support of Φ_0 is the interval $[x_0, x_1]$, the support of Φ_n is the interval $[x_{n-1}, x_n]$ and the support of Φ_i, $i = 1, \ldots, n-1$, is the interval $[x_{i-1}, x_{i+1}]$. In addition, the elements of **g** are given by

$$g_0(t) = \alpha G_1(t), \quad g_n(t) = \alpha G_2(t), \quad g_i(t) = 0, \quad i = 1, \ldots, n-1. \quad (4.3.27)$$

Let us evaluate the mass matrix M. Consider

$$M_{0,0} = \int_{x_0}^{x_1} \Phi_0^2(x)\, dx = \int_0^h \frac{(h-x)^2}{h^2}\, dx = \int_0^h \frac{\xi^2}{h^2}\, d\xi = h/3,$$

as the support of Φ_0 is the interval $[x_0, x_1]$. In addition,

$$M_{n,n} = \int_{x_{n-1}}^{x_n} \Phi_n^2(x)\, dx = \int_{x_{n-1}}^{x_n} \frac{(x - x_{n-1})^2}{h^2}\, dx = \int_0^h \frac{\xi^2}{h^2}\, d\xi = h/3,$$

as the support of Φ_n is the interval $[x_{n-1}, x_n]$. The other elements of the $(n+1)$-by-$(n+1)$ matrix M are shown in Formula (4.3.16). In conclusion:

$$M = \frac{h}{3} \begin{bmatrix} 1 & 1/2 & & & \\ 1/2 & 2 & 1/2 & & \\ & & \cdot & \cdot & \cdot & \\ & & & 1/2 & 2 & 1/2 \\ & & & & 1/2 & 1 \end{bmatrix} \qquad (4.3.28)$$

Let us evaluate the stiffness matrix K. Consider

$$K_{0,0} = \alpha \int_{x_0}^{x_1} \Phi_0'^2(x)\, dx = \alpha/h,$$

as the support of Φ_0 is the interval $[x_0, x_1]$. In addition,

$$K_{n,n} = \alpha \int_{x_{n-1}}^{x_n} \Phi_n'^2(x)\, dx = \alpha/h,$$

as the support of Φ_n is the interval $[x_{n-1}, x_n]$. The other elements of the $(n+1)$-by-$(n+1)$ matrix K are in Formula (4.3.16). In conclusion:

$$K = \frac{\alpha}{h} \begin{bmatrix} 1 & -1 & & & \\ -1 & 2 & -1 & & \\ & & \cdot & \cdot & \cdot & \\ & & & -1 & 2 & -1 \\ & & & & -1 & 1 \end{bmatrix} \qquad (4.3.29)$$

Example 4.3.2 The following listing presents a function that applies the FEM to solve the Neumann problem

$$U_t - U_{xx} = x(x - L) - 2t, \quad 0 < x < L, \quad 0 < t \le T, \qquad (4.3.30)$$

$$U(x, 0) = 0, \quad 0 \le x \le L, \qquad (4.3.31)$$

$$-U_x(0, t) = tL, \quad U_x(L, t) = tL, \quad 0 < t \le T. \qquad (4.3.32)$$

The numerical solution is plotted in Fig. 4.3.2. The listing considers the problem above, but the code can be easily modified for other applications.

```
function u = diffusion_n
% This is the function file diffusion_n.m.
% FEM is applied to the following Neumann problem:
% Ut - alpha Uxx = F, U(x,0) = phi(x), Ux(0,t) = - G1(t), Ux(L,t) = G2(t),
```

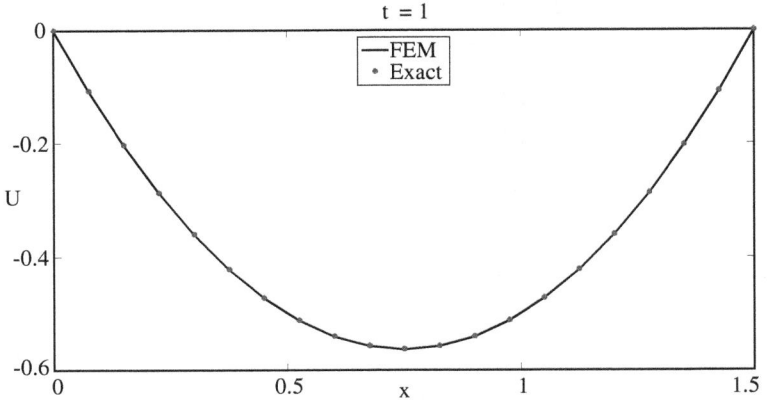

Figure 4.3.2.   Graph of the numerical solution to Problem (4.3.30) to (4.3.32).

```
% where phi(x) = 0, G1(t) = tL, G2(t) = tL, F = x(x - L) - 2t.
% Analytical solution: U = x*(x - L)*t.
% The code can be easily adapted to other applications by modifying the data.
% Initialization
alpha = 1; L = 1.5; T = 1; n = 20;
phi = @(x)        0*x;
G1 = @(t)         t*L;
G2 = @(t)         t*L;
F = @(xi,t)       xi.*(xi - L) - 2*t;
Phil = @(xi, x1, x2)        (xi - x1)/(x2 - x1);
Phir = @(xi, x2, x3)        -(xi - x3)/(x3 - x2);
n = L/n; x = linspace(0,L,n+1);
KK = [-ones(n+1,1)  [1; 2*ones(n-1,1); 1]  -ones(n+1,1)];
K = spdiags(KK, -1:1, n+1, n+1)*alpha/dx;
MM = [ones(n+1,1)/2  [1; 2*ones(n-1,1);1]  ones(n+1,1)/2];
M = spdiags(MM, -1:1, n+1, n+1)*h/3;
B = M^(-1); A = B*K;
u = feval(phi,x');

% FEM
[~,y] = ode15s(@system, [0 T], u, [ ], A, B, alpha, x, n, G1, G2, Phil, Phir, F);
u(1:end,1) = y(end,:);
U = x'.*(x' - L)*T;
plot(x,u,'k',x,U,'r*','LineWidth',2);
```

```
legend('FEM','Exact','Location','North'); xlabel('x'); ylabel('U');
title(['t = ',num2str(T)]);
fprintf( 'Maximum error= %g\n',max(abs(U - u)))
end

% ———— Local function ————
function Du = system(t, u, A, B, alpha, x, n, G1, G2, Phil, Phir, F)
f = zeros(n+1,1);
i = 1; f(i) = integral(@(xi)F(xi,t).*Phir(xi,x(i),x(i+1)), x(i), x(i+1));
for i=2:n
    f(i) = integral(@(xi)F(xi, t).*Phil(xi, x(i-1), x(i)), x(i-1), x(i));
    f(i) = f(i) + integral(@(xi)F(xi, t).*Phir(xi, x(i), x(i+1)), x(i), x(i+1));
end
i = n+1; f(i) = integral(@(xi)F(xi, t).*Phil(xi, x(i-1), x(i)), x(i-1), x(i));
f(1) = f(1) + alpha*G1(t); f(end) = f(end) + alpha*G2(t);
Du = -A*u+B*f;
end
```

4.3.2 *Wave Equation*

Let us discuss FEM for the following wave equation

$$U_{tt} - c^2 U_{xx} = F(x, t) - \lambda U, \quad 0 < x < L, \quad 0 < t \le T. \tag{4.3.33}$$

Consider the Dirichlet problem and assign initial position and velocity

$$U(x, 0) = \varphi(x), \quad U_t(x, 0) = \psi(x), \quad 0 \le x \le L, \tag{4.3.34}$$

and boundary conditions

$$U(0, t) = G_1(t), \quad U(L, t) = G_2(t), \quad 0 < t \le T. \tag{4.3.35}$$

Multiply Eq. (4.3.33) by a smooth test function $v(x)$ and integrate over $[0, L]$

$$\int_0^L [U_{tt}(x, t)v(x) + c^2 U_x(x, t)v'(x) + \lambda U(x, t)v(x)]dx$$

$$= \int_0^L F(x, t)v(x)dx + c^2 [U_x v]_{x=0}^{x=L}. \tag{4.3.36}$$

The FEM considers the weak form above. The approximating solution that satisfies boundary Conditions (4.3.35) is given by

$$u(x,t) = \sum_{j=1}^{n-1} u_j(t)\Phi_j(x) + G_1(t)\Phi_0(x) + G_2(t)\Phi_n(x), \qquad (4.3.37)$$

where $u_j(t)$, $j = 1, ..., n-1$, are unknown functions and $\Phi_j(x)$, $j = 0, ..., n$, are the shape functions introduced in Sec. 4.2.3. Substitute (4.3.37) into (4.3.36) and assume $v = \Phi_i$, $i = 1, ..., n-1$,

$$\sum_{j=1}^{n-1} [\ddot{u}_j(t) + \lambda u_j(t)] \int_0^L \Phi_j(x)\Phi_i(x)dx + \sum_{j=1}^{n-1} u_j(t) \int_0^L c^2 \Phi_j'(x)\Phi_i'(x)dx$$

$$= -[\ddot{G}_1(t) + \lambda G_1(t)] \int_0^L \Phi_0(x)\Phi_i(x)\ dx - c^2 G_1(t) \int_0^L \Phi_0'(x)\Phi_i'(x)\ dx$$

$$-[\ddot{G}_2(t) + \lambda G_2(t)] \int_0^L \Phi_n(x)\Phi_i(x)\ dx - c^2 G_2(t) \int_0^L \Phi_n'(x)\Phi_i'(x)\ dx,$$

$$+ \int_0^L F(x,t)\Phi_i(x)dx, \quad i = 1, ..., n-1,$$

as $\Phi_i(0) = 0$, $\Phi_i(L) = 0$, $i = 1, ..., n-1$. Hence, in matrix notation,

$$\sum_{j=1}^{n-1} M_{ij}[\ddot{u}_j(t) + \lambda u_j(t)] + \sum_{j=1}^{n-1} K_{ij}u_j(t) = f_i(t) - g_i(t), \quad i = 1, ..., n-1,$$

$$(4.3.38)$$

where

$$M_{ij} = \int_0^L \Phi_j(x)\Phi_i(x)dx, \quad K_{ij} = \int_0^L c^2 \Phi_j'(x)\Phi_i'(x)dx, \quad i,j = 1, ..., n-1,$$

$$(4.3.39)$$

$$f_i(t) = \int_0^L F(x,t)\Phi_i(x)dx, \quad g_i(t) = [\ddot{G}_1(t) + \lambda G_1(t)] \int_0^L \Phi_0(x)\Phi_i(x)dx$$

$$+ c^2 G_1(t) \int_0^L \Phi_0'(x)\Phi_i'(x)dx + [\ddot{G}_2(t) + \lambda G_2(t)] \int_0^L \Phi_n(x)\Phi_i(x)dx$$

$$+ c^2 G_2(t) \int_0^L \Phi_n'(x)\Phi_i'(x)dx, \quad i = 1, ..., n-1. \qquad (4.3.40)$$

In compact form, System (4.3.38) is written as

$$M[\ddot{\mathbf{u}}(t) + \lambda \mathbf{u}(t)] + K\mathbf{u}(t) = \mathbf{f}(t) - \mathbf{g}(t), \qquad (4.3.41)$$

where $\mathbf{u}$, $\mathbf{f}$ and $\mathbf{g}$ are column vectors. The integration of the previous second-order system of ordinary differential equations needs the initial conditions

$u_i(0)$ and $\dot{u}_i(0)$. They are derived from (4.3.34):

$$u_i(0) = \varphi(x_i) = \varphi_i, \quad \dot{u}_i(0) = \psi(x_i) = \psi_i, \quad i = 1, \ldots, n. \tag{4.3.42}$$

The first condition was derived in (4.3.11). The second condition is derived similarly. The elements of $\mathbf{f}$ are given by

$$f_i(t) = \int_{x_{i-1}}^{x_{i+1}} F(x,t)\Phi_i(x)dx, \quad i = 1, \ldots, n-1. \tag{4.3.43}$$

Let us evaluate $\mathbf{g}$. Consider

$$g_1(t) = [\ddot{G}_1(t) + \lambda G_1(t)] \int_{x_0}^{x_1} \Phi_0(x)\Phi_1(x) \, dx + c^2 G_1(t) \int_{x_0}^{x_1} \Phi_0'(x)\Phi_1'(x) \, dx,$$

$$g_1(t) = [\ddot{G}_1(t) + \lambda G_1(t)]h/6 - c^2 G_1(t)/h, \tag{4.3.44}$$

as the supports of Φ_0, Φ_1 and Φ_n are the intervals $[x_0, x_1]$, $[x_0, x_2]$ and $[x_{n-1}, x_n]$, respectively. In addition,

$$g_{n-1}(t) = [\ddot{G}_2(t) + \lambda G_2(t)] \int_{x_{n-1}}^{x_n} \Phi_n(x)\Phi_{n-1}(x) \, dx$$

$$+ c^2 G_2(t) \int_{x_{n-1}}^{x_n} \Phi_n'(x)\Phi_{n-1}'(x) \, dx = [\ddot{G}_2(t)+\lambda G_2(t)]\frac{h}{6} - \frac{c^2}{h}G_2(t), \tag{4.3.45}$$

as the supports of Φ_0, Φ_{n-1} and Φ_n are the intervals $[x_0, x_1]$, $[x_{n-2}, x_n]$ and $[x_{n-1}, x_n]$, respectively. The other elements are zero

$$g_i(t) = 0, \quad i = 2, \ldots, n-2. \tag{4.3.46}$$

The matrices M and K are given by (see Formula (4.3.16))

$$M = \frac{h}{3}\begin{bmatrix} 2 & 1/2 & & \\ 1/2 & 2 & 1/2 & \\ & \cdot & \cdot & \cdot \\ & & 1/2 & 2 \end{bmatrix}, \quad K = \frac{c^2}{h}\begin{bmatrix} 2 & -1 & & \\ -1 & 2 & -1 & \\ & \cdot & \cdot & \cdot \\ & & -1 & 2 \end{bmatrix}.$$

Finally, the second-order differential system in (4.3.41) can be converted to a first-order system. Indeed, if the new unknown function

$$\mathbf{w} = \dot{\mathbf{u}}$$

is introduced, we get

$$\begin{bmatrix} \dot{\mathbf{u}} \\ \dot{\mathbf{w}} \end{bmatrix} = \begin{bmatrix} 0 & I \\ -(\lambda I + M^{-1}K) & 0 \end{bmatrix}\begin{bmatrix} \mathbf{u} \\ \mathbf{w} \end{bmatrix} + \begin{bmatrix} 0 \\ M^{-1}(\mathbf{f} - \mathbf{g}) \end{bmatrix}. \tag{4.3.47}$$

Example 4.3.3 Consider the Dirichlet problem

$$U_{tt} - c^2 U_{xx} + \lambda U = 0, \quad 0 < x < L, \quad 0 < t \le T. \tag{4.3.48}$$

$$U(x,0) = \sin(p\pi x/L), \quad U_t(x,0) = 0, \quad 0 < x < L, \tag{4.3.49}$$

$$U(0,t) = 0, \quad U(L,t) = 0, \quad 0 < t \le T. \tag{4.3.50}$$

The following listing presents a function that applies the FEM to solve Problem (4.3.48)–(4.3.50). The numerical solution is plotted in Fig. 4.3.3.

```
function u = wave_d
% This is the function file wave_d.m.
% The FEM is applied to the following Dirichlet problem:
% Utt - c^2 Uxx + lambda U = 0,
% U(x,0) = sin(x*pi*p/L), Ut(x,0) = 0, U(0,t) = 0, U(L,t) = 0.
% Analytical solution:
% U = sin(x*pi*p/L)*cos(t*sqrt(lambda + c^2*pi^2*p^2/L^2) ).
% Initialization
c = 1; L = 1; T = 1; lambda = 20; n = 100; p = 4;
phi = @(x)      sin(x*pi*p/L); psi = @(x)      x*0;
G1 = @(t)      0*t;           G1tt = @(t)      0*t;
G2 = @(t)      0*t;           G2tt = @(t)      0*t;
h = L/n; x = linspace(0,L,n+1);
KK = [-ones(n-1,1)  2*ones(n-1,1)  -ones(n-1,1)];
```

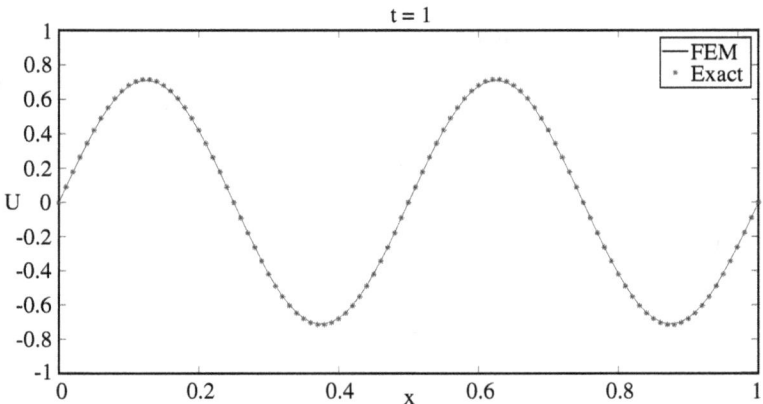

Figure 4.3.3. Graph of the numerical solution to Problem (4.3.48) to (4.3.50).

```
K = spdiags(KK, -1:1, n-1, n-1)*c^2/h;
MM = [ones(n-1,1)/6  2/3*ones(n-1,1)  ones(n-1,1)/6];
M = spdiags(MM, -1:1, n-1, n-1)*h;
B = M^(-1);
A = [zeros(n-1,n-1) eye(n-1,n-1); -lambda*eye(n-1,n-1)-B*K zeros(n-1,n-1)];
u = feval(phi,x'); w = feval(psi,x');
g = zeros(n-1,1);
U = sin(x'*pi*p/L)* cos(T*sqrt(lambda + c^2*pi^2*p^2/L^2) );
% FEM
nt = 40; tt = linspace(0,T,nt+1);
for j=1:nt
    [~,y] = ode45(@system, [tt(j) tt(j+1)], [u(2:n); w(2:n)], [ ], A, B, g,...
            h, c, lambda, G1, G2,G1tt,G2tt);
    u(2:n) = y(end,1:n-1);
    u(1) = G1(tt(j+1)); u(n+1) = G2(tt(j+1));
    w(2:n) = y(end,n:end);
    plot(x,u,'k',x,U,'r*');
    xlabel('x'); ylabel('U'); axis([0 L -1 1]);
    title(['t = ',num2str(tt(j+1))]);
    pause(.1);
end
legend('FEM','Exact');
fprintf( 'Maximum error= %g\n',max(abs(U - u)))
end

% ——— Local function ———
function Du = system(t, uw, A, B, g, h, c, lambda, G1, G2, G1tt, G2tt)
gg = g;
g(1) = (G1tt(t) + lambda*G1(t))*h/6 - c^2*G1(t)/h;
g(end) = (G2tt(t) + lambda*G2(t))*h/6 - c^2*G2(t)/h;
Du = A*uw + [gg; -B*g];
end
```

Another application is suggested in Exercise 4.4.17.

Consider the Neumann problem

$$U_{tt} - c^2 U_{xx} + \lambda U = F(x,t), \quad 0 < x < L, \quad 0 < t \leq T, \qquad (4.3.51)$$

$$U(x,0) = \varphi(x), \quad U_t(x,0) = \psi(x), \quad 0 \leq x \leq L, \qquad (4.3.52)$$

$$-U_x(0,t) = G_1(t), \quad U_x(L,t) = G_2(t), \quad 0 < t \leq T. \qquad (4.3.53)$$

The weak form is given by

$$\int_0^L [U_{tt}(x,t)v(x) + c^2 U_x(x,t)v'(x) + \lambda U(x,t)v(x) - F(x,t)v(x)] \, dx$$

$$= c^2 G_2(t)v(L) + c^2 G_1(t)v(0). \tag{4.3.54}$$

The FEM considers the weak form. The approximating solution to the Neumann problem is expressed as a finite series

$$u(x,t) = \sum_{j=0}^n u_j(t)\Phi_j(x), \tag{4.3.55}$$

where $u_j(t)$, $j = 0, ..., n$, are unknown functions and $\Phi_j(x)$, $j = 0, ..., n$, are the shape functions introduced in Sec. 4.2.3. Substitute $u(x,t)$ into Eq. (4.3.54) and assume $v = \Phi_i$, $i = 0, \ldots, n$,

$$\sum_{j=0}^n [\ddot{u}_j(t) + \lambda u_j(t)] \int_0^L \Phi_j(x)\Phi_i(x)dx + \sum_{j=0}^n u_j(t) \int_0^L c^2 \Phi_j'(x)\Phi_i'(x)dx$$

$$= \int_0^L F(x,t)\Phi_i(x)dx + c^2 G_1(t)\Phi_i(0) + c^2 G_2(t)\Phi_i(L), \ \forall \Phi_i, \quad i = 0, \ldots, n.$$

Hence, using the matrix notation,

$$\sum_{j=0}^n M_{ij}[\ddot{u}_j(t) + \lambda u_j(t)] + \sum_{j=0}^n K_{ij}u_j(t) = f_i(t) + g_i(t), \quad i = 0, \ldots, n, \tag{4.3.56}$$

where

$$M_{ij} = \int_0^L \Phi_j(x)\Phi_i(x)dx, \quad K_{ij} = \int_0^L c^2 \Phi_j'(x)\Phi_i'(x)dx, \quad i, j = 0, \ldots, n,$$

$$f_i(t) = \int_0^L F(x,t)\Phi_i(x)dx, \ g_i(t) = c^2 G_1(t)\Phi_i(0) + c^2 G_2(t)\Phi_i(L), \quad i = 0, \ldots, n.$$

In compact form, System (4.3.56) is written as

$$M[\ddot{\mathbf{u}}(t) + \lambda \mathbf{u}(t)] + K\mathbf{u}(t) = \mathbf{f}(t) + \mathbf{g}(t), \tag{4.3.57}$$

where $\mathbf{u}$, $\mathbf{f}$ and $\mathbf{g}$ are column vectors. The elements of $\mathbf{f}$ are given by

$$f_0(t) = \int_{x_0}^{x_1} F(x,t)\Phi_0(x) \, dx, \quad f_n(t) = \int_{x_{n-1}}^{x_n} F(x,t)\Phi_n(x) \, dx, \tag{4.3.58}$$

$$f_i(t) = \int_{x_{i-1}}^{x_{i+1}} F(x,t)\Phi_i(x) \, dx, \quad i = 1, \ldots, n-1, \tag{4.3.59}$$

as the support of Φ_0 is the interval $[x_0, x_1]$, the support of Φ_n is the interval $[x_{n-1}, x_n]$ and the support of Φ_i, $i = 1, \ldots, n-1$, is the interval $[x_{i-1}, x_{i+1}]$. The vector $\mathbf{g}$ depends on the boundary conditions. Its elements are given by

$$g_0(t) = c^2 G_1(t), \quad g_i(t) = 0, \quad i = 1, \ldots, n-1, \quad g_n(t) = c^2 G_2(t).$$

The matrices M and K were evaluated in Formulas (4.3.28)–(4.3.29) and are given by

$$M = \frac{h}{3} \begin{bmatrix} 1 & 1/2 & & \\ 1/2 & 2 & 1/2 & \\ & . & . & . \\ & & 1/2 & 1 \end{bmatrix}, \quad K = \frac{c^2}{h} \begin{bmatrix} 1 & -1 & & \\ -1 & 2 & -1 & \\ & . & . & . \\ & & -1 & 1 \end{bmatrix}.$$

Finally, note that if the new unknown function

$$\mathbf{w} = \dot{\mathbf{u}},$$

is introduced, then the second-order differential System (4.3.57) is converted to the first-order system

$$\begin{bmatrix} \dot{\mathbf{u}} \\ \dot{\mathbf{w}} \end{bmatrix} = \begin{bmatrix} 0 & I \\ -(\lambda I + M^{-1}K) & 0 \end{bmatrix} \begin{bmatrix} \mathbf{u} \\ \mathbf{w} \end{bmatrix} + \begin{bmatrix} 0 \\ M^{-1}(\mathbf{f} + \mathbf{g}) \end{bmatrix}. \quad (4.3.60)$$

Example 4.3.4 Consider the Neumann problem

$$U_{tt} - U_{xx} = F \sin(\omega t)\delta(x - x_h), \; 0 < x < L, \; 0 < t \leq T, \quad (4.3.61)$$

$$U(x, 0) = 0, \; U_t(x, 0) = 0, \quad 0 < x < L, \quad (4.3.62)$$

$$U_x(0, t) = 0, \quad U_x(L, t) = 0, \quad 0 < t \leq T. \quad (4.3.63)$$

The following listing presents a function that applies the FEM to Problem (4.3.61)–(4.3.63). The solution has a discontinuity for U_x in x_h equal to $F \sin(\omega t)$. Therefore, the load vector (4.3.61)–(4.3.63) is given by

$$f_i = 0, \quad x_i \neq x_h, \quad f_h = F \sin(\omega t), \quad x_i = x_h.$$

The graphs of u and $u(x_h, t)$ are plotted in Fig. 4.3.4.

```
function u = wave_n
% This is the function file wave_n.m.
```

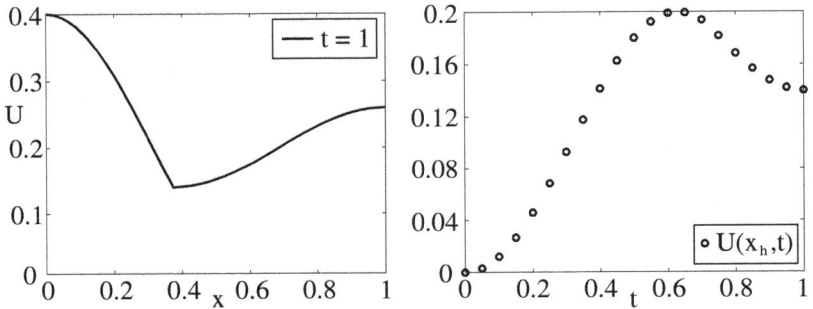

Figure 4.3.4.   Graph of the numerical solution (left); graph of $u(x_h, t)$ (right).

```
% FEM is applied to the following Neumann problem:
% Utt - Uxx = F*delta(x - xh),
% U(x,0) = 0, Ut(x,0) = 0, -Ux(0,t) = 0, Ux(L,t) = 0.
% Initialization
n = 80; L = 1; F = 1; om = 5; T = 1; ih = 31;
h = L/n; x = linspace(0,L,n+1);
KK = [-ones(n+1,1) [1;2*ones(n-1,1);1] -ones(n+1,1)];
K = spdiags(KK, -1:1, n+1, n+1)/h;
MM = [ones(n+1,1)/6 [1/3;2/3*ones(n-1,1);1/3] ones(n+1,1)/6];
M = spdiags(MM, -1:1, n+1, n+1)*h;
C = M^(-1);
B = [zeros(n+1,n+1) eye(n+1,n+1); -C*K zeros(n+1,n+1)];
u = zeros(n+1,1); w = zeros(n+1,1);
g = zeros(n+1,1); f = zeros(n+1,1);
% FEM
nt = 20; tt = linspace(0,T,nt+1); uh = zeros(nt+1,1);
for j = 1:nt % Loop animates the graph.
    [~,y] = ode15s(@system, [tt(j) tt(j+1)], [u(1:n+1); w(1:n+1)], [ ],...
            B, C, g, f, ih, F, om);
    u(1:n+1) = y(end,1:n+1); w(1:n+1)=y(end,n+2:end);
    plot(x,u,'k','LineWidth',2);
    axis([0 L 0 0.4]); xlabel('x'); ylabel('U');
    legend(['t = ',num2str(tt(j+1))],'Location','NorthEast');
    pause(.1);
end
figure(2); plot(tt,uh,'ko','LineWidth',2);
```

```
xlabel('t'); legend('U(x_h, t)','Location','SouthEast');
end

% ———— Local function ————-
function Du = system(t, uw, B, C, g, f, ih, Fh, om)
f(ih)= Fh*sin(om*t);
Du = B*uw + [g; C*f];
end
```

4.4 Exercises

Exercise 4.4.1   Find Eq. (4.1.1) of the parabola $p_2(x)$.

Exercise 4.4.2   Apply integral and int functions to calculate the integrals

$$\int_0^1 \frac{\sin(x)}{x}\, dx, \quad \int_{-\pi}^{\pi} x\cos(|x| + x)\, dx.$$

Exercise 4.4.3   Write the Matlab code that produces Fig. 4.1.3.
Answer.

```
x = linspace(-10, 10, 101);
plot(x, besselj(0,x), 'k', x, besselj(1,x), 'r*:', x, besselj(2,x), 'bo:');
legend('J_0','J_1','J_2');
```

Exercise 4.4.4   Use Matlab to verify that

$$\lim_{z\to 0} J_1(z)/z = .5.$$

Answer. z = sym('z'); limit(besselj(1,z)/z,z,0).

Exercise 4.4.5   Consider the integral_2D_1 listing. Use the local functions to introduce the functions f(x, y), alpha(x), beta(x).
Answer.

```
function integral_2D_2
% This is the function file integral_2D_2.m.
% Integral2 and int functions are called to calculate the integral.
% Local functions are used to define f(x, y), alfa(x), beta(x).
a = 0; b = 4;
Q = integral2(@f, a, b, @alfa, @beta);
x = sym('x'); y = sym('y');
E = int(int(f(x,y), y, alfa(x), beta(x)), x, a, b);
```

```
fprintf('integral2 = %g; int = %g\', Q, double(E))
end
% ———— Local functions ————
function z = f(x, y)
z = x.*y;
end
function y = alfa(x)
y = x.²/4;
end
function y = beta(x)
y = 2*sqrt(x);
end
```

Exercise 4.4.6   Consider a fixed end-fixed end bar in Statics. See Fig. 4.4.1. The bar has length L and is subjected to the axial load $q(x)$. The axial deformation is found by solving the following boundary value problem

$$-AEU''(x) = q(x), \ 0 < x < L, \ U(0) = U(L) = 0.$$

Assume that

$$q(x) = q_A + (q_B - q_A)x/L. \tag{4.4.1}$$

Prove that U and σ are given by

$$U(x) = \frac{1}{AE}[q_A(Lx - x^2)/2 + (q_B - q_A)(L^2x - x^3)/6/L],$$

$$\sigma(x) = \frac{1}{AE}[q_A(L - 2x)/2 + (q_B - q_A)(L^2 - 3x^2)/6/L].$$

Deduce from the last equation that the reactive forces R_A and R_B are given by

$$R_A = -q_AL/2 - (q_B - q_A)L/6, \ \ R_B = -q_AL/2 - 2(q_B - q_A)L/6.$$

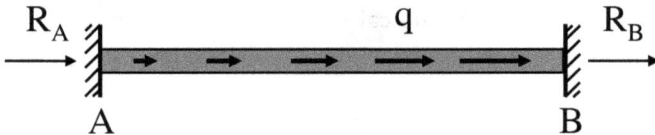

Figure 4.4.1.   Fixed end-fixed end bar subjected to an axial load.

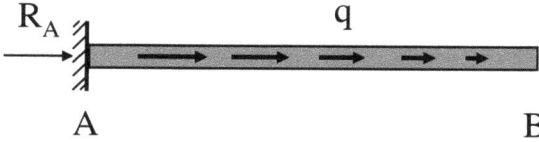

Figure 4.4.2. Fixed end-free end bar subjected to an axial load.

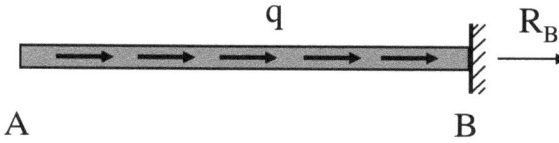

Figure 4.4.3. Free end-fixed end bar subjected to an uniform load.

Note that the trapezoidal load (4.4.1) simplifies to a uniform load for $q_B = q_A$ and to a triangular load for $q_A = 0$.

Exercise 4.4.7 Consider a fixed end-free end bar in Statics. See Fig. 4.4.2. The bar has length L and is subjected to the axial load $q(x)$.

The axial deformation is found by solving the following boundary value problem

$$-AEU''(x) = q(x), \ 0 < x < L, \quad U(0) = 0, \ U'(L) = 0.$$

Assume that $q(x)$ is the triangular load

$$q(x) = q_A(L - x)/L.$$

Prove that U, σ and R_A are given by

$$U(x) = \frac{q_A x^3}{6AEL} - \frac{q_A x^2}{2AE} + \frac{q_A Lx}{2AE}, \ \sigma = \frac{q_A x^2}{2AL} - \frac{q_A x}{A} + \frac{q_A L}{2A}, \ R_A = -\frac{q_A L}{2}.$$

Exercise 4.4.8 Consider a free end-fixed end bar in Statics. See Fig. 4.4.3. The bar has length L and is subjected to the axial uniform load $q(x) = q_A$. Find displacement, stress and reactive force.

Exercise 4.4.9 Consider a fixed end-fixed end bar in Statics. See Fig. 4.4.4. The bar has length L and is subjected to the axial concentrated force

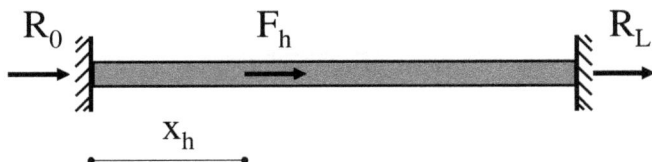

Figure 4.4.4. Fixed end-fixed end bar subjected to an axial concentrated force.

F_h. The axial deformation is found by solving the following boundary value problem

$$-AEU'' = 0, \quad x \in]0, x_h[\, \cup \,]x_h, L[, \tag{4.4.2}$$

$$U(0) = 0, \ \ U(L) = 0, \ \ U(x_h^-) = U(x_h^+), \ \ A\sigma(x_h^+) - A\sigma(x_h^-) = -F_h. \tag{4.4.3}$$

Find the displacement, stress and reactive forces.

Answer. U is linear on both intervals $(0, x_h)$ and (x_h, L) because of (4.4.2). In addition, U must satisfy the first two conditions in (4.4.3). Therefore,

$$U(x) = \begin{cases} C_1 x, & \text{if } \ 0 < x < x_h, \\ C_2(L - x), & \text{if } \ x_h < x < L. \end{cases} \tag{4.4.4}$$

The constants C_1 and C_2 are founded by using the last two conditions in (4.4.3):

$$C_1 x_h = C_2(L - x_h), \quad -AEC_2 - AEC_1 = -F_h.$$

Solving with respect to C_1, C_2 yields

$$C_1 = (L - x_h)F_h/(AEL), \quad C_2 = F_h x_h/(AEL).$$

Substituting the last result into (4.4.4), one arrives at the expression of U

$$U(x) = \begin{cases} (L - x_h)F_h x/(AEL), & \text{if } \ 0 < x < x_h, \\ (L - x)F_h x_h/(AEL), & \text{if } \ x_h < x < L. \end{cases}$$

Hence, stress and reactive forces are

$$\sigma(x) = EU'(x) = \begin{cases} (L - x_h)F_h/(AL), & \text{if } \ 0 < x < x_h, \\ -F_h x_h/(AL), & \text{if } \ x_h < x < L, \end{cases}$$

$$R_0 = -(L - x_h)F_h/L, \quad R_L = -F_h x_h/L.$$

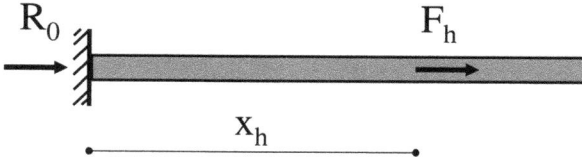

Figure 4.4.5.   Fixed end-free end bar subjected to an axial concentrated force

Exercise 4.4.10   Consider a fixed end-free end bar in Statics. See Fig. 4.4.5. The bar has length L and is subjected to the axial concentrated force F_h. The axial deformation is found by solving the following boundary value problem

$$-EAU'' = 0, \quad x \in]0, x_h[\cup]x_h, L[,$$

$$U(0) = 0, \ U'(L) = 0, \quad U(x_h^-) = U(x_h^+), \quad A\sigma(x_h^+) - A\sigma(x_h^-) = -F_h.$$

Prove that

$$U(x) = \begin{cases} F_h x/(AE), & \text{if } 0 < x < x_h, \\ F_h x_h/(AE), & \text{if } x_h < x < L, \end{cases}$$

$$\sigma(x) = \begin{cases} F_h/A, & \text{if } 0 < x < x_h, \\ 0, & \text{if } x_h < x < L, \end{cases} \quad R_0 = -F_h.$$

Exercise 4.4.11   Write a listing that calls the fem_dd function and solves the boundary value problem

$$-U'' = cx^2, \quad 0 < x < L, \quad U(0) = U_0, \quad U(L) = U_L.$$

Exercise 4.4.12   Write a function, say fem_nd, that solves the Neumann–Dirichlet problem

$$-U'' = F(x), \quad 0 < x < L, \quad U'(0) = U_0', \quad U(L) = U_L.$$

Exercise 4.4.13   Write a function that applies the FEM to calculate the displacement, stress and reactive force for the bar in Example 4.2.6.

Answer.

```
function [u, sigma] = stress_2
% This is the function file stress_2.m.
% FEM is applied to solve the following problem:
% -AE*U"(x) = q(x), q(x) = qA + (L - x)/L, U(0) = 0, U'(L) = 0.
% Analytical solution:
% U = qA*x^3/6/E/A/L - qA*x^2/2/E/A + qA*L*x/2/A/E.
% Initialization
n = 30; u0 = 0; uxL = 0;
L = 1000; % mm
E = 90000; % N/mm^2
A = 1000; % mm^2
qA = 8; % N/mm
Phil = @(xi, x1, x2)        (xi-x1)/(x2-x1);
Phir = @(xi, x2, x3)        -(xi-x3)/(x3-x2);
F = @(xi)        qA/A/E/L*(L - xi);
h = L/n; x = linspace(0,L,n+1);
u = zeros(n+1,1); sigma = zeros(n+1,1); f = zeros(n-1,1);
U = qA*x'.^3/6/E/A/L - qA*x'.^2/2/E/A + qA*L*x'/2/A/E;
S = qA/A/L/2*x'.^2 - qA/A*x' + qA*L/2/A;
% FEM
KK = [-ones(n,1)  [2*ones(n-1,1);1]  -ones(n,1)];
K = spdiags(KK, -1:1, n, n)/h;
for i = 2:n
    f(i-1) =integral(@(xi)F(xi).*Phil(xi,x(i-1),x(i)), x(i-1), x(i));
    f(i-1) = f(i-1) + integral(@(xi)F(xi).*Phir(xi,x(i),x(i+1)), x(i), x(i+1));
end
f(n) = integral(@(xi)F(xi).*Phil(xi,x(n),x(n+1)), x(n), x(n+1))+uxL;
f(1) = f(1) + u0/h;
u(2:end) = K\f; u(1) = u0;
for i=1:n
    sigma(i,1) = E*integral(@(xi)F(xi).*Phir(xi,x(i),x(i+1)), x(i), x(i+1))...
    - u(i)*E/h + u(i+1)*E/h;
end
sigma(n+1) = 0;
rA = -A*sigma(1);
plot(x,u,'k',x,U,'r*','LineWidth',2);
legend('FEM','Exact','Location','SouthEast');
xlabel('x'); ylabel('U'); title('Displacement');
```

```
figure(2); plot(x,sigma,'k',x,S,'r*','LineWidth',2);
legend('FEM','Exact'); xlabel('x'); ylabel('\sigma');title('Stress');
fprintf('RA=%f\n',rA)
fprintf( 'Maximum error for displacement = %g\n',max(abs(U - u)))
fprintf( 'Maximum error for stress= %g\n',max(abs(S - sigma)))
end
```

Exercise 4.4.14 Apply the FEM to discuss the bar in Fig. 4.4.6, which is being subjected to an axial uniform load q and axial concentrated force (x_h, F_h).

Exercise 4.4.15 Calculate the elements of the matrix M shown in Formula (4.3.16).

Answer. The matrix M is symmetric. Consider the main diagonal

$$M_{i,i} = \int_{x_{i-1}}^{x_{i+1}} \Phi_i^2(x) \; dx = \int_{x_{i-1}}^{x_i} \frac{(x - x_{i-1})^2}{h^2} \; dx + \int_{x_i}^{x_{i+1}} \frac{(x_{i+1} - x)^2}{h^2} \; dx,$$

as the support of Φ_i is the interval $[x_{i-1}, x_{i+1}]$. Hence,

$$M_{i,i} = \int_0^h \frac{\xi^2}{h^2} \; d\xi + \int_0^h \frac{\xi^2}{h^2} \; d\xi = 2h/3, \quad i = 1, \ldots, n - 1.$$

Consider $M_{i,i+1}$

$$M_{i,i+1} = \int_{x_i}^{x_{i+1}} \Phi_i(x)\Phi_{i+1}(x) \; dx = -\int_{x_i}^{x_{i+1}} \frac{(x - x_{i+1})(x - x_i)}{h^2} \; dx,$$

as the support of $\Phi_i\Phi_{i+1}$ is the interval $[x_i, x_{i+1}]$. Hence,

$$M_{i,i+1} = \int_0^h \frac{(h - \xi)\xi}{h^2} \; d\xi = \frac{h}{6}, \quad i = 1, \ldots, n - 2.$$

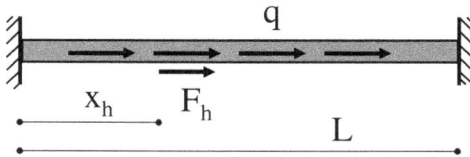

Figure 4.4.6. Fixed end-fixed end bar subjected to an axial uniform load and a concentrated force.

The remaining elements of M are zero. For example, consider

$$M_{1,3} = \int_0^L \Phi_1(x)\Phi_3(x)\,dx.$$

The supports of Φ_1 and Φ_3 are the intervals $[x_0, x_2]$ and $[x_2, x_4]$, respectively. Their intersection is void and $M_{1,3} = 0$.

Exercise 4.4.16 Write a function, say diffusion_d_A, that applies the FEM to solve the following Dirichlet problem

$$U_t - U_{xx} = 0, \quad 0 < x < L, \quad 0 < t \le T,$$

$$U(x,0) = x^2, \quad 0 \le x \le L,$$

$$U(0,t) = 2t, \quad U(L,t) = 2t + L^2, \quad 0 < t \le T.$$

Exercise 4.4.17 Write a function, say wave_d_A, that applies the FEM to the simple Dirichlet problem

$$U_{tt} - U_{xx} = 0, \quad 0 < x < L, \quad 0 < t \le T,$$

$$U(x,0) = x^2, \quad U_t(x,0) = 0, \quad 0 < x < L,$$

$$U(0,t) = t^2, \quad U(L,t) = t^2 + L^2, \quad 0 < t \le T.$$

Chapter 5

Introduction to the Finite Element Method in Two Spatial Dimensions

The Finite Element Method (FEM) in two spatial dimensions is presented. Since the method is applied to elliptic partial differential equations, the section below is devoted to this topic, where the uniqueness of the solutions of the main boundary value problems is discussed. In addition, Green's identities are introduced, as they are related to the weak formulation.

The FEM in two spatial dimensions is introduced in Sec. 5.2 and the shape functions on the triangle element are defined. A number of examples and exercises will help the reader to familiarize themselves with the new concept. The weak formulation is discussed and applied to the boundary value problems. Engineering applications are provided.

The last section is devoted to the Finite Difference Method for elliptic partial differential equations Knabner and Angermann (2003). The Five-Point Method is presented and illustrated with examples and applications.

5.1 Elliptic Partial Differential Equations

5.1.1 *Green's Identities*

Consider the divergence theorem

$$\int_\Omega \nabla \cdot \mathbf{q} \, d\Omega = \int_{\partial\Omega} \mathbf{q} \cdot \mathbf{n} \, dS, \tag{5.1.1}$$

where Ω is a bounded domain, $\mathbf{q}$ is a smooth vector function and $\mathbf{n}$ is the outward unit normal vector. Some important identities can be derived from (5.1.1). If it is assumed that $\mathbf{q} = \nabla u$, then from (5.1.1), it follows

$$\int_\Omega \Delta u \, d\Omega = \int_{\partial\Omega} \frac{\partial u}{\partial n} \, dS, \tag{5.1.2}$$

where $\partial u / \partial n = \nabla u \cdot \mathbf{n}$ is the outward normal derivative to $\partial\Omega$. Moreover, if $\mathbf{q} = v\nabla u$, then (5.1.1) gives

$$\int_\Omega (v\Delta u + \nabla v \cdot \nabla u)\, d\Omega = \int_{\partial\Omega} v\frac{\partial u}{\partial n}\, dS. \qquad (5.1.3)$$

Similarly, if $\mathbf{q} = u\nabla v$, then

$$\int_\Omega (u\Delta v + \nabla u \cdot \nabla v)\, d\Omega = \int_{\partial\Omega} u\frac{\partial v}{\partial n}\, dS.$$

Subtracting the last equation to (5.1.3) yields

$$\int_\Omega (v\Delta u - u\Delta v)\, d\Omega = \int_{\partial\Omega} \left(v\frac{\partial u}{\partial n} - u\frac{\partial v}{\partial n}\right) dS. \qquad (5.1.4)$$

Integral Relationships (5.1.2) to (5.1.4) are named *Green's*[1] *identities*. They are used to discuss the uniqueness of the solutions to boundary value problems for the Poisson[2] equation in Sec. 5.1.2. In addition, Green's identities will be applied to derive the weak form of the Poisson equation in Sec. 5.2.2.

5.1.2 *Boundary Value Problems*

Consider the Dirichlet problem for the Poisson equation

$$\begin{cases} \Delta U = F(x,y,z), & (x,y,z) \in \Omega, \\ U = g(x,y,z), & (x,y,z) \in \partial\Omega. \end{cases} \qquad (5.1.5)$$

When $F = 0$, the Poisson's equation is named Laplace's equation. Let U_1 and U_2 be two solutions to Problem (5.1.5), and V be their difference

$$V = U_1 - U_2.$$

The function V is the solution of the homogeneous problem

$$\begin{cases} \Delta V(x,y,z) = 0, & (x,y,z) \in \Omega, \\ V(x,y,z) = 0, & (x,y,z) \in \partial\Omega. \end{cases} \qquad (5.1.6)$$

---

[1] George Green, a British scientist, 1793–1841. He made fundamental scientific contributions in Electromagnetism in *An Essay on the Application of Mathematical Analysis to the Theories of Electricity and Magnetism* (1828).
[2] Siméon Denis Poisson, a French scientist, 1781–1840. He was Professor at École Polytechnique in Paris. He made scientific contributions in Electromagnetism and Analytical Mechanics.

Consider Identity (5.1.3) for $u = v = V$,

$$\int_{\Omega} (V \Delta V + \nabla V \cdot \nabla V) \, d\Omega = \int_{\partial \Omega} V \frac{\partial V}{\partial n} \, dS.$$

Using (5.1.6) we have

$$\int_{\Omega} (V_x^2 + V_y^2 + V_z^2) \, d\Omega = 0.$$

Hence,

$$V_x^2 + V_y^2 + V_z^2 = 0, \quad \text{on } \Omega \Rightarrow V_x = V_y = V_z = 0, \quad \text{on } \Omega.$$

Therefore,

$$V = U_1 - U_2 = c, \quad \text{on } \Omega,$$

where the constant c must be zero because of $(5.1.6)_2$. In conclusion, it is $U_1 = U_2$, proving that the solution to the Dirichlet Problem (5.1.5) is unique.

Consider the Neumann problem for the Poisson equation

$$\begin{cases} \Delta U = F(x, y, z), & (x, y, z) \in \Omega, \\ \partial U / \partial n = g(x, y, z), & (x, y, z) \in \partial \Omega. \end{cases} \tag{5.1.7}$$

If U_1 and U_2 are two solutions to Problem (5.1.7), then their difference $V = U_1 - U_2$ is the solution of the homogeneous problem

$$\begin{cases} \Delta V(x, y, z) = 0, & (x, y, z) \in \Omega, \\ \partial V(x, y, z) / \partial n = 0, & (x, y, z) \in \partial \Omega. \end{cases}$$

By using the same reasoning as in the previous problem, one arrives at

$$V = U_1 - U_2 = c, \quad \text{on } \Omega.$$

Nothing else can be deduced, as we cannot know the value of V on the boundary in this case. In addition, consider Identity (5.1.2) for $u = U$,

$$\int_{\Omega} \Delta U \, d\Omega = \int_{\partial \Omega} \frac{\partial U}{\partial n} \, dS,$$

and use (5.1.7)

$$\int_{\Omega} F \, d\Omega = \int_{\partial \Omega} g \, dS. \tag{5.1.8}$$

Relationship (5.1.8) is a necessary condition for the existence of solutions to Problem (5.1.7). If the data F and g are not assigned in agreement with (5.1.8), then Problem (5.1.7) has no solution. In conclusion, the Neumann problem has solutions only when compatibility Condition (5.1.8) is satisfied. In this case, the problem has infinite solutions that differ by a constant.

Let us consider the Dirichlet–Neumann problem

$$\begin{cases} \Delta U = F(x,y,z), & (x,y,z) \in \Omega, \\ U = g_1(x,y,z), & (x,y,z) \in \partial\Omega_1, \\ \partial U/\partial n = g_2(x,y,z), & (x,y,z) \in \partial\Omega_2, \end{cases} \qquad (5.1.9)$$

where $\partial\Omega_1 \cup \partial\Omega_2 = \partial\Omega$ and $\partial\Omega_1 \cap \partial\Omega_2 = \emptyset$. From the previous reasoning, it is easily realized that Problem (5.1.9) has a unique solution.

Let us consider the Robin problem

$$\begin{cases} \Delta U = F(x,y,z), & (x,y,z) \in \Omega, \\ \partial U/\partial n + \alpha U = g(x,y,z), & (x,y,z) \in \partial\Omega, \end{cases} \qquad (5.1.10)$$

where α is a positive function

$$\alpha(x,y,z) > 0, \quad (x,y,z) \in \partial\Omega$$

The difference $V = U_1 - U_2$ of two solutions to Problem (5.1.10) satisfies the homogeneous problem

$$\begin{cases} \Delta V = 0, & (x,y,z) \in \Omega, \\ \partial V/\partial n + \alpha V = 0, & (x,y,z) \in \partial\Omega. \end{cases} \qquad (5.1.11)$$

Consider Identity (5.1.3) for $u = v = V$,

$$\int_\Omega (V\Delta V + \nabla V \cdot \nabla V) \, d\Omega = \int_{\partial\Omega} V \frac{\partial V}{\partial n} \, dS.$$

Using (5.1.11), we have

$$\int_\Omega (V_x^2 + V_y^2 + V_z^2) \, d\Omega + \int_{\partial\Omega} \alpha V^2 \, dS = 0.$$

This relationship implies

$$V = 0, \quad \text{on } \partial\Omega, \qquad (5.1.12)$$

and

$$V_x^2 + V_y^2 + V_z^2 = 0, \quad \text{on } \Omega.$$

As in the previous problems, the last equation leads to

$$V = U_1 - U_2 = c, \quad \text{on } \Omega,$$

where the constant c must be zero because of (5.1.12). The Robin problem has a unique solution.

5.2 Finite Element Method in Two Spatial Dimensions

5.2.1 *Shape Functions*

The FEM in two spatial dimensions considers the triangle element. The 2D domain is subdivided into triangles. Examples of simple *triangulations* are illustrated in Figs. 5.2.1 and 5.2.3. Some Matlab functions able to create and plot a triangulation will be provided in the next examples.

Example 5.2.1   The following listing creates the triangulation in Fig. 5.2.1, and produces the plots in Fig. 5.2.2.

```
function triangulation_plot1
% This is the function file triangulation_plot1.m.
% A triangulation is created manually and by calling the delaunay
% function. The triangulation is plotted.
h = 1; k = 1;
x = [h; 2*h; h; 2*h; h; 2*h; h; 2*h; 3*h; 3*h; 3*h; 2*h; h; 0; 0; 0];
y = [k; k; 2*k; 2*k; 3*k; 3*k; 0; 0; k; 2*k; 3*k; 4*k; 4*k; 3*k; 2*k; k];
```

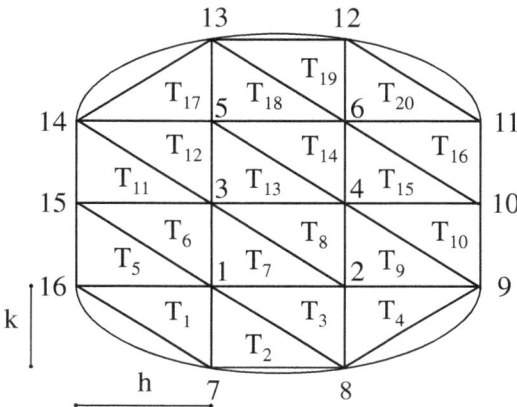

Figure 5.2.1.   Example of triangulation.

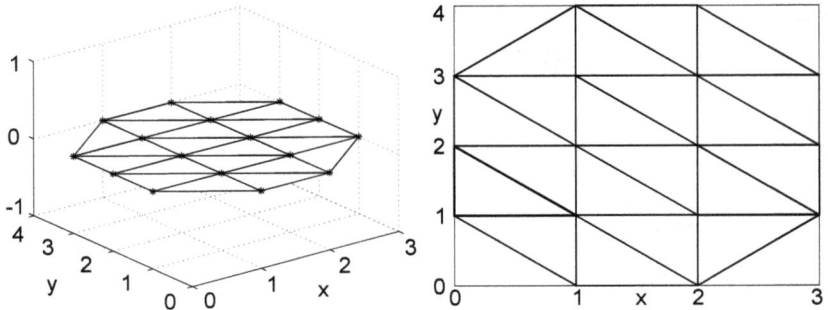

Figure 5.2.2.   Triangulation plots.

```
T= [1 16 7; 1 7 8; 1 8 2; 2 8 9; 1 15 16;% Triangulation created manually.
1 3 15; 1 2 3; 2 4 3; 2 9 4; 9 10 4; 3 14 15; 3 5 14; 3 4 5; 4 6 5; 4 10 6;
10 11 6; 5 13 14; 5 6 13; 6 12 13; 6 11 12];
z = zeros(length(x), 1);
trimesh(T, x, y, z,'LineWidth',1, 'edgecolor', 'k', 'Marker','*');
xlabel('x'); ylabel('y');
Td = delaunay(x,y); % Triangulation created by calling the delaunay function.
figure(2);
triplot(Td, x, y, 'k','LineWidth',1); xlabel('x'); ylabel('y');
end
```

Example 5.2.2   The following listing creates the triangulation in Fig. 5.2.3. The undesirable triangulation is shown in Fig. 5.2.4.

```
function triangulation_plot2
% This is the function file triangulation_plot2.m.
% A triangulation is created by calling the delaunay function.
% The triangulation is plotted.
n = 4; m = 8; h = 1;
x1(1:n*m,1) = repmat([0; h; 2*h; 3*h], 8, 1);
for k=1:m
     y1(1+(k-1)*n:k*n,1) = (k-1)*h*ones(n,1);
end
x2 = [4*h; 4*h; 4*h]; y2 =[0; h; 2*h]; x = [x1; x2]; y = [y1; y2];
Tw = delaunay(x,y); triplot(Tw, x, y, 'k','LineWidth',1); axis('equal');
     % The delaunay function can generate undesirable triangulations when
     % the domain is nonconvex. In this situation, constrained triangulations
```

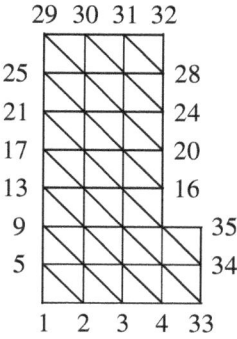

Figure 5.2.3.   Example of triangulation.

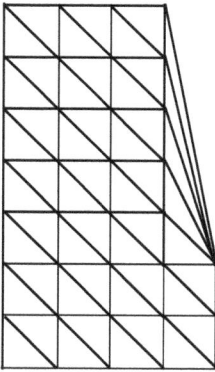

Figure 5.2.4.   Undesirable triangulation.

```
% should be created and the delaunayTriangulation class should be used.
% However, in this simple example, the triangulation is corrected manually.
T1 = delaunay(x1,y1);
T2 = [4 33 8; 33 34 8; 8 34 12; 34 35 12]; T = [T1; T2];
figure(2); triplot(T, x, y, 'k','LineWidth',1); axis('equal');
end
```

Consider the generic triangle element. Its vertices, named *nodes* in this contest, are locally numbered 1, 2 and 3, counterclockwise. The coordinates of a node are referred to a global reference system and indicated with (x_1, y_1), (x_2, y_2) and (x_3, y_3), as shown in Fig. 5.2.5. Consider the linear

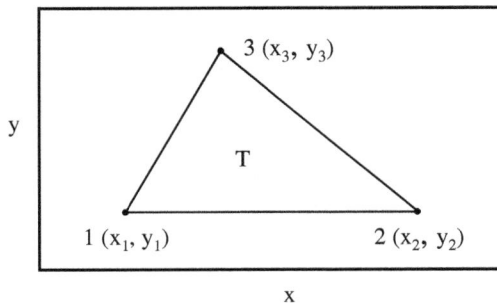

Figure 5.2.5.   Triangle element.

functions on triangle T

$$\Phi_i(x,y) = a_i x + b_i y + c_i, \quad i = 1,2,3, \quad (x,y) \in T, \tag{5.2.1}$$

defined by the Kronecker δ functions

$$\Phi_i(x_j, y_j) = \delta_{ij} = \begin{cases} 1, & i = j, \\ 0, & i \neq j, \end{cases} \quad i,j = 1,2,3. \tag{5.2.2}$$

The explicit expression of Φ_i is derived from Property (5.2.2). Indeed, consider Φ_1 and note that

$$\begin{array}{ll} \Phi_1(x_1, y_1) = 1, & a_1 x_1 + b_1 y_1 + c_1 = 1, \\ \Phi_1(x_2, y_2) = 0, \quad \Rightarrow & a_1 x_2 + b_1 y_2 + c_1 = 0, \\ \Phi_1(x_3, y_3) = 0, & a_1 x_3 + b_1 y_3 + c_1 = 0. \end{array}$$

Hence,

$$\begin{bmatrix} x_1 & y_1 & 1 \\ x_2 & y_2 & 1 \\ x_3 & y_3 & 1 \end{bmatrix} \begin{bmatrix} a_1 \\ b_1 \\ c_1 \end{bmatrix} = \begin{bmatrix} 1 \\ 0 \\ 0 \end{bmatrix}. \tag{5.2.3}$$

Solving System (5.2.3) with respect to a_1, b_1 and c_1, and the result substituted into (5.2.1) yields (see Exercise 5.4.1)

$$\Phi_1(x,y) = [(y_2 - y_3)x - (x_2 - x_3)y + x_2 y_3 - x_3 y_2]/(2A), \tag{5.2.4}$$

where A is the area of the triangle

$$A = \frac{1}{2} \begin{vmatrix} x_1 & y_1 & 1 \\ x_2 & y_2 & 1 \\ x_3 & y_3 & 1 \end{vmatrix}. \tag{5.2.5}$$

The functions Φ_2 and Φ_3 are derived similarly

$$\Phi_2(x,y) = [(y_3 - y_1)x - (x_3 - x_1)y + x_3 y_1 - x_1 y_3]/(2A),$$
$$\Phi_3(x,y) = [(y_1 - y_2)x - (x_3 - x_2)y + x_1 y_2 - x_2 y_1]/(2A). \tag{5.2.6}$$

The gradient of Φ_i

$$\nabla \Phi_i(x,y) = \left(\frac{\partial \Phi_i}{\partial x}, \frac{\partial \Phi_i}{\partial y} \right), \quad (x,y) \in T, \tag{5.2.7}$$

is a constant vector function, as Φ_i is linear. The partial derivatives are the same as the coefficients of x and y of the linear function Φ_i.

The functions Φ_i, $i = 1, 2, 3$, are a basis for the linear functions defined on T. Indeed, any linear function $u(x,y)$ defined on T can be expressed as a linear combination of Φ_1, Φ_2 and Φ_3,

$$u(x,y) = u_1 \Phi_1(x,y) + u_2 \Phi_2(x,y) + u_3 \Phi_3(x,y), \quad (x,y) \in T, \tag{5.2.8}$$

where $u_i = u(x_i, y_i)$. Formula (5.2.8) follows from the straightforward calculations shown in Exercise 5.4.2. Moreover,

$$u_1 \Phi_1(x,y) + u_2 \Phi_2(x,y) + u_3 \Phi_3(x,y) = 0, \quad (x,y) \in T, \tag{5.2.9}$$

if and only if

$$u_i = 0, \quad i = 1, 2, 3. \tag{5.2.10}$$

It must be proved that $(5.2.9) \Rightarrow (5.2.10)$, as $\Leftarrow$ is obvious. From (5.2.9) and Property (5.2.2), it follows

$$0 = u_1 \Phi_1(x_i, y_i) + u_2 \Phi_2(x_i, y_i) + u_3 \Phi_3(x_i, y_i) = u_i \Phi_i(x_i, y_i) = u_i,$$

that is the desired result.

Consider a triangulation with m triangles and n nodes, for example, see Fig. 5.2.1. From the previous discussion related to a single element, it follows that the linear functions definined by the Kronecker δ functions

$$\Phi_i(x_j, y_j) = \delta_{ij} = \begin{cases} 1, i = j, \\ 0, i \neq j, \end{cases} \quad i, j = 1, \ldots, n, \tag{5.2.11}$$

are a global basis for the continuous functions u that are linear on each element, i.e., the piecewise linear continuous functions u defined on the triangulated domain. Therefore, u can be expressed as

$$u(x, y) = \sum_{i=1}^{n} u_i \Phi_i(x, y), \qquad (5.2.12)$$

where $u_i = u(x_i, y_i)$. The support of Φ_i related to node i is composed of the triangles that have a vertex that is the same as node i. The expression of Φ_i on each triangle of the support is deduced from the general Formulas (5.2.4) and (5.2.6).

Example 5.2.3 Let us determine Φ_i and $\nabla\Phi_i$ for the triangulation in Fig. 5.2.1. Consider the node 1. The support of Φ_1 is $\{T_1, T_2, T_3, T_7, T_6, T_5\}$. See Fig. 5.2.6. Let us add the local numbering 1, 2 and 3 to the nodes of the triangles belonging to the support, as shown in Fig. 5.2.7 for the triangles T_1, T_2 and T_3. In this way, it is easy to apply Formula (5.2.4), and

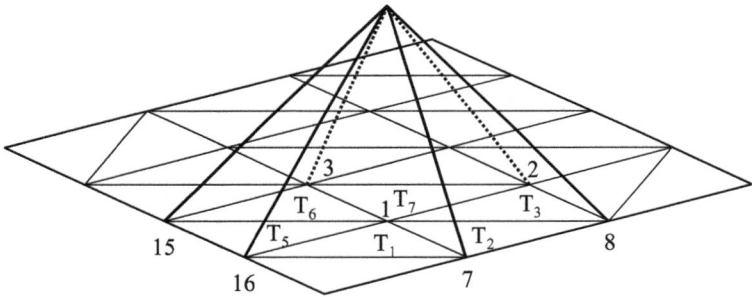

Figure 5.2.6.   Support of Φ_1.

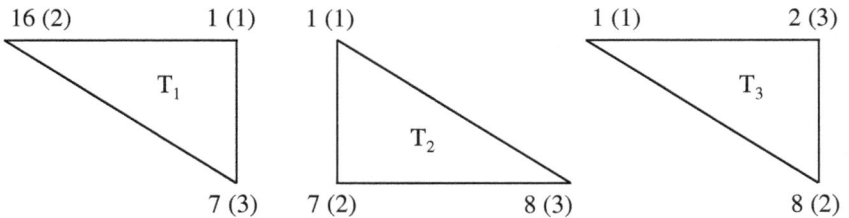

Figure 5.2.7. T_1, T_2, T_3: global and local numbering.

for node 1, we get

$$\Phi_1 = \begin{cases} x/h + y/k - 1, & \text{on } T_1, \\ y/k, & \text{on } T_2, \\ -x/h + 2, & \text{on } T_3, \\ -x/h - y/k + 3, & \text{on } T_7, \\ -y/k + 2, & \text{on } T_6, \\ x/h, & \text{on } T_5 \\ 0, & \text{otherwise,} \end{cases} \qquad \nabla\Phi_1 = \begin{cases} (1/h, 1/k), & \text{on } T_1, \\ (0, 1/k), & \text{on } T_2, \\ (-1/h, 0), & \text{on } T_3, \\ (-1/h, -1/k), & \text{on } T_7, \\ (0, -1/k), & \text{on } T_6, \\ (1/h, 0), & \text{on } T_5, \\ 0, & \text{otherwise.} \end{cases}$$

Consider node 2. The support of Φ_2 is $\{T_3, T_4, T_9, T_8, T_7\}$. Applying Formula (5.2.4) yields

$$\Phi_2 = \begin{cases} x/h + y/k - 2, & \text{on } T_3, \\ -x/h + y/k + 2, & \text{on } T_4, \\ -x/h - y/k + 4, & \text{on } T_9, \\ -y/k + 2, & \text{on } T_8, \\ x/h - 1, & \text{on } T_7, \\ 0, & \text{otherwise,} \end{cases} \qquad \nabla\Phi_2 = \begin{cases} (1/h, 1/k), & \text{on } T_3, \\ (-1/h, 1/k), & \text{on } T_4, \\ (-1/h, -1/k), & \text{on } T_9, \\ (0, -1/k), & \text{on } T_8, \\ (1/h, 0), & \text{on } T_7, \\ 0, & \text{otherwise.} \end{cases}$$

The support of Φ_3 is $\{T_6, T_7, T_8, T_{13}, T_{12}, T_{11}\}$. Applying Formula (5.2.4) yields

$$\Phi_3 = \begin{cases} x/h + y/k - 2, & \text{on } T_6, \\ y/k - 1, & \text{on } T_7, \\ -x/h + 2, & \text{on } T_8, \\ -x/h - y/k + 4, & \text{on } T_{13}, \\ -y/k + 3, & \text{on } T_{12}, \\ x/h, & \text{on } T_{11}, \\ 0, & \text{otherwise,} \end{cases} \qquad \nabla\Phi_3 = \begin{cases} (1/h, 1/k), & \text{on } T_6, \\ (0, 1/k), & \text{on } T_7, \\ (-1/h, 0), & \text{on } T_8, \\ (-1/h, -1/k), & \text{on } T_{13}, \\ (0, -1/k), & \text{on } T_{12}, \\ (1/h, 0), & \text{on } T_{11}, \\ 0, & \text{otherwise.} \end{cases}$$

See Exercise 5.4.3 for the remaining nodes.

Example 5.2.4 The following listing presents a function that calculates Φ_i and $\nabla\Phi_i$ for a given triangulation.

```
function [Phi, nablaPhi] = pyramid(T, x, y, node)
% This is the function file pyramid.m.
% The input arguments are: triangulation T, coordinates x, y of the nodes
% and the node related to Φ to calculate. The function returns two
% matrices with the coefficients of Φ and ∇Φ. The number of rows
% in the matrices is the same as the number of triangles in the support of Φ.
```

% For example, consider that triangulation in Example 5.2.1 is passed
% and node $= 1$. If it is assumed $h = k = 1$, then the function returns

$$
\%\Phi_1 = \begin{bmatrix} 1 & 1 & -1 & 1 \\ 0 & 1 & 0 & 2 \\ -1 & 0 & 2 & 3 \\ 1 & 0 & 0 & 5 \\ 0 & -1 & 2 & 6 \\ -1 & -1 & 3 & 7 \end{bmatrix}, \quad
\nabla\Phi_1 = \begin{bmatrix} 1 & 1 & 1 \\ 0 & 1 & 2 \\ -1 & 0 & 3 \\ 1 & 0 & 5 \\ 0 & -1 & 6 \\ -1 & -1 & 7 \end{bmatrix}.
$$

% The first three elements of a row in Φ_1 are the coefficients of the linear
% function Φ_1 related to the triangle specified by the fourth element in the
% row. The first two elements of a row in $\nabla\Phi_1$ are the components of the
% gradient of Φ_1 related to the triangle specified by the third element in
% the row.

```
[support, index] = find(T == node);
mt = length(support);
Phi = zeros(mt, 4); nablaPhi = zeros(mt, 3);
for i=1:mt
      local = local_base(T, support(i), index(i));
             % Local base for support(i) with node = 1
      area= det([ x(T(support(i),:)') y(T(support(i),:)') [1;1;1]])/2;
             % area of support(i)
      a = (y(local(2)) - y(local(3)))/2/area;
      b = -(x(local(2)) - x(local(3)))/2/area;
      c = (x(local(2))*y(local(3))-x(local(3))*y(local(2)) )/2/area;
      Phi(i,:) = [a b c support(i)]; nablaPhi(i,:) = [a b support(i)];
end
end
% ———— Local function ————
function first = local_base(T, triangle, index)
first = T(triangle, :); % The nodes of 'triangle' are saved in 'first'.
if index > 1 % Local base (counterclockwise).
      if index > 2
             first(1) = T(triangle, 3);
             first(2) = T(triangle, 1);
             first(3) = T(triangle, 2);
      else
             first(1) = T(triangle, 2);
```

```
        first(2) = T(triangle, 3);
        first(3) = T(triangle, 1);
    end
end
end
```

Example 5.2.5   The following listing applies the pyramid function to the domain in Fig. 5.2.1. A specific Φ is returned and plotted. See Fig. 5.2.8.

```
function [Phi, nablaPhi] = pyramid_phi
% This is the function file pyramid_phi.m.
% Pyramid function is called to determine the function Phi related to
% the node specified by the User. Phi is plotted.
h = 1; k = 1;
x = [h; 2*h; h; 2*h; h; 2*h; h; 2*h; 3*h; 3*h; 3*h; 2*h; h; 0; 0; 0];
y = [k; k; 2*k; 2*k; 3*k; 3*k; 0; 0; k; 2*k; 3*k; 4*k; 4*k; 3*k; 2*k; k];
T= [1 16 7; 1 7 8; 1 8 2; 2 8 9; 1 15 16; 1 3 15; 1 2 3;
    2 4 3; 2 9 4; 9 10 4; 3 14 15; 3 5 14; 3 4 5; 4 6 5;
    4 10 6; 10 11 6; 5 13 14; 5 6 13; 6 12 13; 6 11 12];
node = 4; % Specify the node here.
[Phi, nablaPhi] = pyramid(T, x, y, node);
z = zeros(length(x), 1); z(node) = 1; trimesh(T, x, y, z, 'edgecolor', 'k');
xlabel('x'); ylabel('y'); zlabel('\Phi');
hidden off;
end
```

5.2.2   *Weak Form of the Poisson Equation*

Consider the Dirichlet problem for the Poisson's equation in two dimensions

$$-\Delta U(x, y) = F(x, y), \quad (x, y) \in \Omega, \tag{5.2.13}$$

$$U(x, y) = g(x, y), \quad (x, y) \in \partial\Omega. \tag{5.2.14}$$

The minus sign in Eq. (5.2.13) could be eliminated and included in F. However, Eq. (5.2.13) is usually written like this in the FEM context. Multiply Eq. (5.2.13) by the smooth test function $v(x, y)$ and integrate over Ω

$$-\int_\Omega v\Delta U \ d\Omega = \int_\Omega Fv \ d\Omega. \tag{5.2.15}$$

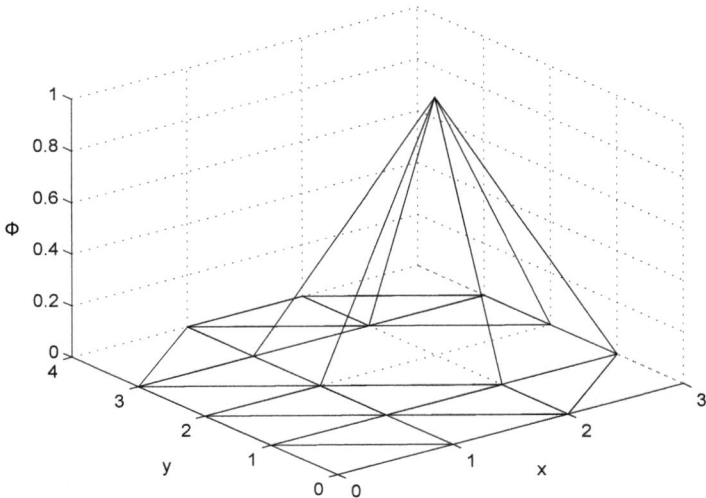

Figure 5.2.8. $\Phi(4)$.

Let us use Green's identity (5.1.3) in the previous equation and get

$$\int_\Omega \nabla v \cdot \nabla U \; d\Omega = \int_{\partial\Omega} v \frac{\partial U}{\partial n} \; ds + \int_\Omega F v \; d\Omega, \qquad (5.2.16)$$

where $\partial U/\partial n$ is the outward normal derivative to $\partial\Omega$. The last equation is the *weak form* of Eq. (5.2.13) that is named the *strong form* by contrast. The FEM considers Eq. (5.2.16). The approximating solution to the Dirichlet problem is expressed as a finite series

$$u(x,y) = \sum_{j=1}^n u_j \Phi_j(x,y), \qquad (5.2.17)$$

where u_j, $j = 1, ..., n$ are unknown coefficients and $\Phi_j(x,y)$ are given shape functions. Substituting (5.2.17) into (5.2.16) and assuming $v = \Phi_i$, $i = 1, \ldots, n$, yields

$$\sum_{j=1}^n u_j \int_\Omega \nabla\Phi_i \cdot \nabla\Phi_j \; d\Omega = \int_{\partial\Omega} \Phi_i \frac{\partial u}{\partial n} \; ds + \int_\Omega F\Phi_i \; d\Omega, \; i = 1, \ldots, n. \;\; (5.2.18)$$

Hence,

$$\sum_{j=1}^n K_{ij} u_j = f_i, \quad i = 1, \ldots, n, \qquad (5.2.19)$$

where

$$K_{ij} = \int_\Omega \nabla\Phi_i \cdot \nabla\Phi_j \, d\Omega, \quad f_i = \int_{\partial\Omega} \Phi_i \frac{\partial u}{\partial n} \, ds + \int_\Omega F\Phi_i \, d\Omega, \quad i = j = 1, \ldots, n.$$

$$(5.2.20)$$

In compact form, System (5.2.19) is written as

$$K\mathbf{u} = \mathbf{f},$$

where K is the stiffness matrix and $\mathbf{f}$ is the load vector. The nodes where the solution is known (as given by the boundary conditions) are named *constrained nodes*. By contrast, the nodes where the solution is unknown are named *free nodes*. For example, consider the Dirichlet problem. The nodes internal to the domain are free and the nodes on the boundary are constrained.

Example 5.2.6   Let us discuss the following Dirichlet problem

$$-\Delta U(x, y) = 0, \quad (x, y) \in \Omega, \quad\quad\quad (5.2.21)$$

$$U(x, y) = x^2 - y^2, \quad (x, y) \in \partial\Omega, \quad\quad\quad (5.2.22)$$

where Ω is the domain in Fig. 5.2.1. The free nodes are $n_f = 6$, namely, the nodes from 1 to 6. The remaining nodes from 7 to 15 are constrained. The solution on these nodes is given by boundary Condition (5.2.22)

$$\begin{aligned}
&u_7 = h^2, &u_8 = 4h^2, &\quad u_9 = 9h^2 - k^2, \\
&u_{10} = 9h^2 - 4k^2, \; u_{11} = 9h^2 - 9k^2, \; &u_{12} = 4h^2 - 16k^2, \\
&u_{13} = h^2 - 16k^2, \; u_{14} = -9k^2, &u_{15} = -4k^2, \\
&u_{16} = -k^2.
\end{aligned}$$

$$(5.2.23)$$

System (5.2.19) for Problem (5.2.21) and (5.2.22) has $n_f = 6$ unknowns $u_1, \ldots, u_6$. Therefore, it simplifies to

$$\sum_{j=1}^{n_f} u_j \int_\Omega \nabla\Phi_i \cdot \nabla\Phi_j \, d\Omega = - \sum_{j=n_f+1}^{n} u_j \int_\Omega \nabla\Phi_i \cdot \nabla\Phi_j \, d\Omega, i = 1, \ldots, n_f,$$

$$(5.2.24)$$

as $F = 0$ and $\Phi_i(x, y) = 0$, $i = 1, \ldots, n_f$ on the boundary. System (5.2.24) can be written in matrix notation as follows

$$\sum_{j=1}^{n_f} K_{ij} u_j = g_i, \quad i = 1, \ldots, n_f, \quad\quad\quad (5.2.25)$$

where

$$g_i = - \sum_{j=n_f+1}^{n} u_j \int_{\Omega} \nabla \Phi_i \cdot \nabla \Phi_j \, d\Omega = - \sum_{j=n_f+1}^{n} b_{ij} u_j, \quad i = 1, \ldots, n_f.$$

(5.2.26)

A listing that applies the FEM to solve Problem (5.2.21) and (5.2.22) will be illustrated in Example 5.2.7. Now, the integrals in Formulas (5.2.25) and (5.2.26) are calculated manually in order to better understand the code in the listing. Note that the integrals over Ω in (5.2.21) and (5.2.22) are calculated on smaller domains. Indeed, if the integrand function is $\nabla \Phi_i \cdot \nabla \Phi_i$, then the integral is calculated over the support of Φ_i. If the integrand function is $\nabla \Phi_i \cdot \nabla \Phi_j$, then the integral is calculated over the intersection of the two supports and it is zero if the intersection is void. Consider $i = 1$ and use the results provided in Example 5.2.3 and Exercise 5.4.4. Therefore, we get

$$K_{11} = \int_{T_1 \cup T_2 \cup T_3 \cup T_7 \cup T_6 \cup T_5} \nabla \Phi_1 \cdot \nabla \Phi_1 = 2hk \left[\frac{1}{h^2} + \frac{1}{k^2} \right],$$

$$K_{12} = \int_{T_3 \cup T_7} \nabla \Phi_1 \cdot \nabla \Phi_2 = -hk \frac{1}{h^2}, \quad K_{13} = \int_{T_6 \cup T_7} \nabla \Phi_1 \cdot \nabla \Phi_3 = -hk \frac{1}{k^2},$$

$$K_{14} = K_{15} = K_{16} = 0,$$

$$b_{17} = \int_{T_1 \cup T_2} \nabla \Phi_1 \cdot \nabla \Phi_7 = -hk \frac{1}{k^2}, \quad b_{18} = b_{19} = b_{1,10} = b_{1,11} = 0,$$

$$b_{1,12} = b_{1,13} = b_{1,14} = b_{1,15} = 0, \quad b_{1,16} = \int_{T_1 \cup T_5} \nabla \Phi_1 \cdot \nabla \Phi_{16} = -hk \frac{1}{h^2}.$$

Lastly, consider boundary Values (5.2.23) and obtain

$$g_1 = \frac{h}{k} u_7 + \frac{k}{h} u_{16} = \frac{h}{k} h^2 - \frac{k}{h} k^2.$$

The other integrals are calculated in Exercise 5.4.4.

Example 5.2.7   The following listing applies the FEM to solve Problem (5.2.21) and (5.2.22). If it is assumed $h = k = 1$, then the following values are obtained for the unknown coefficients:

$$u_1 = 0.0, \quad u_2 = 3.0, \quad u_3 = -3.0, \quad u_4 = 0.0, \quad u_5 = -8.0, \quad u_6 = -5.0.$$

The numerical solution is very accurate as it can be immediately verified by considering the analytical solution of the problem: $U(x, y) = x^2 - y^2$. The graph of the numerical solution is shown in Fig 5.2.9.

```
function u = laplace
% This is the function file laplace.m.
% FEM is applied to solve the Dirichlet problem in Example 5.2.7, Sec. 5.2.2.
h = 1; k = 1;
x = [h; 2*h; h; 2*h; h; 2*h; h; 2*h; 3*h; 3*h; 3*h; 2*h; h; 0; 0; 0];
y = [k; k; 2*k; 2*k; 3*k; 3*k; 0; 0; k; 2*k; 3*k; 4*k; 4*k; 3*k; 2*k; k];
T= [1 16 7; 1 7 8; 1 8 2; 2 8 9; 1 15 16; 1 3 15; 1 2 3; 2 4 3; 2 9 4; 9 10 4;
    3 14 15; 3 5 14; 3 4 5; 4 6 5; 4 10 6; 10 11 6; 5 13 14; 5 6 13; 6 12 13;
    6 11 12];
n = length(x); % Number of nodes.
nf = 6;% Free nodes.
K = zeros(nf,nf);% Stiffness matrix.
b = zeros(nf,n-nf);
ub = [h²; 4*h²; 9*h²-k²; 9*h²-4*k²; 9*h²-9*k²; 4*h²-16*k²;
      h²-16*k²; -9*k²; -4*k²; -k²]; % Boundary conditions.
```

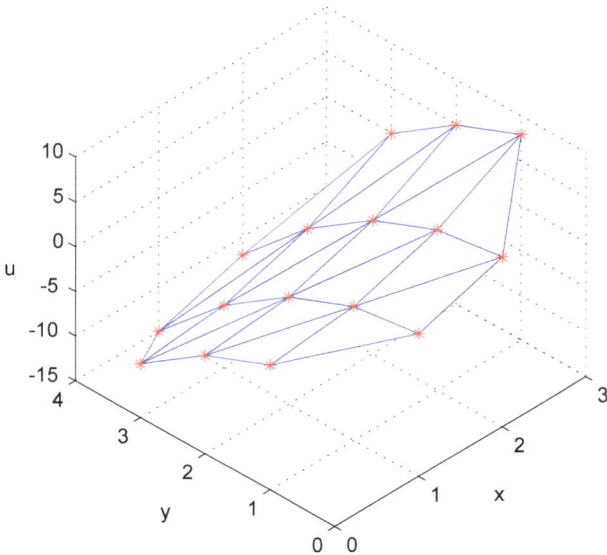

Figure 5.2.9.   Graph of the numerical solution to Problem (5.2.21) and (5.2.22).

```
% FEM
for i=1:nf
    [~,nablaPhii] = pyramid(T, x, y, i);
    support_i = nablaPhii(:,3);
    for j=i:n
        [~,nablaPhij] = pyramid(T, x, y, j);
        support_j = nablaPhij(:,3);
        mt = length(support_j);
        for r=1:mt
          if ismember(support_j(r), support_i)==1
            index = find(support_i == support_j(r));
            area=det([x(T(support_j(r),:)',:) y(T(support_j(r),:)',:) [1;1;1]])/2;
            % Area of support_j(r).
            if j <= nf % If j is a free node, matrix K is updated.
                K(i,j)=K(i,j)+nablaPhij(r,1:2)*nablaPhii(index,1:2)'*area;
                K(j,i)=K(i,j);
            else % If j is a constrained node, matrix b is updated.
                b(i,j-nf)=b(i,j-nf)+nablaPhij(r,1:2)*nablaPhii(index,1:2)'*area;
            end
          end
        end
    end
end
g = -b * ub; u = K\g; u = [u; ub];
U = [h^2-k^2; 4*h^2-k^2; h^2-4*k^2; 4*h^2-4*k^2; h^2-9*k^2; 4*h^2-9*k^2;ub];
    % Analytical solution.
fprintf('Maximum error = %g\n',max( abs(U-u) ) )
trimesh(T, x, y, u, 'edgecolor', 'b', 'Marker','*'); view(-46,43); hold on;
trimesh(T, x, y, U, 'edgecolor', 'b', 'Marker', '*' ,'MarkerEdgecolor', 'r');
xlabel('x'); ylabel('y'); zlabel('u'); hold off
end
```

5.2.3 *Dirichlet–Neumann Problem*

Consider the Dirichlet–Neumann problem for the Poisson equation in two dimensions

$$-\Delta U(x,y) = F(x,y), \quad (x,y) \in \Omega. \qquad (5.2.27)$$

The function U is assigned on $\partial\Omega_1 \subset \partial\Omega$

$$U(x,y) = G_1(x,y), \quad (x,y) \in \partial\Omega_1, \tag{5.2.28}$$

and the normal derivative $\partial U/\partial n$ is assigned on $\partial\Omega_2 \subset \partial\Omega$

$$\frac{\partial U}{\partial n}(x,y) = G_2(x,y), \quad (x,y) \in \partial\Omega_2, \tag{5.2.29}$$

where $\partial\Omega_1 \cup \partial\Omega_2 = \partial\Omega$ and $\partial\Omega_1 \cap \partial\Omega_2 = \emptyset$. Multiply Eq. (5.2.27) by the smooth test function $v(x,y)$ and integrate over Ω

$$-\int_\Omega v\Delta U \; d\Omega = \int_\Omega Fv \; d\Omega.$$

Let us use Green's identity (5.1.3) in the previous equation and get

$$\int_\Omega \nabla v \cdot \nabla U \; d\Omega = \int_{\partial\Omega_1} v\frac{\partial U}{\partial n} \; ds + \int_{\partial\Omega_2} vG_2 \; ds + \int_\Omega Fv \; d\Omega, \tag{5.2.30}$$

where boundary Condition (5.2.29) was used. The last equation is the *weak form* of Eq. (5.2.27) that is named the *strong form* by contrast. The FEM considers the weak form (5.2.30) and the approximating solution is expressed as a finite series

$$u(x,y) = \sum_{j=1}^n u_j \Phi_j(x,y),$$

where u_j, $j = 1, ..., n$ are unknown coefficients and $\Phi_j(x,y)$ are given shape functions.

Example 5.2.8   Consider the following Dirichlet–Neumann problem

$$-\Delta U(x,y) = 0, \quad (x,y) \in \Omega, \tag{5.2.31}$$

$$U = 0 \quad \text{on segment } x = y, \; (\partial\Omega_1), \tag{5.2.32}$$

$$\frac{\partial U}{\partial n} = \begin{cases} 0 & \text{on segment } y = 0, \; (\partial\Omega_2), \\ 4h & \text{on segment } x = 2h, \; (\partial\Omega_2), \end{cases} \tag{5.2.33}$$

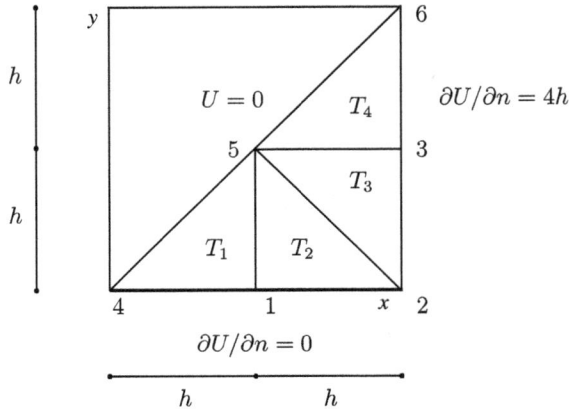

Figure 5.2.10.   Dirichlet–Neumann Problem (5.2.31) to (5.2.33).

where Ω is the domain in Fig. 5.2.10. By considering boundary Condition (5.2.32), the approximating solution to Problem (5.2.31)–(5.2.33) simplifies to

$$u(x,y) = \sum_{j=1}^{3} u_j \Phi_j(x,y). \tag{5.2.34}$$

Substituting (5.2.34) into (5.2.30) and assuming $v = \Phi_i$, $i = 1, 2, 3$, yields

$$\sum_{j=1}^{3} u_j \int_{\Omega} \nabla \Phi_j \cdot \nabla \Phi_i d\Omega = \int_{\partial \Omega_2} \Phi_i \frac{\partial u}{\partial n}\, ds, \quad i = 1, 2, 3,$$

as $F = 0$ and $\Phi_i = 0$, $i = 1, 2, 3$, on segment $x = y$ ($\partial \Omega_1$). By considering boundary Condition (5.2.33), it follows from the previous equation that

$$\sum_{j=1}^{3} K_{ij} u_j = 4h \int_{y_2}^{y_6} \Phi_i(2h, y)dy, \quad i = 1, 2, 3. \tag{5.2.35}$$

The supports of Φ_1, Φ_2 and Φ_3 are $\{T_1, T_2\}$, $\{T_2, T_3\}$ and $\{T_3, T_4\}$, respectively. In addition, the functions Φ_1, Φ_2 and Φ_3 are given by (see Formula (5.2.4))

$$\Phi_1 = \begin{cases} x/h - y/h, & \text{on } T_1, \\ -x/h - y/h + 2/h, & \text{on } T_2, \end{cases} \qquad \Phi_2 = \begin{cases} x/h - 1, & \text{on } T_2, \\ -y/h + 1, & \text{on } T_3, \end{cases}$$

$$\Phi_3 = \begin{cases} x/h + y/h - 2, & \text{on } T_3, \\ x/h - y/h, & \text{on } T_4. \end{cases}$$

By using the previous formulas, we can calculate the stiffness matrix

$$K_{11} = 2, \ K_{12} = -0.5, \ K_{13} = 0, \ K_{22} = 1, \ K_{23} = -0.5, \ K_{33} = 2. \quad (5.2.36)$$

Calculate the integrals in the known term of System (5.2.35). For $i = 1$

$$4h \int_{y_2}^{y_6} \Phi_1(2h, y)dy = 0, \qquad (5.2.37)$$

as $\Phi_1 = 0$ on $\{x = 2h, \ 0 < y < 2h\}$. Consider $i = 2$

$$4h \int_{y_2}^{y_6} \Phi_2(2h, y)dy = 4h \int_0^h (1 - y/h)dy = 2h^2, \qquad (5.2.38)$$

as $\Phi_2 = 1 - y/h$ on $\{x = 2h, \ 0 < y < h\}$ and $\Phi_2 = 0$ on $\{x = 2h, \ h < y < 2h\}$. Lastly, for $i = 3$

$$4h \int_{y_2}^{y_6} \Phi_3(2h, y)dy = 4h \int_0^h y/h \ dy + 4h \int_h^{2h} (2 - y/h)dy = 4h^2, \quad (5.2.39)$$

as $\Phi_3 = y/h$ on $\{x = 2h, \ 0 < y < h\}$ and $\Phi_3 = 2 - y/h$ on $\{x = 2h, \ h < y < 2h\}$. Substituting (5.2.36) to (5.2.39) into (5.2.35), one arrives at the system

$$\begin{cases} 2u_1 - 0.5u_2 = 0, \\ -0.5u_1 + u_2 - 0.5u_3 = 2h^2, \\ -0.5u_2 + 2u_3 = 4h^2. \end{cases}$$

Solving the previous system yields

$$u_1 = h^2, \quad u_2 = 4h^2, \quad u_3 = 3h^2.$$

The solution provided by the FEM is very accurate, as it can be verified by using the analytical solution of Problem (5.2.31)–(5.2.33): $U = x^2 - y^2$.

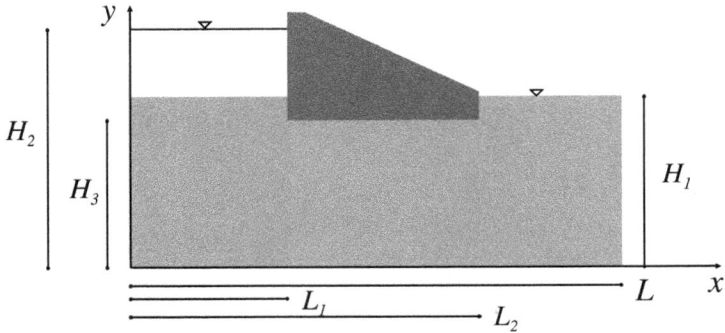

Figure 5.2.11.   Simple model of a dam.

5.2.4 *Applications to the Dam and Sheet Pile Wall*

Consider the simple model of a dam in Fig. 5.2.11. The steady state seepage motions are governed by the following Dirichlet–Neumann problem for the piezometric head $h = h(x, y)$

$$\Delta h(x, y) = 0, \quad (x, y) \in \Omega, \tag{5.2.40}$$

$$h = H_1 \text{ on } \partial\Omega_1, \quad h = H_2 \text{ on } \partial\Omega_2, \tag{5.2.41}$$

$$h_x = 0 \text{ on } \partial\Omega_3 \cup \partial\Omega_4 \cup \partial\Omega_5 \cup \partial\Omega_6, \quad h_y = 0 \text{ on } \partial\Omega_7 \cup \partial\Omega_8, \tag{5.2.42}$$

where

$$\partial\Omega_1 = \{L_2 \leq x \leq L, \ y = H_1\}, \quad \partial\Omega_2 = \{0 \leq x \leq L_1, \ y = H_1\},$$

$$\partial\Omega_3 = \{x = 0, \ 0 < y < H_1\}, \quad \partial\Omega_4 = \{x = L, \ 0 < y < H_1\},$$

$$\partial\Omega_5 = \{x = L_1, \ H_3 < y < H_1\}, \quad \partial\Omega_6 = \{x = L_2, \ H_3 < y < H_1\},$$

$$\partial\Omega_7 = \{0 \leq x \leq L, \ y = 0\}, \quad \partial\Omega_8 = \{L_1 \leq x \leq L_2, \ y = H_3\}.$$

Consider the FEM. The approximating solution h is given by the finite series

$$h(x, y) = \sum_{j=1}^{n_f} h_j \Phi_j(x, y) + \sum_{j=n_f+1}^{n} h_j \Phi_j(x, y), \tag{5.2.43}$$

where n is the total number of nodes and n_f is the number of free nodes. Substituting (5.2.43) into (5.2.30) and assuming $v = \Phi_i$, $i = 1, ..., n_f$, yields

$$\sum_{j=1}^{n_f} K_{ij} h_j = - \sum_{j=n_f+1}^{n} K_{ij} h_j, \quad i = 1, \ldots, n_f, \qquad (5.2.44)$$

where the parameters h_j for the constrained nodes $j = n_f + 1, \ldots, n$, are given by boundary Conditions (5.2.41).

Example 5.2.9   The following listing presents a function that applies the FEM to Problem (5.2.40) to (5.2.42). The numerical solution is plotted in Fig. 5.2.12. The streamlines and the equipotential lines are illustrated in Fig. 5.2.13. Lastly, Fig. 5.2.14 shows the simple triangulation used in the listing.

```
function [u, hh] = dam
% This is the function file dam.m.
% The FEM is applied to solve the Dirichlet–Neumann problem in Example
% 5.2.9. The function returns the matrix and the vector expressions of the
% numerical solution.
```

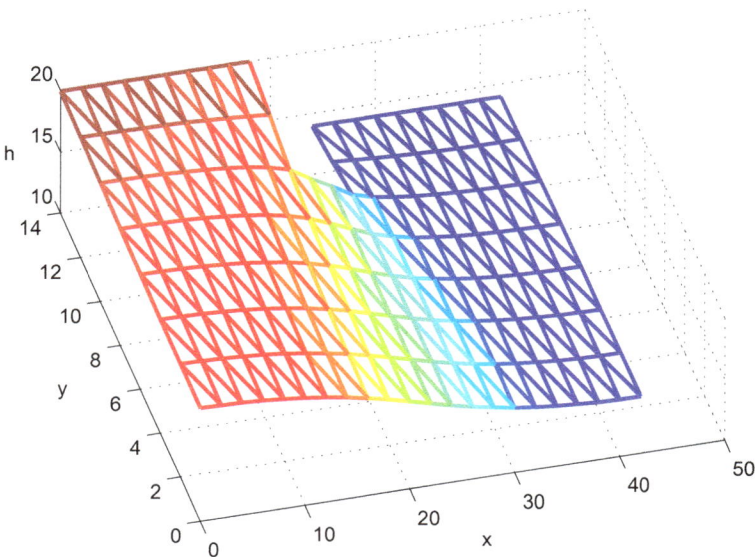

Figure 5.2.12.   Numerical solution to Problem (5.2.40) to (5.2.42).

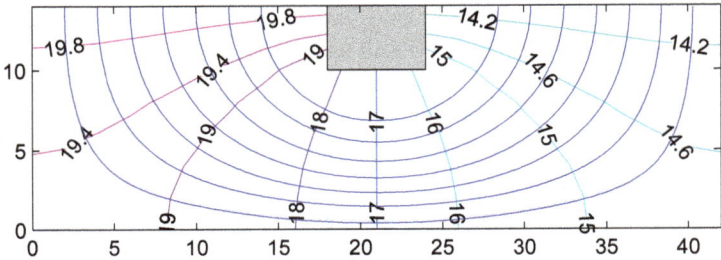

Figure 5.2.13.   Streamlines and equipotential lines.

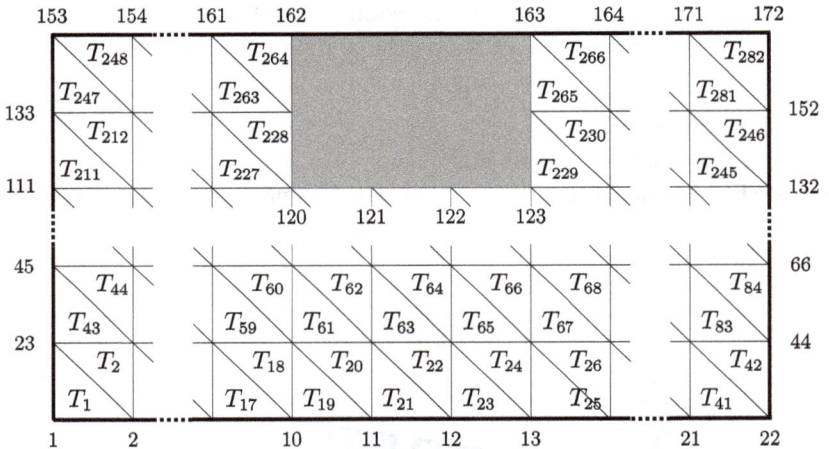

Figure 5.2.14.   Triangulation.

```
% Initialization
L = 42; H1 = 14; H2 = 20; d = 2; nx = L/d; ny = H1/d;
% L = 42; H1 = 14; H2 = H1; d = 2; nx = L/d; ny = H1/d; % test
x = linspace(0,L,nx+1); y = linspace(0,H1,ny+1);
k1 = 6; i1 = 10; i2 = 13;
H3 = y(k1); L1 = x(i1); L2 = x(i2); % H3 = 10; L1 = 18; L2 = 24;
P = Points(x, y, nx+1, ny+1, i1, k1); xt = P(:,1); yt = P(:,2);
n = length(xt); hb = [H2*ones(i1,1); H1*ones(nx + 2 - i2,1)];
nf = n - length(hb); K = zeros(nf,nf); b = zeros(nf,n-nf);
T = Triangulation(nx, nx+1, ny+1, i1, i2, k1);
% FEM
for i=1:nf
```

```
    [~,nablaPhii] = pyramid(T, xt, yt, i);
    support_i = nablaPhii(:,3);
    for j=i:n
        [~,nablaPhij] = pyramid(T, xt, yt, j);
        support_j = nablaPhij(:,3); mt = length(support_j);
        for r=1:mt
          if ismember(support_j(r), support_i) == 1
            index = find(support_i == support_j(r));
            area = det([xt(T(support_j(r),:)',:) yt(T(support_j(r),:)',:)...
                [1;1;1]])/2;
            if j<nf+1
                K(i,j)=K(i,j)+ nablaPhij(r,1:2)*nablaPhii(index,1:2)'*area;
                K(j,i)=K(i,j);
            else
                b(i,j-nf)=b(i,j-nf)+nablaPhij(r,1:2)*nablaPhii(index,1:2)'*area;
            end
          end
        end
    end
end
f = -b * hb; v = K\f; hh = [v;hb];
H = [v(1:k1*(nx+1)+i1); nan*ones(i2-i1-1,1); v(k1*(nx+1)+i1+1:nf);...
    H2*ones(i1,1); nan*ones(i2-i1-1,1); H1*ones(nx + 2 - i2,1)];
u = reshape(H,nx+1,ny+1);
% Plot
trimesh(T, xt, yt, hh,'LineWidth',2); view(-15,69);
xlabel('x'); ylabel('y'); zlabel('h');
[hx, hy] = gradient(u',d,d);
KK =10^(-8); qx = -KK*hx; qy = -KK*hy;
figure(2);
[~,cc] = contour(x,y,u',[14.2,14.6,15,16,17,18,19,19.4,19.8]);
set(cc,'ShowText','on'); colormap cool;
axis('equal'); axis([0 L 0 H1]);
rectangle('position',[0, 0, L, H1])
rectangle('position',[L1, H3 ,L2-L1 ,H1-H3],...
    'FaceColor',[0.8 0.8 0.8],'LineWidth',1);
```

```matlab
hold on;
streamline(x,y,qx,qy,x(2:8),y(ny+1)*ones(1,length(x(2:8))));
end

% ———— Local functions ————
function P = Points(x, y, n1, n2, i1, k1)
P = zeros(n1*n2,2);
P(1:n1*n2,1)= repmat(x',n2,1);
for k=1:n2
    P(1+(k-1)*n1:k*n1,2)= y(k)*ones(n1,1);
end
np1 = k1*n1+i1; np2 = (k1+1)*n1+i1;
P = P([1:np1,np1+3:np2,np2+3:n1*n2],:);
end

function T = Triangulation(nx, n1, n2, i1, i2, k1)
m = nx*2*(n2-1)-(i2-i1)*2*(n2-k1); T = zeros(m,3);
for k=1:k1-
    for i=1:nx
        T((k-1)*2*nx+2*i-1,:)= [(k-1)*n1+i (k-1)*n1+i+1 k*n1+i];
        T((k-1)*2*nx+2*i,:) = [(k-1)*n1+i+1 k*n1+i+1 k*n1+i];
    end
end
lt = nx*2*(k1-1); lp1 = n1*(k1-1); lp2 = n1*k1;
for i=1:i1-1
    T(lt+2*i-1,:)= [lp1+i lp1+i+1 lp2+i];
    T(lt+2*i,:) = [lp1+i+1 lp2+i+1 lp2+i];
end
lt = lt+2*(i1-1); lp1 = lp1+i1+2; lp2 = lp2+i1;
for i=1:n1-i2
    T(lt+2*i-1,:)= [lp1+i lp1+i+1 lp2+i];
    T(lt+2*i,:) = [lp1+i+1 lp2+i+1 lp2+i];
end
lt = lt+2*(n1-i2); lp1 = lp1+n1-i2+1; lp2 = lp2+n1-i2+1;
for i=1:i1-1
    T(lt+2*i-1,:)= [lp1+i lp1+i+1 lp2+i];
    T(lt+2*i,:) = [lp1+i+1 lp2+i+1 lp2+i];
end
```

It = It+2*(i1-1); lp1 = lp1+i1; lp2 = lp2+i1;
for i=1:n1-i2
    T(It+2*i-1,:)= [lp1+i lp1+i+1 lp2+i];
    T(It+2*i,:) = [lp1+i+1 lp2+i+1 lp2+i];
end
end

See Exercise 5.4.5.

Example 5.2.10   Consider the simple model of a sheet pile wall in Fig. 5.2.15. The steady state seepage analysis leads to considering the following Dirichlet–Neumann problem for the piezometric head $h = h(x, y)$

$$\Delta h(x, y) = 0, \quad (x, y) \in \Omega, \tag{5.2.45}$$

$$h = \begin{cases} 16 \text{ on } \{y = 12,\ 0 \le x \le x_1\}, \\ 13 \text{ on } \{y = 12,\ x_2 \le x \le 22\}, \end{cases} \tag{5.2.46}$$

$$\partial h/\partial n = 0 \text{ on} \begin{cases} \{x = 0,\ 0 \le y < 12\},\ \{x = 22, 0 \le y < 12\}, \\ \{y = 0,\ 0 \le x \le 22\},\ \{y = 6,\ x_1 \le x \le x_2\}, \\ \{x = x_1,\ 6 \le y \le 12\}, \{x = x_2,\ 6 \le y \le 12\}, \end{cases} \tag{5.2.47}$$

where x_1 is the abscissa of the left side of the sheet pile wall and x_2 is the abscissa of the right side. Problem (5.2.45) to (5.2.47) is similar to the problem discussed in Example 5.2.9. A function that applies the FEM to Problem (5.2.45) to (5.2.47) is suggested in Exercise 5.4.6.

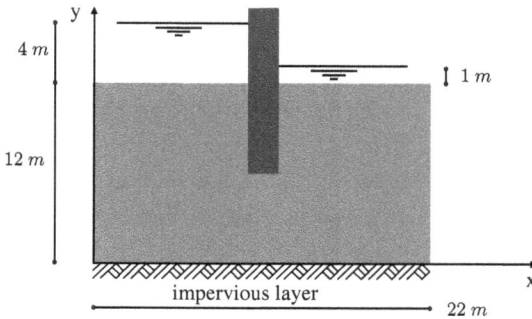

Figure 5.2.15.   Sheet pile wall.

5.3 Finite Difference Method

5.3.1 *Five-Point Method*

Consider the Dirichlet problem for Laplace's equation in two dimensions

$$\Delta U(x,y) = 0, \quad (x,y) \in \Omega, \tag{5.3.1}$$

$$U(x,y) = g(x,y), \quad (x,y) \in \partial\Omega. \tag{5.3.2}$$

When the domain Ω is a rectangle

$$\Omega = \{0 < x < L, \ 0 < y < H\},$$

boundary Condition (5.3.2) is written as

$$\begin{cases} U(0,y) = g_1(y), & U(L,y) = g_3(y), \quad 0 \le y \le H, \\ U(x,0) = g_2(x), & U(x,H) = g_4(x), \quad 0 \le x \le L. \end{cases} \tag{5.3.3}$$

Let us apply the central approximation to the derivatives in Eq. (5.3.1)

$$(u_{i+1,k} - 2u_{i,k} + u_{i-1,k})/(\Delta x)^2 + (u_{i,k+1} - 2u_{i,k} + u_{i,k-1})/(\Delta y)^2 = 0.$$

The previous finite-difference equation holds for all points within the domain. Moreover, setting

$$\sigma = \Delta y/\Delta x, \quad s = -2(1 + \sigma^2), \tag{5.3.4}$$

we get

$$\begin{cases} su_{i,k} + \sigma^2(u_{i-1,k} + u_{i+1,k}) + u_{i,k-1} + u_{i,k+1} = 0, \\ i = 1, \ldots, n-1, \ k = 1, \ldots, m-1, \ n = L/\Delta x, \ m = H/\Delta y. \end{cases} \tag{5.3.5}$$

The finite-difference equation in (5.3.5) is named the *Five-Point Method* (Fig. 5.3.1). As outlined in (5.3.5), the Five-Point Method provides a linear algebraic system of $(n-1)(m-1)$ equations in $(n-1)(m-1)$ unknowns. For example, for the mesh in Fig. 5.3.1, the equations and unknowns are 15. For $k = 1$, System (5.3.5) can be written in matrix form as follows

$$\begin{bmatrix} s & \sigma^2 & & \\ \sigma^2 & s & \sigma^2 & \\ & \cdot & \cdot & \cdot \\ & & \sigma^2 & s \end{bmatrix} \begin{bmatrix} u_{1,1} \\ u_{2,1} \\ \cdot \\ u_{n-1,1} \end{bmatrix} + \begin{bmatrix} u_{1,2} \\ u_{2,2} \\ \cdot \\ u_{n-1,2} \end{bmatrix} + \begin{bmatrix} u_{1,0} \\ u_{2,0} \\ \cdot \\ u_{n-1,0} \end{bmatrix} + \sigma^2 \begin{bmatrix} u_{0,1} \\ 0 \\ \cdot \\ u_{n,1} \end{bmatrix} = \mathbf{0}.$$

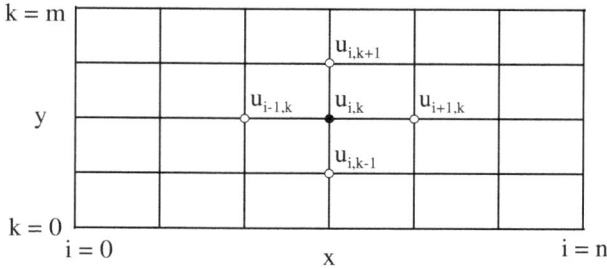

Figure 5.3.1. Five-Point Method.

Hence,

$$Bu_1 + u_2 + b_1 = 0, \tag{5.3.6}$$

where

$$B = \begin{bmatrix} s & \sigma^2 & & \\ \sigma^2 & s & \sigma^2 & \\ & \cdot & \cdot & \cdot \\ & & \sigma^2 & s \end{bmatrix}, \quad u_1 = \begin{bmatrix} u_{1,1} \\ u_{2,1} \\ \cdot \\ u_{n-1,1} \end{bmatrix}, \quad u_2 = \begin{bmatrix} u_{1,2} \\ u_{2,2} \\ \cdot \\ u_{n-1,2} \end{bmatrix},$$

$$b_1 = \begin{bmatrix} u_{1,0} \\ u_{2,0} \\ \cdot \\ u_{n-1,0} \end{bmatrix} + \sigma^2 \begin{bmatrix} u_{0,1} \\ 0 \\ \cdot \\ u_{n,1} \end{bmatrix} = \begin{bmatrix} g_{2,1} \\ g_{2,2} \\ \cdot \\ g_{2,n-1} \end{bmatrix} + \sigma^2 \begin{bmatrix} g_{1,1} \\ 0 \\ \cdot \\ g_{3,1} \end{bmatrix}.$$

Note that b_1 is known, as expressed in terms of boundary Conditions (5.3.3). For $k = 2, \ldots, m - 2$, System (5.3.5) implies

$$\begin{bmatrix} u_{1,k-1} \\ u_{2,k-1} \\ \cdot \\ u_{n-1,k-1} \end{bmatrix} + \begin{bmatrix} s & \sigma^2 & & \\ \sigma^2 & s & \sigma^2 & \\ & \cdot & \cdot & \cdot \\ & & \sigma^2 & s \end{bmatrix} \begin{bmatrix} u_{1,k} \\ u_{2,k} \\ \cdot \\ u_{n-1,k} \end{bmatrix} + \begin{bmatrix} u_{1,k+1} \\ u_{2,k+1} \\ \cdot \\ u_{n-1,k+1} \end{bmatrix} + \sigma^2 \begin{bmatrix} u_{0,k} \\ 0 \\ \cdot \\ u_{n,k} \end{bmatrix} = 0,$$

$$u_{k-1} + Bu_k + u_{k+1} + b_k = 0, \quad k = 2, \ldots, m - 2, \tag{5.3.7}$$

where b_k is a known term, as $u_{0,k} = g_{1,k}$ and $u_{n,k} = g_{3,k}$. Lastly, for $k = m - 1$, System (5.3.5) implies

$$\begin{bmatrix} u_{1,m-2} \\ u_{2,m-2} \\ \cdot \\ u_{n-1,m-2} \end{bmatrix} + \begin{bmatrix} s & \sigma^2 & & \\ \sigma^2 & s & \sigma^2 & \\ & \cdot & \cdot & \cdot \\ & & \sigma^2 & s \end{bmatrix} \begin{bmatrix} u_{1,m-1} \\ u_{2,m-1} \\ \cdot \\ u_{n-1,m-1} \end{bmatrix}$$

$$+ \begin{bmatrix} u_{1,m} \\ u_{2,m} \\ \cdot \\ u_{n-1,m} \end{bmatrix} + \sigma^2 \begin{bmatrix} u_{0,m-1} \\ 0 \\ \cdot \\ u_{n,m-1} \end{bmatrix} = \mathbf{0},$$

$$\mathbf{u}_{m-2} + B\mathbf{u}_{m-1} + \mathbf{b}_{m-1} = \mathbf{0}, \tag{5.3.8}$$

where the known term $\mathbf{b}_{m-1}$ is given by

$$\mathbf{b}_{m-1} = \begin{bmatrix} u_{1,m} \\ u_{2,m} \\ \cdot \\ u_{n-1,m} \end{bmatrix} + \sigma^2 \begin{bmatrix} u_{0,m-1} \\ 0 \\ \cdot \\ u_{n,m-1} \end{bmatrix} = \begin{bmatrix} g_{4,1} \\ g_{4,2} \\ \cdot \\ g_{4,n-1} \end{bmatrix} + \sigma^2 \begin{bmatrix} g_{1,m-1} \\ 0 \\ \cdot \\ g_{3,m-1} \end{bmatrix}.$$

Equations (5.3.6) to (5.3.8) can be combined in a single equation by using the block matrices

$$\begin{bmatrix} B & I_{n-1} & 0_{n-1} & \\ I_{n-1} & B & I_{n-1} & \\ & \cdot & \cdot & \cdot \\ & & 0_{n-1} & I_{n-1} & B \end{bmatrix} \begin{bmatrix} \mathbf{u}_1 \\ \mathbf{u}_2 \\ \cdot \\ \mathbf{u}_{m-1} \end{bmatrix} + \begin{bmatrix} \mathbf{b}_1 \\ \mathbf{b}_2 \\ \cdot \\ \mathbf{b}_{m-1} \end{bmatrix} = \mathbf{0},$$

where I_{n-1} is the $(n-1)$-by-$(n-1)$ unit matrix and 0_{n-1} is the zero matrix. Hence, with the clear meaning of the symbols,

$$A\mathbf{u} + \mathbf{b} = \mathbf{0}. \tag{5.3.9}$$

It is a linear system of $(n-1)(m-1)$ equations in $(n-1)(m-1)$ unknowns.

Example 5.3.1 Consider the special Dirichlet problem

$$\Delta U(x,y) = 0, \quad 0 < x < L, \quad 0 < y < H, \tag{5.3.10}$$

$$\begin{cases} U(0,y) = g_1(y) = -y^2, \quad U(L,y) = g_3(y) = L^2 - y^2, \quad 0 \le y \le H, \\ U(x,0) = g_2(x) = x^2, \quad U(x,H) = g_4(x) = x^2 - H^2, \quad 0 \le x \le L. \end{cases}$$

$$\tag{5.3.11}$$

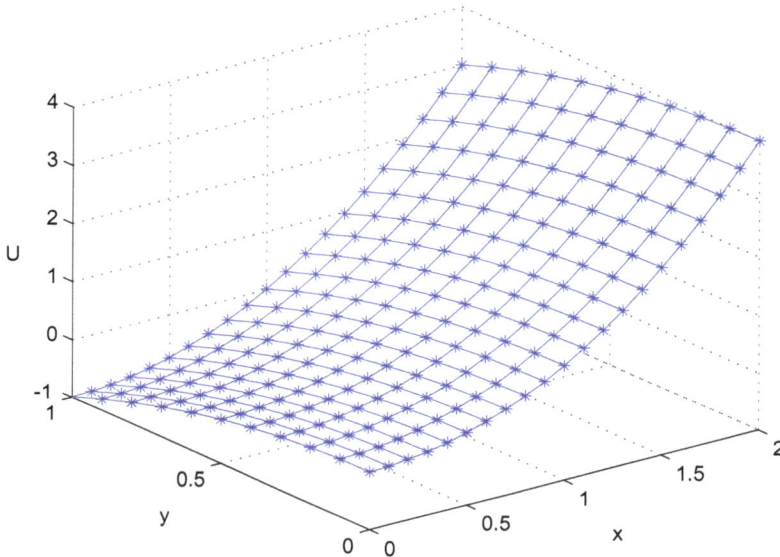

Figure 5.3.2.   Numerical solution to Problem (5.3.10) to (5.3.11).

The following listing presents a function that applies the Five-Point Method to Problem (5.3.10) and (5.3.11). The graph of the numerical solution is shown in Fig. 5.3.2.

```
u = five_point
% This is the function file five_point.m.
% Five-Point Method is applied to solve the following Dirichlet problem for
% the Laplace equation: Uyy + Uxx = 0,
% U(0,y) = -y.^2, U(x,0) = x^2, U(L,y) = L^2-y^2, U(x,H) = x^2-H^2.
% Initialization
L = 2; H = 1; n = 20; m = 10;
dx = L/n; dy = H/m; sigma = dy/dx;
x = linspace(0,L,n+1); y = linspace(0,H,m+1);
b = zeros((n-1)*(m-1),1); u = zeros(n+1, m+1);
g1 = @(y) -y.^2;
g2 = @(x)  x.^2;
g3 = @(y)  L^2 - y.^2;
g4 = @(x)  x.^2 - H^2;
AA = repmat([ones(n-1,1)  [sigma^2*ones(n-2,1);0]...
```

```
-2*(1+sigma²)*ones(n-1,1)[0;sigma²*ones(n-2,1)] ones(n-1,1)], (m-1),1);
A = spdiags(AA, [-(n-1) -1:1 (n-1)], (n-1)*(m-1), (n-1)*(m-1));
% Five-Point Method
u(1,:) = g1(y); % Boundary values.
u(:,1) = g2(x); u(n+1,:) = g3(y); u(:,m+1) = g4(x);
b(1:n-1) = u(2:n,1); % Known term.
b( (n-1)*(m-2)+1 :(m-1)*(n-1) ) = u(2:n,m+1);
for k = 2:m
    b((k-2)*(n-1)+1) = b((k-2)*(n-1)+1) + sigma²*u(1,k);
    b((k-1)*(n-1)) = b((k-1)*(n-1)) + sigma²*u(n+1,k);
end
v = -A\b;
u(2:n,2:m) = reshape(v,n-1,m-1); u = u';
U = @(x,y) x.²-y.²; U = feval(U, x, y');% Analytical solution.
mesh(x,y,u,'edgecolor', 'b'); mesh(x,y,U,'edgecolor', 'b', 'Marker','*');
xlabel('x'); ylabel('y'); zlabel('U');
fprintf('Maximum error = %g\n',max(max(abs(U -u ))))
end
```

See Exercise 5.4.7.

Consider the Dirichlet problem for the Poisson equation

$$\Delta U(x, y) = F(x, y), \quad (x, y) \in \Omega,$$

$$U(x, y) = g(x, y), \quad (x, y) \in \partial\Omega.$$

The Five-Point Method for the previous problem is given by

$$\frac{u_{i+1,k} - 2u_{i,k} + u_{i-1,k}}{(\Delta x)^2} + \frac{u_{i,k+1} - 2u_{i,k} + u_{i,k-1}}{(\Delta y)^2} = f_{i,k},$$

$$su_{i,k} + \sigma^2(u_{i-1,k} + u_{i+1,k}) + u_{i,k-1} + u_{i,k+1} = f_{i,k}(\Delta y)^2. \qquad (5.3.12)$$

When the domain is a rectangle $\Omega = \{0 < x < L, \ 0 < y < H\}$, System (5.3.9) is generalized to

$$
\begin{bmatrix}
B & I_{n-1} & 0_{n-1} & & \\
I_{n-1} & B & I_{n-1} & & \\
& \cdot & \cdot & \cdot & \\
& & 0_{n-1} & I_{n-1} & B
\end{bmatrix}
\begin{bmatrix}
u_1 \\ u_2 \\ \cdot \\ u_{m-1}
\end{bmatrix}
+
\begin{bmatrix}
b_1 \\ b_2 \\ \cdot \\ b_{m-1}
\end{bmatrix}
=
\begin{bmatrix}
f_1 \\ f_2 \\ \cdot \\ f_{m-1}
\end{bmatrix},
$$

where

$$\mathbf{f}_k = (\Delta y)^2 \begin{bmatrix} f_{1,k} \\ f_{2,k} \\ \cdot \\ f_{n-1,k} \end{bmatrix}, \quad k = 1, ..., m-1.$$

Example 5.3.2   Consider the special Dirichlet problem

$$\Delta U(x,y) = -\sin x - \cos y, \quad 0 < x < L, \quad 0 < y < H, \qquad (5.3.13)$$

$$\begin{cases} U(0,y) = g_1(y) = \cos y, \\ U(L,y) = g_3(y) = \sin L + \cos y, \end{cases} \quad 0 \le y \le H, \qquad (5.3.14)$$

$$\begin{cases} U(x,0) = g_2(x) = \sin x + 1, \\ U(x,H) = g_4(x) = \sin x + \cos H, \end{cases} \quad 0 \le x \le L. \qquad (5.3.15)$$

A function that applies the Five-Point Method to Problem (5.3.13) to (5.3.15) is provided below. The numerical solution is shown in Fig. 5.3.3.

u = five_point_f1

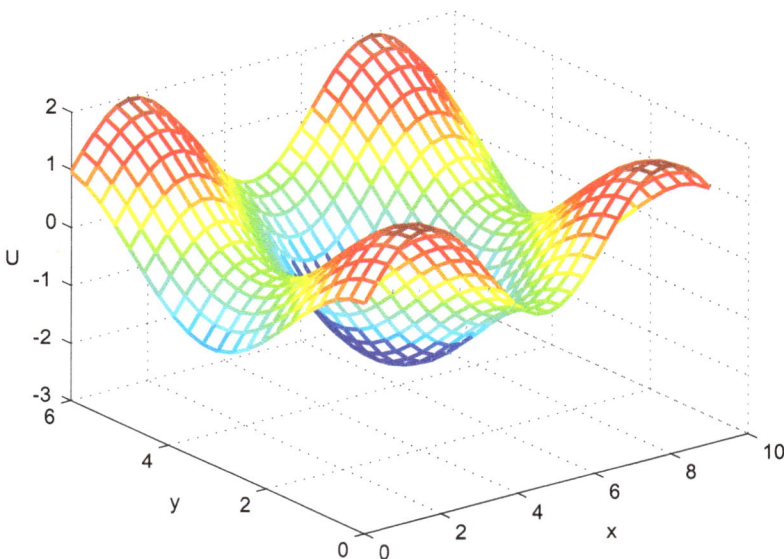

Figure 5.3.3.   Numerical solution to Problem (5.3.13) to (5.3.15).

```
% This is the function file five_point_f1.m.
% The Five-Point Method is applied to solve the following Dirichlet problem
% for the Poisson equation: Uyy + Uxx = - sin x - cos y,
% U(0,y) = cos y, U(L,y) = sin L + cos y,
% U(x,0) = sin x + 1, U(x,H) = sin x + cos H.
% Analytical solution: U = sin x + cos y.
% Initialization
L = 9; H = 6; n = 30; m = 20;
dx = L/n; dy = H/m; sigma = dy/dx;
x = linspace(0,L,n+1); y = linspace(0,H,m+1);
b = zeros((n-1)*(m-1),1); u = zeros(n+1,m+1);
g1 = @(y)  cos(y);
g2 = @(x)  sin(x) + 1;
g3 = @(y)  sin(L) + cos(y);
g4 = @(x)  sin(x) + cos(H);
F = @(x,y)  - sin(x) - cos(y);
f = zeros((n-1)*(m-1),1);
for k=2:m
    f((k-2)*(n-1)+1:(k-1)*(n-1),1) = F(x(2:n)',y(k))*dy^2;
end
       Same code as five_point function
v = A\(f-b);
u(2:n,2:m) = reshape(v,n-1,m-1); u = u';
mesh(x,y,u,'LineWidth',2); xlabel('x'); ylabel('y'); zlabel('U');
U = @(x,y) sin(x) + cos(y); U = feval(U,x,y');
fprintf('Maximum error = %g\n',max(max(abs(U -u ))))
end
```

Another application is suggested in Exercise 5.4.8.

Consider the 3D Poisson equation

$$\Delta U(x, y, z) = F(x, y, z), \quad (x, y, z) \in \Omega. \qquad (5.3.16)$$

The generalization of the Five-Point Method to Eq. (5.3.16) is named the *Seven-Point Method*. See Exercise 5.4.9.

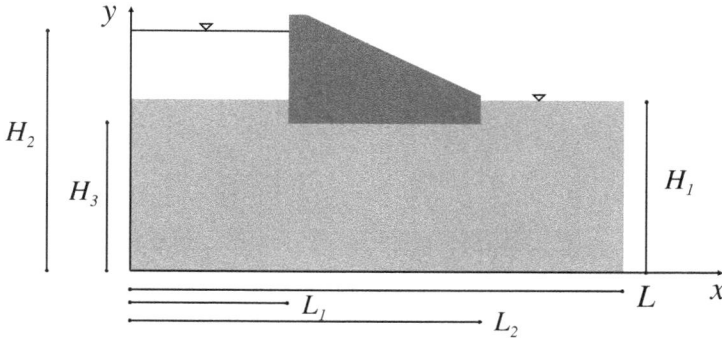

Figure 5.3.4.   Simple model of dam.

5.3.2 *Model of a Dam*

Consider the simple model of a dam in Fig. 5.3.4. The steady state seepage motions are governed by the following Dirichlet–Neumann problem for the piezometric head $h = h(x, y)$

$$\Delta h(x, y) = 0, \quad (x, y) \in \Omega, \tag{5.3.17}$$

$$h = H_1 \text{ on } \partial\Omega_1, \quad h = H_2 \text{ on } \partial\Omega_2, \tag{5.3.18}$$

$$h_x = 0 \text{ on } \partial\Omega_3 \cup \partial\Omega_4 \cup \partial\Omega_5 \cup \partial\Omega_6, \quad h_y = 0 \text{ on } \partial\Omega_7 \cup \partial\Omega_8, \tag{5.3.19}$$

where

$$\partial\Omega_1 = \{L_2 \leq x \leq L, \ y = H_1\}, \quad \partial\Omega_2 = \{0 \leq x \leq L_1, \ y = H_1\},$$

$$\partial\Omega_3 = \{x = 0, \ 0 < y < H_1\}, \quad \partial\Omega_4 = \{x = L, \ 0 < y < H_1\},$$

$$\partial\Omega_5 = \{x = L_1, \ H_3 < y < H_1\}, \quad \partial\Omega_6 = \{x = L_2, \ H_3 < y < H_1\},$$

$$\partial\Omega_7 = \{0 \leq x \leq L, \ y = 0\}, \quad \partial\Omega_8 = \{L_1 \leq x \leq L_2, \ y = H_3\}.$$

Problem (5.3.17)–(5.3.19) was discussed with the FEM in Sec. 5.2.4. Now, the Five-Point Method is applied. The numerical solution is indicated with

$$h_{i,k} = h(x_i, y_k), \ i = 1, ..., n+1, \ k = 1, ..., m+1.$$

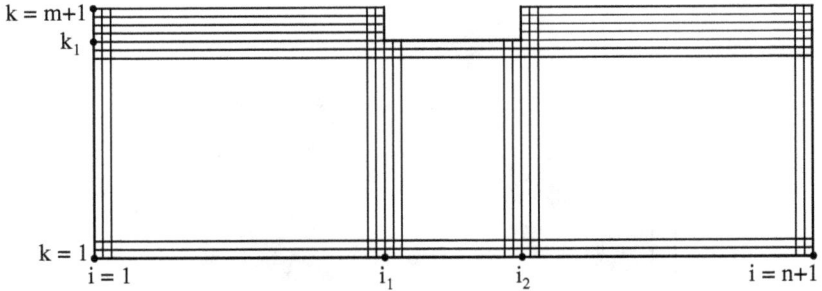

Figure 5.3.5.   Mesh for the dam problem.

The mesh is shown in Fig. 5.3.5. Note that $x_1 = 0$, $x_{n+1} = L$, $x_{i_1} = L_1$, $x_{i_2} = L_2$, $y_1 = 0$, $y_{m+1} = H_1$ and $y_{k_1} = H_3$. The Neumann boundary conditions are approximated by using the forward (f) and backward (b) approximations:

$$\{x = 0,\ 0 < y < H_1\},\quad h_{2,k} = h_{1,k},\quad \text{(f)},$$
$$\{x = L,\ 0 < y < H_1\},\quad h_{n+1,k} = h_{n,k},\quad \text{(b)},\qquad 2 < k < m, \qquad (5.3.20)$$

$$\{x = L_1,\ H_3 < y < H_1\},\quad h_{i_1,k} = h_{i_1-1,k},\quad \text{(b)},$$
$$\{x = L_2,\ H_3 < y < H_1\},\quad h_{i_2+1,k} = h_{i_2,k},\quad \text{(f)},\qquad k_1 < k < m, \qquad (5.3.21)$$

$$\{0 \le x \le L,\ y = 0\},\quad h_{i,2} = h_{i,1},\quad \text{(f)},\quad 1 \le i \le n+1, \qquad (5.3.22)$$

$$\{L_1 \le x \le L_2,\ y = H_3\},\quad h_{i,k_1} = h_{i,k_1-1},\quad \text{(b)},\quad i_1 \le i \le i_2. \qquad (5.3.23)$$

The linear system provided by the Five-Point Method is illustrated in Fig. 5.3.6. Note that the vectors $\mathbf{h}_k$, $k < k_1$, have a greater length than the vectors $\mathbf{h}_k^*$, $k \ge k_1$. Attention is needed for $k = k_1$. Consider the Five-Point Method in matrix form

$$\begin{bmatrix} h_{2,k-1} \\ h_{3,k-1} \\ . \\ h_{n,k-1} \end{bmatrix} + \begin{bmatrix} s & \sigma^2 & & \\ \sigma^2 & s & \sigma^2 & \\ & . & . & . \\ & & \sigma^2 & s \end{bmatrix} \begin{bmatrix} h_{2,k} \\ h_{3,k} \\ . \\ h_{n,k} \end{bmatrix} + \begin{bmatrix} h_{2,k+1} \\ h_{3,k+1} \\ . \\ h_{n,k+1} \end{bmatrix} + \sigma^2 \begin{bmatrix} h_{1,k} \\ 0 \\ . \\ h_{n+1,k} \end{bmatrix} = \mathbf{0}.$$

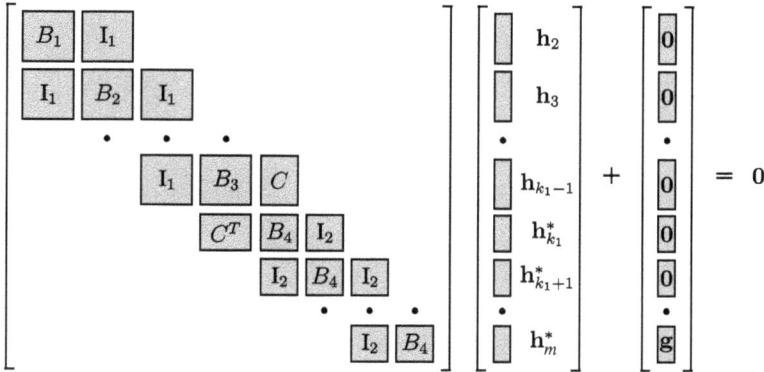

Figure 5.3.6.   Graphical representation of the linear system.

Using (5.3.20) in the previous formula yields

$$\begin{bmatrix} h_{2,k-1} \\ h_{3,k-1} \\ \cdot \\ h_{n,k-1} \end{bmatrix} + \begin{bmatrix} s+\sigma^2 & \sigma^2 \\ \sigma^2 & s & \sigma^2 \\ & \cdot & \cdot & \cdot \\ & & \sigma^2 & s+\sigma^2 \end{bmatrix} \begin{bmatrix} h_{2,k} \\ h_{3,k} \\ \cdot \\ h_{n,k} \end{bmatrix} + \begin{bmatrix} h_{2,k+1} \\ h_{3,k+1} \\ \cdot \\ h_{n,k+1} \end{bmatrix} = 0. \quad (5.3.24)$$

If $k = 2$, then $h_{i,2} = h_{i,1}$ because of (5.3.22), hence Formula (5.3.24) is written as

$$\begin{bmatrix} s+1+\sigma^2 & \sigma^2 \\ \sigma^2 & s+1 & \sigma^2 \\ & \cdot & \cdot \\ & & \sigma^2 & s+1+\sigma^2 \end{bmatrix} \begin{bmatrix} h_{2,2} \\ h_{3,2} \\ \cdot \\ h_{n,2} \end{bmatrix} + \begin{bmatrix} h_{2,3} \\ h_{3,3} \\ \cdot \\ h_{n,3} \end{bmatrix} = 0,$$

$$B_1 \mathbf{h}_2 + \mathbf{h}_3 = 0. \quad (5.3.25)$$

In addition, for $3 \le k \le k_1 - 2$, Formula (5.3.24) is written as

$$\mathbf{h}_{k-1} + B_2 \mathbf{h}_k + \mathbf{h}_{k+1} = 0, \quad 3 \le k \le k_1 - 2. \quad (5.3.26)$$

For $k = k_1 - 1$, Formula (5.3.24) implies

$$
\begin{bmatrix} h_{2,k_1-2} \\ h_{3,k_1-2} \\ . \\ h_{i_1,k_1-2} \\ . \\ h_{i_2,k_1-2} \\ . \\ h_{n,k_1-2} \end{bmatrix} + \begin{bmatrix} s+\sigma^2 & \sigma^2 & & & & \\ \sigma^2 & s & \sigma^2 & & & \\ & . & . & . & & \\ & & \sigma^2 & s & \sigma^2 & \\ & & & . & . & . \\ & & & & \sigma^2 & s & \sigma^2 \\ & & & & & . & . & . \\ & & & & & & \sigma^2 & s+\sigma^2 \end{bmatrix} \begin{bmatrix} h_{2,k_1-1} \\ h_{3,k_1-1} \\ . \\ h_{i_1,k_1-1} \\ . \\ h_{i_2,k_1-1} \\ . \\ h_{n,k_1-1} \end{bmatrix} + \begin{bmatrix} h_{2,k_1} \\ h_{3,k_1} \\ . \\ h_{i_1,k_1} \\ . \\ h_{i_2,k_1} \\ . \\ h_{n,k_1} \end{bmatrix} = \mathbf{0}.
$$

The elements $h_{i_1,k_1} \ldots h_{i_2,k_1}$ of the last column vector are equal to $h_{i_1,k_1-1} \ldots h_{i_2,k_1-1}$, respectively, because of (5.3.23). Therefore, the mentioned elements must be moved to the preceding column vector. The resulting vector $\mathbf{h}_{k_1}^*$ has length $i_1 - 2 + n - i_2$, less than the other vectors in the same formula, and cannot be added to those. Therefore, we introduce the $(n-1)$-by-$(i_1-2+n-i_2)$ matrix C such that the product $C\mathbf{h}_{k_1}^*$ is equal to the vector $[h_{2,k_1}, \ldots, h_{i_1-1,k_1}, 0, \ldots, 0, h_{i_2+1,k_1}, \ldots, h_{n_x,k_1}]^T$ that has same length as the other vectors. Considering this in the previous formula, one arrives at

$$
\begin{bmatrix} s+\sigma^2 & \sigma^2 & & & & & \\ & . & . & & & & \\ \sigma^2 & s & \sigma^2 & & & & \\ & \sigma^2 & s+1 & \sigma^2 & & & \\ & & . & . & . & & \\ & & & \sigma^2 & s+1 & \sigma^2 & \\ & & & & \sigma^2 & s & \sigma^2 \\ & & & & & . & . & . \\ & & & & & & \sigma^2 & s+\sigma^2 \end{bmatrix} \begin{bmatrix} h_{2,k_1-1} \\ . \\ h_{i_1-1,k_1-1} \\ h_{i_1,k_1-1} \\ . \\ h_{i_2,k_1-1} \\ h_{i_2+1,k_1-1} \\ . \\ h_{n,k_1-1} \end{bmatrix}
$$

$$
+ \begin{bmatrix} 1 & & & & & \\ & . & & & & \\ & & 1 & & & \\ & & & 0 & & \\ & & & & . & \\ & & & & & 0 \\ & & & & & & 1 \\ & & & & & & & . \\ & & & & & & & & 1 \end{bmatrix} \begin{bmatrix} h_{2,k_1} \\ . \\ h_{i_1-1,k_1} \\ h_{i_2+1,k_1} \\ . \\ h_{n,k_1} \end{bmatrix} + \begin{bmatrix} h_{2,k_1-2} \\ . \\ h_{i_1-1,k_1-2} \\ h_{i_1,k_1-2} \\ . \\ h_{i_2,k_1-2} \\ h_{i_2+1,k_1-2} \\ . \\ h_{n,k_1-2} \end{bmatrix} = \mathbf{0},
$$

$$
\mathbf{h}_{k_1-2} + B_3 \mathbf{h}_{k_1-1} + C \mathbf{h}_{k_1}^* = \mathbf{0}. \tag{5.3.27}
$$

Consider the Five-Point Method for $k = k_1$ and $i = 2, \ldots, i_1 - 1$,

$$\begin{bmatrix} h_{2,k_1-1} \\ h_{3,k_1-1} \\ \cdot \\ h_{i_1-1,k_1-1} \end{bmatrix} + \begin{bmatrix} s & \sigma^2 & & \\ \sigma^2 & s & \sigma^2 & \\ & \cdot & \cdot & \cdot \\ & & \sigma^2 & s \end{bmatrix} \begin{bmatrix} h_{2,k_1} \\ h_{3,k_1} \\ \cdot \\ h_{i_1-1,k_1} \end{bmatrix} + \begin{bmatrix} h_{2,k_1+1} \\ h_{3,k_1+1} \\ \cdot \\ h_{i_1-1,k_1+1} \end{bmatrix} + \sigma^2 \begin{bmatrix} h_{1,k_1} \\ 0 \\ \cdot \\ h_{i_1,k_1} \end{bmatrix} = 0,$$

and use Formulas (5.3.20) and (5.3.21),

$$\begin{bmatrix} h_{2,k_1-1} \\ h_{3,k_1-1} \\ \cdot \\ h_{i_1-1,k_1-1} \end{bmatrix} + \begin{bmatrix} s+\sigma^2 & \sigma^2 & & \\ \sigma^2 & s & \sigma^2 & \\ & \cdot & \cdot & \cdot \\ & & \sigma^2 & s+\sigma^2 \end{bmatrix} \begin{bmatrix} h_{2,k_1} \\ h_{3,k_1} \\ \cdot \\ h_{i_1-1,k_1} \end{bmatrix} + \begin{bmatrix} h_{2,k_1+1} \\ h_{3,k_1+1} \\ \cdot \\ h_{i_1-1,k_1+1} \end{bmatrix} = 0.$$

Similarly, for $k = k_1$ and $i = i_2 + 1, \ldots, n$,

$$\begin{bmatrix} h_{i_2+1,k_1-1} \\ h_{i_2+2,k_1-1} \\ \cdot \\ h_{n,k_1-1} \end{bmatrix} + \begin{bmatrix} s+\sigma^2 & \sigma^2 & & \\ \sigma^2 & s & \sigma^2 & \\ & \cdot & \cdot & \cdot \\ & & \sigma^2 & s+\sigma^2 \end{bmatrix} \begin{bmatrix} h_{i_2+1,k_1} \\ h_{i_2+2,k_1} \\ \cdot \\ h_{n,k_1} \end{bmatrix} + \begin{bmatrix} h_{i_2+1,k_1+1} \\ h_{i_2+2,k_1+1} \\ \cdot \\ h_{n,k_1+1} \end{bmatrix} = 0.$$

By combining the two previous formulas into a single matrix equation, we get

$$\begin{bmatrix} s+\sigma^2 & \sigma^2 & & & & & \\ \sigma^2 & s & \sigma^2 & & & & \\ & \cdot & \cdot & & & & \\ & & s+\sigma^2 & \sigma^2 & & & \\ & & & s+\sigma^2 & \sigma^2 & & \\ & & & \sigma^2 & s & \sigma^2 & \\ & & & & \cdot & \cdot & \cdot \\ & & & & & \sigma^2 & s+\sigma^2 \end{bmatrix} \begin{bmatrix} h_{2,k_1} \\ h_{3,k_1} \\ \cdot \\ h_{i_1-1,k_1} \\ h_{i_2+1,k_1} \\ h_{i_2+2,k_1} \\ \cdot \\ h_{n,k_1} \end{bmatrix}$$

$$+ \begin{bmatrix} h_{2,k_1-1} \\ h_{3,k_1-1} \\ \cdot \\ h_{i_1-1,k_1-1} \\ h_{i_2+1,k_1-1} \\ h_{i_2+2,k_1-1} \\ \cdot \\ h_{n,k_1-1} \end{bmatrix} + \begin{bmatrix} h_{2,k_1+1} \\ h_{3,k_1+1} \\ \cdot \\ h_{i_1-1,k_1+1} \\ h_{i_2+1,k_1+1} \\ h_{i_2+2,k_1+1} \\ \cdot \\ h_{n,k_1+1} \end{bmatrix} = 0.$$

The vector in the first part of the previous formula is $\mathbf{h}^*_{k_1}$. The last vector is $\mathbf{h}^*_{k_1+1}$. The remaining vector is a subset of $\mathbf{h}_{k_1-1}$. Since it can be written as

$$
\begin{bmatrix}
h_{2,k_1-1} \\
h_{3,k_1-1} \\
\cdot \\
h_{i_1-1,k_1-1} \\
h_{i_2+1,k_1-1} \\
h_{i_2+2,k_1-1} \\
\cdot \\
h_{n,k_1-1}
\end{bmatrix}
= C^T \mathbf{h}_{k_1-1},
$$

we obtain

$$
C^T \mathbf{h}_{k_1-1} + B_4 \mathbf{h}^*_{k_1} + \mathbf{h}^*_{k_1+1} = \mathbf{0}. \tag{5.3.28}
$$

For $k_1 + 1 \le k \le m - 1$ and $i = 1, \dots, i_1$, $i = i_2, \dots, n+1$, the Five-Point Method gives

$$
\begin{bmatrix}
s+\sigma^2 & \sigma^2 & & & & & & \\
\sigma^2 & s & \sigma^2 & & & & & \\
& & \cdot & \cdot & & & & \\
& & & s+\sigma^2 & \sigma^2 & & & \\
& & & & s+\sigma^2 & \sigma^2 & & \\
& & & & \sigma^2 & s & \sigma^2 & \\
& & & & & & \cdot & \cdot \\
& & & & & & \sigma^2 & s+\sigma^2
\end{bmatrix}
\begin{bmatrix}
h_{2,k} \\
h_{3,k} \\
\cdot \\
h_{i_1-1,k} \\
h_{i_2+1,k} \\
h_{i_2+2,k} \\
\cdot \\
h_{n,k}
\end{bmatrix}
$$

$$
+
\begin{bmatrix}
h_{2,k-1} \\
h_{3,k-1} \\
\cdot \\
h_{i_1-1,k-1} \\
h_{i_2+1,k-1} \\
h_{i_2+2,k-1} \\
\cdot \\
h_{n,k-1}
\end{bmatrix}
+
\begin{bmatrix}
h_{2,k+1} \\
h_{3,k+1} \\
\cdot \\
h_{i_1-1,k+1} \\
h_{i_2+1,k+1} \\
h_{i_2+2,k+1} \\
\cdot \\
h_{n,k+1}
\end{bmatrix}
= \mathbf{0},
$$

where Formulas (5.3.20)–(5.3.21) were applied. Hence,

$$
\mathbf{h}^*_{k-1} + B_4 \mathbf{h}^*_k + \mathbf{h}^*_{k+1} = \mathbf{0}, \quad k_1 + 1 \le k \le m - 1. \tag{5.3.29}
$$

Lastly, for $k = m$,

$$
\begin{bmatrix}
s+\sigma^2 & \sigma^2 & & & & & \\
\sigma^2 & s & \sigma^2 & & & & \\
& & \cdot & \cdot & \cdot & & \\
& & s+\sigma^2 & \sigma^2 & & & \\
& & & s+\sigma^2 & \sigma^2 & & \\
& & & \sigma^2 & s & \sigma^2 & \\
& & & & \cdot & \cdot & \cdot \\
& & & & & \sigma^2 & s+\sigma^2
\end{bmatrix}
\begin{bmatrix}
h_{2,m} \\
h_{3,m} \\
\cdot \\
h_{i_1-1,m} \\
h_{i_2+1,m} \\
h_{i_2+2,m} \\
\cdot \\
h_{n,m}
\end{bmatrix}
$$

$$
+
\begin{bmatrix}
h_{2,m-1} \\
h_{3,m-1} \\
\cdot \\
h_{i_1-1,m-1} \\
h_{i_2+1,m-1} \\
h_{i_2+2,m-1} \\
\cdot \\
h_{n,m-1}
\end{bmatrix}
+
\begin{bmatrix}
H_2 \\
H_2 \\
\cdot \\
H_2 \\
H_1 \\
H_1 \\
\cdot \\
H_1
\end{bmatrix}
= \mathbf{0}.
$$

Hence,

$$\mathbf{h}^*_{m-1} + B_4 \mathbf{h}^*_m + \mathbf{g} = \mathbf{0}. \tag{5.3.30}$$

Equations (5.3.25)–(5.3.30) can be assembled by using a block matrix

$$
\begin{bmatrix}
B_1 & I_1 & & & & & \\
I_1 & B_2 & I_1 & & & & \\
& & \cdot & \cdot & \cdot & & \\
& & I_1 & B_3 & C & & \\
& & & C^T & B_4 & I_2 & \\
& & & & I_2 & B_4 & I_2 \\
& & & & & \cdot & \cdot & \cdot \\
& & & & & & I_2 & B_4
\end{bmatrix}
\begin{bmatrix}
\mathbf{h}_2 \\
\mathbf{h}_3 \\
\cdot \\
\mathbf{h}_{k_1-1} \\
\mathbf{h}^*_{k_1} \\
\mathbf{h}^*_{k_1+1} \\
\cdot \\
\mathbf{h}^*_m
\end{bmatrix}
+
\begin{bmatrix}
\mathbf{0} \\
\cdot \\
\\
\\
\\
\mathbf{0} \\
\mathbf{g}
\end{bmatrix}
= \mathbf{0},
$$

where I_1 and I_2 are the $n-1$-by-$n-1$ and $i_1 - 2 + n - i_2$-by-$i_1 - 2 + n - i_2$ unit matrices, respectively. Hence, the desired system

$$A\mathbf{h} + \mathbf{a} = \mathbf{0}. \tag{5.3.31}$$

Example 5.3.3 The following listing presents a function that applies the Five-Point Method to Problem (5.3.17) to (5.3.19). The graph of the numerical solution is shown in Fig. 5.3.7. The equipotential lines and the stream lines are illustrated in Fig. 5.3.8.

```
function H = dam_5p
% This is the function file dam_5p.m.
% Five-Point Method is applied to the dam problem in Sec. 5.3.2.

% Initialization
L = 42; H1 = 14; H2 = 20; dx = .5; dy = dx; n = 84; m = 28;
x = linspace(0,L,n+1); y = linspace(0,H1,m+1);
k1 = 21; i1 = 37; i2 = 49; % H3 = y(k1)=10; L1 = x(i1)=18; L2 = x(i2)=24;
sigma = dy/dx; s = -2*(1+sigma^2);
p1 = i1 - 2; p2 = n - i2; p = p1+ p2;
q = (n - 1)*(k1-2) + (m-k1+1)*p; % Number of unknowns.
H = zeros(n+1,m+1);
A = zeros(q,q); % Block matrix.
a = [zeros(q-p,1); H2*ones(p1,1); H1*ones(p2,1)]; % Known term.
% Matrices B1, B2, B3, B4, C.
BB1=[sigma^2*ones(n-1,1)  [s/2;(s+1)*ones(n-3,1);s/2]  sigma^2*ones(n-1,1)];
B1 = spdiags(BB1,-1:1,n-1,n-1);
a2 = [sigma^2+s; s*ones(n - 3,1); sigma^2+s];
```

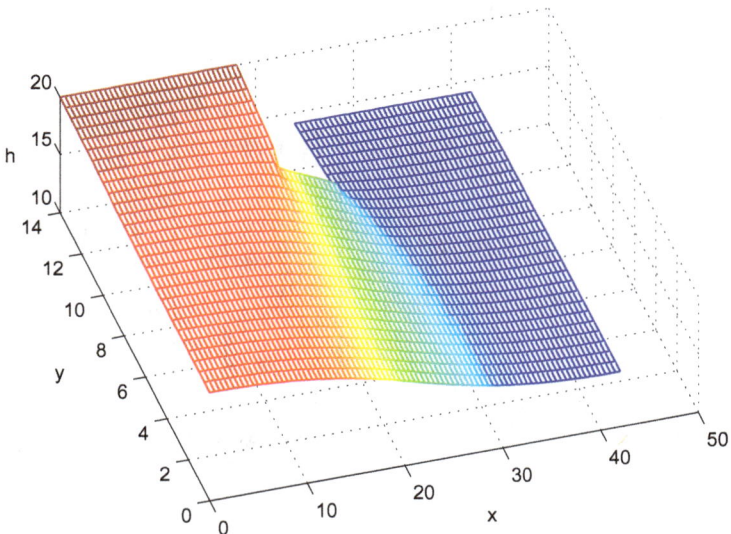

Figure 5.3.7.   Graph of the numerical solution to Problem (5.3.17) to (5.3.19).

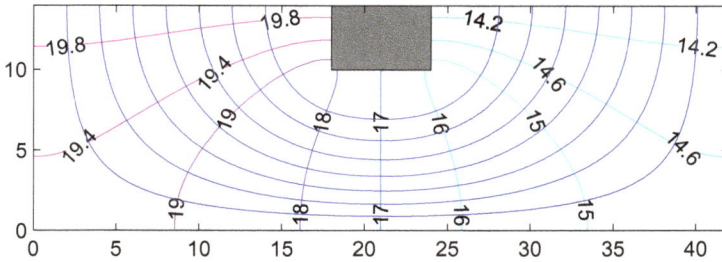

Figure 5.3.8.   Equipotential lines and the stream lines.

```
BB2 = [sigma²*ones(n - 1,1)  a2  sigma²*ones(n - 1,1)];
B2 = spdiags(BB2,-1:1,n - 1,n - 1);
a3 = [s+sigma²;s*ones(i1-3,1);(s+1)*ones(i2-i1+1,1);s*ones(n-i2-1,1);...
    s+sigma²];
BB3 = [sigma²*ones(n - 1,1)  a3  sigma²*ones(n - 1,1)];
B3 = spdiags(BB3,-1:1,n - 1,n - 1);
bb41 = [sigma²*ones(p1,1)  [sigma²+s;s*ones(p1-2,1);sigma²+s]...
    sigma²*ones(p1,1)];
b41 = spdiags(bb41,-1:1,p1,p1);
bb42 = [sigma²*ones(p2,1)  [sigma²+s;s*ones(p2-2,1);sigma²+s]...
    sigma²*ones(p2,1)];
b42 = spdiags(bb42,-1:1,p2,p2);
B4(1:p1,1:p1) = b41; B4(p1+1:p,p1+1:p) = b42;
C = zeros(n - 1,p);
C(1:i1-2,1:i1-2) = eye(i1-2); C(i2:n - 1,i1-2+1:p) = eye(n - 1-(i2-1));

% Matrix A
A(1:n - 1,1:n - 1) = B1; A(1:n - 1,n:2*(n - 1)) = eye(n - 1);
for k=3:k1-2
    A((k-2)*(n - 1)+1:(k-1)*(n - 1),(k-3)*(n - 1)+1:(k-2)*(n - 1)) = eye(n-1);
    A((k-2)*(n - 1)+1:(k-1)*(n - 1),(k-2)*(n - 1)+1:(k-1)*(n - 1)) = B2;
    A((k-2)*(n - 1)+1:(k-1)*(n - 1),(k-1)*(n - 1)+1:k*(n - 1)) = eye(n - 1);
end
k = k1 - 1;
A((k-2)*(n - 1)+1:(k-1)*(n - 1),(k-3)*(n - 1)+1:(k-2)*(n - 1)) = eye(n - 1);
A((k-2)*(n - 1)+1:(k-1)*(n - 1),(k-2)*(n - 1)+1:(k-1)*(n - 1)) = B3;
A((k-2)*(n - 1)+1:(k-1)*(n - 1),(k-1)*(n - 1)+1:(k-1)*(n -1)+p) = C;
k = k1; k2 = (k1-2)*(n - 1);
A(k2+1:k2+p,(k-3)*(n - 1)+1:(k-2)*(n - 1)) = C';
```

```
A(k2+1:k2+p,k2+1:k2+p) = B4; A(k2+1:k2+p,k2+p+1:k2+2*p) = eye(p);
for k=k1+1:m-1
    h = k - k1;
    A(k2+h*p+1:k2+h*p+p,k2+h*p-p+1:k2+h*p) = eye(p);
    A(k2+h*p+1:k2+h*p+p,k2+h*p+1:k2+h*p+p) = B4;
    A(k2+h*p+1:k2+h*p+p,k2+h*p+p+1:k2+h*p+2*p) = eye(p);
end
k = m; h = k - k1;
A(k2+h*p+1:k2+h*p+p,k2+h*p-p+1:k2+h*p) = eye(p);
A(k2+h*p+1:k2+h*p+p,k2+h*p+1:k2+h*p+p) = B4;

% Five-Point Method
v = -A\a;
for k=2:k1-1
    H(2:n,k) = v((k-2)*(n - 1)+1:(k-1)*(n - 1));
end
for k=k1:m
    h = k-k1;
    H(2:i1-1,k) = v(k2+h*p+1:k2+h*p+p1);
    H(i2+1:n,k) = v(k2+h*p+p1+1:k2+h*p+p);
    H(i1+1:i2-1,k+1) = nan; % nan means Not A Number.
end
H(2:i1-1,m+1)=H2*ones(p1,1); % Boundary values (Bv) for k=m+1, i=2:i1-1
H(i2+1:n,m+1) = H1*ones(p2,1); % Bv for k = m + 1, i = i2+1:n
H(i1,k1+1:m+1) = H(i1-1,k1+1:m+1); % Bv for i = i1, k = k1+2:m+1
H(i2,k1+1:m+1) = H(i2+1,k1+1:m+1); % Bv for i = i2, k = k1+2:m+1
H(i1:i2,k1) = H(i1:i2,k1-1); % Bv for k = k1, i = i1:i2
H(2:n,1) = H(2:n,2); % Bv for k = 1, i = 2:n
H(1,1:m+1) = H(2,1:m+1); % Bv for i = 1, k = 1:m+1
H(n+1,1:m+1) = H(n,1:m+1); % Bv for i = n+1, k=1:m+1

%Plot
mesh(x,y,H','LineWidth',1); view(-17,69);
xlabel('x'); ylabel('y'); zlabel('h');
[hx, hy] = gradient(H',dx,dy);
K = 10^(-8); qx = -K*hx; qy = - K*hy;
figure(2)
[~,cc] = contour(x,y,H',[14.2,14.6,15,16,17,18,19,19.4,19.8]);
axis('equal'); axis([0  L  0  H1]);
set(cc,'ShowText','on'); colormap cool;
```

rectangle('position',[x(i1),y(k1),x(i2)-x(i1),H1-y(k1)],...
   'Facecolor', [0.5 0.5 0.5]);
hold on;
streamline(x, y, qx, qy, x(5:4:i1-8), y(m+1)*ones(1,length(x(5:4:i1-8))));
end

See Exercises 5.4.10 and 5.4.11.

5.4 Exercises

Exercise 5.4.1 Solve Algebraic System (5.2.3).

Answer.

$$\begin{bmatrix} x_1 & y_1 & 1 \\ x_2 & y_2 & 1 \\ x_3 & y_3 & 1 \end{bmatrix} \begin{bmatrix} a_1 \\ b_1 \\ c_1 \end{bmatrix} = \begin{bmatrix} 1 \\ 0 \\ 0 \end{bmatrix} \Rightarrow B \begin{bmatrix} a_1 \\ b_1 \\ c_1 \end{bmatrix} = \begin{bmatrix} 1 \\ 0 \\ 0 \end{bmatrix} \Rightarrow \begin{bmatrix} a_1 \\ b_1 \\ c_1 \end{bmatrix} = B^{-1} \begin{bmatrix} 1 \\ 0 \\ 0 \end{bmatrix}.$$

A simple calculation shows that

$$B^{-1} = \frac{1}{\det(B)} \begin{bmatrix} y_2 - y_3 & y_3 - y_1 & y_1 - y_2 \\ -(x_2 - x_3) & -(x_3 - x_1) & -(x_1 - x_2) \\ x_2 y_3 - x_3 y_2 & x_3 y_1 - x_1 y_3 & x_1 y_2 - x_2 y_1 \end{bmatrix}.$$

Hence,

$$\begin{bmatrix} a_1 \\ b_1 \\ c_1 \end{bmatrix} = \frac{1}{\det(B)} \begin{bmatrix} y_2 - y_3 \\ -(x_2 - x_3) \\ x_2 y_3 - x_3 y_2 \end{bmatrix},$$

that is the desired result.

Exercise 5.4.2 Prove Formula (5.2.8).

Answer. The function $u(x, y)$ is linear

$$u(x, y) = ax + by + c, \qquad (5.4.1)$$

and such that $u_i = u(x_i, y_i)$, $i = 1, 2, 3$. Therefore,

$$\begin{cases} ax_1 + by_1 + c = u_1 \\ ax_2 + by_2 + c = u_2 \\ ax_3 + by_3 + c = u_3 \end{cases} \Rightarrow \begin{bmatrix} x_1 & y_1 & 1 \\ x_2 & y_2 & 1 \\ x_3 & y_3 & 1 \end{bmatrix} \begin{bmatrix} a \\ b \\ c \end{bmatrix} = \begin{bmatrix} u_1 \\ u_2 \\ u_3 \end{bmatrix} \Rightarrow B \begin{bmatrix} a \\ b \\ c \end{bmatrix} = \begin{bmatrix} u_1 \\ u_2 \\ u_3 \end{bmatrix},$$

$$
\begin{bmatrix} a \\ b \\ c \end{bmatrix} = B^{-1} \begin{bmatrix} u_1 \\ u_2 \\ u_3 \end{bmatrix} \Rightarrow \begin{cases} a = (B^{-1})_{1,l}u_1 + (B^{-1})_{1,2}u_2 + (B^{-1})_{1,3}u_3 \\ b = (B^{-1})_{2,l}u_1 + (B^{-1})_{2,2}u_2 + (B^{-1})_{2,3}u_3 \\ c = (B^{-1})_{3,l}u_1 + (B^{-1})_{3,2}u_2 + (B^{-1})_{3,3}u_3. \end{cases}
$$

Using the values of $(B^{-1})_{i,j}$ provided in Exercise 5.4.1 and substituting a, b and c into (5.4.1) yields the desired result.

Exercise 5.4.3   Determine all $\Phi_i(x, y)$ for the triangulation in Fig. 5.2.1.

Answer. The support of Φ_4 is $\{T_8, T_9, T_{10}, T_{15}, T_{14}, T_{13}\}$ and the support of Φ_5 is $\{T_{12}, T_{13}, T_{14}, T_{18}, T_{17}\}$. Applying Formula (5.2.4) yields

$$
\Phi_4 = \begin{cases} x/h + y/k - 3, & \text{on } T_8, \\ y/k - 1, & \text{on } T_9, \\ -x/h + 2, & \text{on } T_{10}, \\ -x/h - y/k + 5, & \text{on } T_{15}, \\ -y/k + 3, & \text{on } T_{14}, \\ x/h - 1, & \text{on } T_{13}, \\ 0, & \text{otherwise,} \end{cases}
\qquad
\Phi_5 = \begin{cases} x/h + y/k - 3, & \text{on } T_{12}, \\ y/k - 2, & \text{on } T_{13}, \\ -x/h + 2, & \text{on } T_{14}, \\ -x/h - y/k + 5, & \text{on } T_{18}, \\ x/h - y/k + 3, & \text{on } T_{17}, \\ 0, & \text{otherwise.} \end{cases}
$$

The support of Φ_6 is $\{T_{14}, T_{15}, T_{16}, T_{20}, T_{19}, T_{18}\}$ and the support of Φ_7 is $\{T_1, T_2\}$. Applying Formula (5.2.4) yields

$$
\Phi_6 = \begin{cases} x/h + y/k - 4, & \text{on } T_{14}, \\ y/k - 2, & \text{on } T_{15}, \\ -x/h + 3, & \text{on } T_{16}, \\ -x/h - y/k + 6, & \text{on } T_{20}, \\ -y/k + 4, & \text{on } T_{19}, \\ x/h - 1, & \text{on } T_{18}, \\ 0, & \text{otherwise,} \end{cases}
\qquad
\Phi_7 = \begin{cases} -y/k + 1, & \text{on } T_1, \\ -x/h - y/k + 2, & \text{on } T_2, \\ 0, & \text{otherwise.} \end{cases}
$$

The support of Φ_8 is $\{T_2, T_3, T_4\}$ and the support of Φ_9 is $\{T_4, T_9, T_{10}\}$. Applying Formula (5.2.4) yields

$$
\Phi_8 = \begin{cases} x/h - 1, & \text{on } T_2, \\ -y/k + 1, & \text{on } T_3, \\ -y/k + 1, & \text{on } T_4, \\ 0, & \text{otherwise,} \end{cases}
\qquad
\Phi_9 = \begin{cases} x/h - 2, & \text{on } T_4, \\ x/h - 2, & \text{on } T_9, \\ -y/k + 2, & \text{on } T_{10}, \\ 0, & \text{otherwise.} \end{cases}
$$

The support of Φ_{10} is $\{T_{10}, T_{15}, T_{16}\}$ and the support of Φ_{11} is $\{T_{16}, T_{20}\}$. Applying Formula (5.2.4) yields

$$\Phi_{10} = \begin{cases} x/h + y/k - 4, & \text{on } T_{10}, \\ x/h - 2, & \text{on } T_{15}, \\ -y/k + 3, & \text{on } T_{16}, \\ 0, & \text{otherwise}, \end{cases} \qquad \Phi_{11} = \begin{cases} x/h + y/k - 5, & \text{on } T_{16}, \\ x/h - 2, & \text{on } T_{20}, \\ 0, & \text{otherwise}. \end{cases}$$

The support of Φ_{12} is $\{T_{19}, T_{20}\}$ and the support of Φ_{13} is $\{T_{17}, T_{18}, T_{19}\}$. Applying Formula (5.2.4) yields

$$\Phi_{12} = \begin{cases} x/h + y/k - 5, & \text{on } T_{19}, \\ y/k - 3, & \text{on } T_{20}, \\ 0, & \text{otherwise}, \end{cases} \qquad \Phi_{13} = \begin{cases} y/k - 3, & \text{on } T_{17}, \\ y/k + 3, & \text{on } T_{18}, \\ -x/h + 2, & \text{on } T_{19}, \\ 0, & \text{otherwise}. \end{cases}$$

The support of Φ_{14} is $\{T_{11}, T_{12}, T_{17}\}$ and the support of Φ_{16} is $\{T_1, T_5\}$. Applying Formula (5.2.4) yields

$$\Phi_{14} = \begin{cases} y/k - 2, & \text{on } T_{11}, \\ -x/h + 1, & \text{on } T_{12}, \\ -x/h + 1, & \text{on } T_{17}, \\ 0, & \text{otherwise}, \end{cases} \qquad \Phi_{16} = \begin{cases} -x/h + 1, & \text{on } T_1, \\ -x/h - y/k + 2, & \text{on } T_5, \\ 0, & \text{otherwise}. \end{cases}$$

The support of Φ_{15} is $\{T_5, T_6, T_{11}\}$ (Fig. 5.4.1). Applying Formula (5.2.4) yields

$$\Phi_{15} = \begin{cases} y/k - 1, & \text{on } T_5, \\ -x/h + 1, & \text{on } T_6, \\ -x/h - y/k + 3, & \text{on } T_{11}, \\ 0, & \text{otherwise}. \end{cases}$$

Exercise 5.4.4   Calculate the integrals in Formulas (5.2.21) and (5.2.22) for $i = 2, \ldots, 6$.

   Answer. For $i = 2$, it is

$$K_{21} = K_{12} = -hk\frac{1}{h^2}, \quad K_{22} = \int_{T_3 \cup T_4 \cup T_7 \cup T_8 \cup T_9} \nabla\Phi_2 \cdot \nabla\Phi_2 = 2hk\left[\frac{1}{h^2} + \frac{1}{k^2}\right],$$

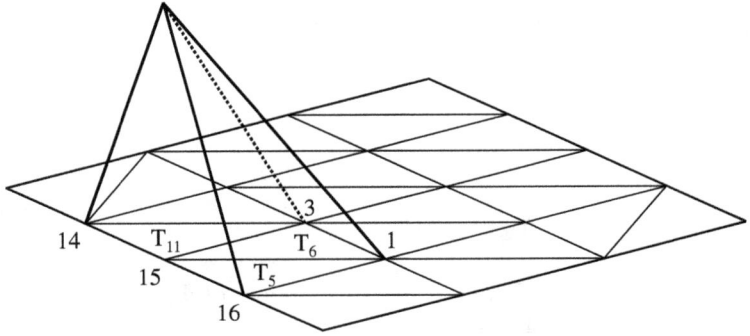

Figure 5.4.1. Φ_{15}.

$$K_{23} = 0; \quad K_{24} = \int_{T_8 \cup T_9} \nabla \Phi_2 \cdot \nabla \Phi_4 = -hk\frac{1}{k^2}, \quad K_{25} = K_{26} = 0,$$

$$b_{27} = 0; \quad b_{28} = \int_{T_3 \cup T_4} \nabla \Phi_2 \cdot \nabla \Phi_8 = -hk\frac{1}{k^2},$$

$$b_{29} = \int_{T_4 \cup T_9} \nabla \Phi_2 \cdot \nabla \Phi_9 = -hk\frac{1}{h^2},$$

$$b_{2,10} = b_{2,11} = b_{2,12} = b_{2,13} = b_{2,14} = b_{2,15} = b_{2,16} = 0,$$

$$g_2 = \frac{h}{k}u_8 + \frac{k}{h}u_9 = \frac{h}{k}4h^2 + \frac{k}{h}(9h^2 - k^2).$$

For $i = 3$, it is

$$K_{31} = K_{13} = -hk\frac{1}{k^2}, \quad K_{32} = K_{23} = 0,$$

$$K_{33} = \int_{T_6 \cup T_7 \cup T_8 \cup T_{11} \cup T_{12} \cup T_{13}} \nabla \Phi_3 \cdot \nabla \Phi_3 = 2hk\left[\frac{1}{h^2} + \frac{1}{k^2}\right],$$

$$K_{34} = \int_{T_8 \cup T_{13}} \nabla \Phi_3 \cdot \nabla \Phi_4 = -hk\frac{1}{h^2},$$

$$K_{35} = \int_{T_{12} \cup T_{13}} \nabla \Phi_3 \cdot \nabla \Phi_5 = -hk\frac{1}{k^2}, \quad K_{36} = 0,$$

$$b_{37} = b_{38} = b_{39} = b_{3,10} = b_{3,11} = b_{3,12} = b_{3,13} = b_{3,14} = 0,$$

$$b_{3,15} = \int_{T_6 \cup T_{11}} \nabla \Phi_3 \cdot \nabla \Phi_{15} = -hk\frac{1}{h^2}, \quad b_{3,16} = 0,$$

$$g_3 = \frac{k}{h}u_{15} = -\frac{k}{h}4k^2.$$

For $i = 4$, it is

$$K_{41} = K_{14} = 0, \quad K_{42} = K_{24} = -hk\frac{1}{k^2}, \quad K_{43} = K_{34} = -hk\frac{1}{h^2},$$

$$K_{44} = \int_{T_8 \cup T_9 \cup T_{10} \cup T_{11} \cup T_{12} \cup T_{13}} \nabla\Phi_4 \cdot \nabla\Phi_4 = 2hk\left[\frac{1}{h^2} + \frac{1}{k^2}\right],$$

$$K_{45} = 0, \quad K_{46} = \int_{T_{14} \cup T_{15}} \nabla\Phi_4 \cdot \nabla\Phi_6 = -hk\frac{1}{k^2},$$

$$b_{47} = b_{48} = b_{49} = 0, \quad b_{4,10} = \int_{T_{10} \cup T_{15}} \nabla\Phi_4 \cdot \nabla\Phi_{10} = -hk\frac{1}{h^2},$$

$$b_{4,11} = b_{4,12} = b_{4,13} = b_{4,14} = b_{4,15} = b_{4,16} = 0,$$

$$g_4 = \frac{k}{h}u_{10} = \frac{k}{h}(9h^2 - 4k^2).$$

For $i = 5$, it is

$$K_{51} = K_{15} = 0, \quad K_{52} = K_{25} = 0, \quad K_{53} = K_{35} = -hk\frac{1}{k^2}, \quad K_{54} = K_{45} = 0,$$

$$K_{55} = \int_{T_{12} \cup T_{13} \cup T_{14} \cup T_{17} \cup T_{18}} \nabla\Phi_5 \cdot \nabla\Phi_5 = 2hk\left[\frac{1}{h^2} + \frac{1}{k^2}\right],$$

$$K_{56} = \int_{T_{14} \cup T_{18}} \nabla\Phi_5 \cdot \nabla\Phi_6 = -\frac{hk}{2}\left[\frac{1}{h^2} + \frac{1}{k^2}\right], \quad b_{57} = b_{58} = b_{59} = 0,$$

$$b_{5,10} = b_{5,11} = b_{5,12} = 0, \quad b_{5,13} = \int_{T_{17} \cup T_{18}} \nabla\Phi_5 \cdot \nabla\Phi_{13} = -hk\frac{1}{k^2},$$

$$b_{5,14} = \int_{T_{12} \cup T_{17}} \nabla\Phi_5 \cdot \nabla\Phi_{14} = -hk\frac{1}{h^2}, \quad b_{5,15} = b_{5,16} = 0,$$

$$g_5 = \frac{h}{k}u_{13} + \frac{k}{h}u_{14} = \frac{h}{k}(h^2 - 16k^2) - \frac{k}{h}9k^2.$$

For $i = 6$, it is

$$K_{61} = K_{16} = 0, \quad K_{62} = K_{26} = 0, \quad K_{63} = K_{36} = 0,$$

$$K_{64} = K_{46} = -hk\frac{1}{k^2}, \quad K_{65} = K_{56} = -\frac{hk}{2}\left[\frac{1}{h^2} + \frac{1}{k^2}\right],$$

$$K_{66} = \int_{T_{14} \cup T_{15} \cup T_{16} \cup T_{18} \cup T_{19} \cup T_{20}} \nabla \Phi_6 \cdot \nabla \Phi_6 = 2hk \left[\frac{1}{h^2} + \frac{1}{k^2} \right],$$

$$b_{67} = b_{68} = b_{69} = b_{6,10} = 0, \quad b_{6,11} = \int_{T_{16} \cup T_{20}} \nabla \Phi_6 \cdot \nabla \Phi_{11} = -hk\frac{1}{h^2},$$

$$b_{6,12} = \int_{T_{19} \cup T_{20}} \nabla \Phi_6 \cdot \nabla \Phi_{12} = -hk\frac{1}{k^2}, \quad b_{6,13} = b_{6,14} = b_{6,15} = b_{6,16} = 0,$$

$$g_6 = \frac{k}{h}u_{11} + \frac{h}{k}u_{12} = \frac{k}{h}(9h^2 - 9k^2) + \frac{h}{k}(4h^2 - 16k^2).$$

Exercise 5.4.5   Consider the dam function in Example 5.2.9. Replace the code line

   L = 42; H1 = 14; H2 = 20; d = 2; nx = L/d; ny = H1/d;

with

   L = 42; H1 = 14; H2 = H1; d = 2; nx = L/d; ny = H1/d;

What happens?

Answer. The numerical solution is to $H = 14$, which is the same as the analytical solution. This result indirectly shows the accuracy of the numerical method.

Exercise 5.4.6   Write a function, say sheet_pile_wall, that applies the FEM to Problem (5.2.45) to (5.2.47) in Example 5.2.10.

Hint. The following listing considers the simple triangulation in Fig. 5.4.2. The numerical solution is plotted in Fig. 5.4.3.

```
function [u, hh] = sheet_pile_wall
% This is the function file sheet_pile_wall.m.
% The FEM is applied to solve the Dirichlet–Neumann problem in Example
% 5.2.10. The function returns the matrix and the vector expressions of
% the numerical solution.

% Initialization
Lx = 22; Ly = 12; d = 2; nx = Lx/d; ny = Ly/d;
n = (nx+1)*(ny+1);% Number of nodes.
x = linspace(0,Lx,nx+1); y = linspace(0,Ly,ny+1);
P = Nodes(x, y, nx+1, ny+1); xt = P(:,1); yt = P(:,2);
T = Triangulation(nx, nx+1, ny);
```

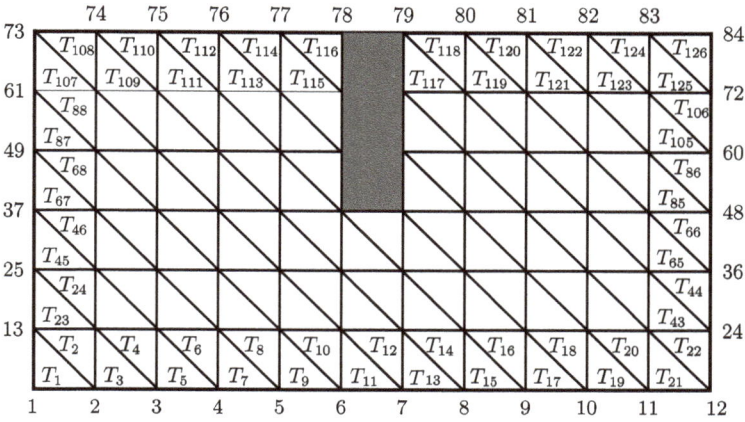

Figure 5.4.2. Triangulation for Problem (5.2.45) to (5.2.47).

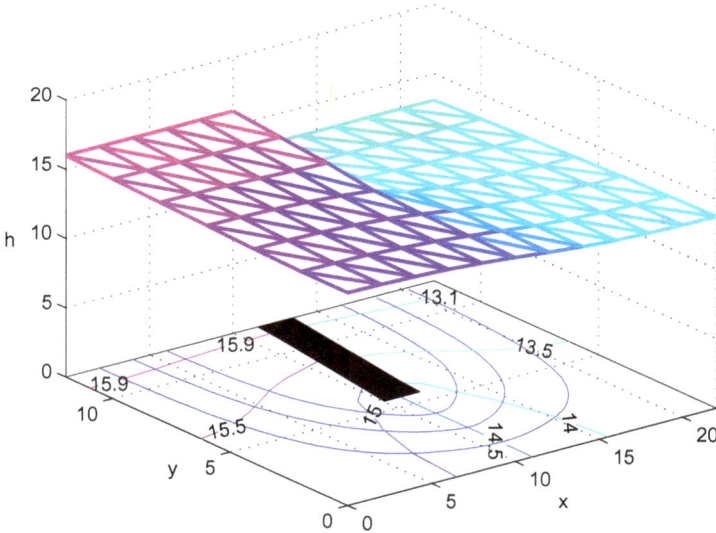

Figure 5.4.3. Numerical solution to Problem (5.2.45) to (5.2.47).

```
hb = [16*ones(6,1); 13*ones(6,1)];% Boundary conditions.
nf = n - length(hb);% Free nodes.
K = zeros(nf,nf);% Stiffness matrix.
b = zeros(nf,n-nf);
```

```
% FEM
for i=1:nf
    same code as dam function
end
f = -b * hb; v = K\f; hh = [v;hb];
u = reshape(hh,nx+1,ny+1);
% Plot
trimesh(T, xt, yt, hh,'LineWidth',2); view(-6,55);
xlabel('x'); ylabel('y'); zlabel('h');
[hx, hy] = gradient(u',d,d);
KK =10^(-8); qx = -KK*hx; qy = -KK*hy;
hold on;
[~,cc] = contour(x,y,u',[13.1,13.5,14,14.5,15,15.5,15.9]);
set(cc,'ShowText','on'); colormap cool;
axis([0  Lx  0  Ly]); rectangle('position',[0,  0,  Lx,  Ly])
rectangle('position',[x(6),y(4),d,3*d ],'FaceColor','black','LineWidth',1)
hold on;
streamline(x,y,qx,qy,x(2:4),y(ny+1)*ones(1,3));
end
% ———— Local functions ————-
function P = Nodes(x, y, n1, n2)
P = zeros(n1*n2,2);
P(1:n1*n2,1) = repmat(x',n2,1);
for jj=1:n2
    P(1+(jj-1)*n1:jj*n1,2) = y(jj)*ones(n1,1);
end
end
function T = Triangulation(nx, n1, ny)
T = zeros(2*nx*ny,3);
for k=1:ny
    for i=1:nx
        T((k-1)*2*nx+2*i-1,:)= [(k-1)*n1+i (k-1)*n1+i+1 k*n1+i];
        T((k-1)*2*nx+2*i,:) = [(k-1)*n1+i+1 k*n1+i+1 k*n1+i];
    end
end
T = T([1:76, 79:98, 101:120, 123:2*nx*ny],:);
end
```

Exercise 5.4.7 Consider the five_point function in Example 5.3.1. Verify that matrix A was introduced with the following code

```
AA = repmat([ones(n-1,1)  [sigma²*ones(n-2,1);0]...
   -2*(1+sigma²)*ones(n-1,1) [0;sigma²*ones(n-2,1)]  ones(n-1,1)], (m-1),1);
A = spdiags(AA, [-(n-1)  -1:1  (n-1)], (n-1)*(m-1), (n-1)*(m-1));
```

can also be introduced with the following loop

```
B1 = sigma²*ones(n-2,1); B2 = -2*(1+sigma²)*ones(n-1,1);
B = diag(B1,-1) + diag(B2) + diag(B1,1);
A(1:n-1,1:n-1) = B; A(1:n-1,n:2*(n-1)) = eye(n-1);
for k=2:m-2
    A((k-1)*(n-1)+1:k*(n-1),(k-1)*(n-1)+1:k*(n-1)) = B;
    A((k-1)*(n-1)+1:k*(n-1),k*(n-1)+1:(k+1)*(n-1)) = eye(n-1);
    A((k-1)*(n-1)+1:k*(n-1),(k-2)*(n-1)+1:(k-1)*(n-1)) = eye(n-1);
end
A((m-2)*(n-1)+1:(m-1)*(n-1),(m-2)*(n-1)+1:(m-1)*(n-1)) = B;
A((m-2)*(n-1)+1:(m-1)*(n-1),(m-3)*(n-1)+1:(m-2)*(n-1)) = eye(n-1);
```

Exercise 5.4.8 Consider the following Dirichlet problem

$$\Delta U(x,y) = x + y, \quad 0 < x < L, \quad 0 < y < H, \tag{5.4.2}$$

$$\begin{cases} U(0,y) = g_1(y) = y^3/6, \\ U(L,y) = g_3(y) = (L^3 + y^3)/6, \end{cases} \quad 0 \le y \le H, \tag{5.4.3}$$

$$\begin{cases} U(x,0) = g_2(x) = x^3/6, \\ U(x,H) = g_4(x) = (x^3 - H^3)/6, \end{cases} \quad 0 \le x \le L. \tag{5.4.4}$$

Write a function, say five_point_f2, that applies the Five-Point Method to Problem (5.4.2) to (5.4.4).

Exercise 5.4.9 Write the Seven-Point Method for Poisson's Eq. (5.3.16). *Hint.* Consider Fig. 5.4.4.

Exercise 5.4.10 Consider the dam_5p function, Example 5.3.3. Check the accuracy of the numerical result. *Hint.* See Exercise 5.4.5.

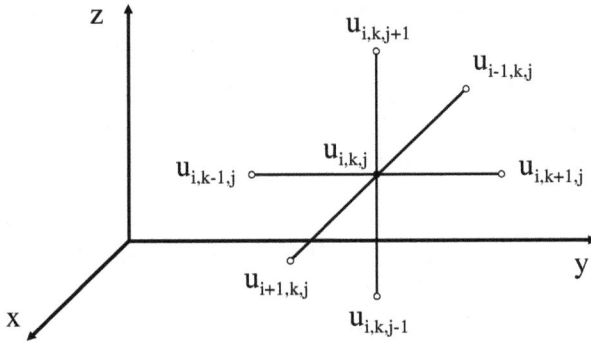

Figure 5.4.4. Seven-Point Method.

Exercise 5.4.11 The dam problem was discussed with the FEM in Sec. 5.2.4 and with the Five-Point Method in Sec. 5.3.2. Compare the results provided by the two methods.

Hint. Let us consider the matrices u and H returned by dam and dam_5p functions, respectively. The matrices u and H have different sizes. Consider the matrix $H\_f = H(1:4:n+1,1:4:m+1)$ that has the same size as u. The elements of the matrices u and H_f refer to the same points and can be compared.

Chapter 6

The Euler–Bernoulli Beam

The equation governing the lateral vibrations of a beam is derived in the framework of the Euler–Bernoull theory. The weak formulation is presented and the Finite Element Method (FEM) is introduced Fenner (2005); Rao (2005). The application of the method is illustrated with problems in Statics. The Matlab codes consider beams subjected to several constraints and loads. A section is devoted to beams subjected to concentrated forces. The weak form for the related equation is introduced and Matlab applications are illustrated in this new situation.

6.1 Finite Element Method

6.1.1 *Euler–Bernoulli Beam Equation*

Consider the lateral vibrations $U(x,t)$ of a beam subjected to the distributed load $F(x,t)$ (Fig. 6.1.1). The governing equation is derived in the framework of the Euler–Bernoull beam theory, where it is assumed that plain cross-sections remain plain and deformed beam angles (slopes) are small. Consider Newton's Second Law

$$\int_{x_1}^{x_2} \rho \frac{\partial^2 U}{\partial t^2}\, dx = \int_{x_1}^{x_2} F dx + V(x_1,t) - V(x_2,t), \qquad (6.1.1)$$

$$\int_{x_1}^{x_2} \left(F - \frac{\partial V}{\partial x} - \rho \frac{\partial^2 U}{\partial t^2} \right) dx = 0, \qquad (6.1.2)$$

where ρ is the density per unit length and V is the shear force. Consider the moment equation about x_1 (neglecting rotary effects)

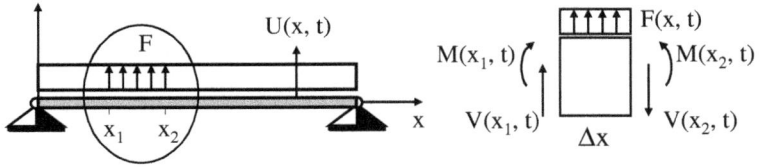

Figure 6.1.1.   Lateral vibrations of a beam.

$$\int_{x_1}^{x_2} (x - x_1)(F - \rho U_{tt})dx - M(x_1, t) + M(x_2, t) - (x_2 - x_1)V(x_2, t) = 0.$$

$$(6.1.3)$$

$$\int_{x_1}^{x_2} \left[(x - x_1)(F - \rho U_{tt}) + \frac{\partial M}{\partial x} - \frac{\partial (x - x_1)V}{\partial x} \right] dx = 0. \qquad (6.1.4)$$

Equation (6.1.2) holds for any $\Delta x = x_2 - x_1$. Therefore, it implies

$$\rho U_{tt} = F - V_x. \qquad (6.1.5)$$

Similarly, from (6.1.4), it follows that

$$(x - x_1)(F - \rho U_{tt} - V_x) + M_x - V = 0,$$

that simplifies to

$$V = M_x. \qquad (6.1.6)$$

because of Eq. (6.1.5). Consider the relationship between the bending moment M and (approximate) curvature

$$M = EIU_{xx}, \qquad (6.1.7)$$

where E is the Young modulus and I is the beam's second moment of area. Substitute (6.1.7) into (6.1.6) and, then, into (6.1.5) to obtain the equation for the lateral motions of an elastic beam

$$\rho U_{tt} + (EIU_{xx})_{xx} = F. \qquad (6.1.8)$$

Often, the product EI is constant and we have

$$\rho U_{tt} + EIU_{xxxx} = F. \qquad (6.1.9)$$

Equation (6.1.8) is named the *Euler–Lagrange equation*. In Statics, Eq. (6.1.8) simplifies to the *Euler–Bernoulli equation*.

$$(EIU_{xx})_{xx} = F, \qquad (6.1.10)$$

Figure 6.1.2.   Pinned-pinned beam (left), fixed-free beam (right).

Two initial conditions are prescribed to Eq. (6.1.8)

$$U(x,0) = \varphi(x), \quad U_t(x,0) = \psi(x), \tag{6.1.11}$$

corresponding to the beam initial position and velocity. In addtion, four boundary conditions must be assigned. For example, for the pinned-pinned beam in Fig. 6.1.2 (left), we have

$$U(x_A,t) = 0, \quad U_{xx}(x_A,t) = 0, \quad U(x_B,t) = 0, \quad U_{xx}(x_B,t) = 0, \tag{6.1.12}$$

where the second and fourth conditions are equivalent to $M(x_A,t) = M(x_B,t) = 0$, because of (6.1.7). For the fixed end-free end beam in Fig. 6.1.2 (right), we have

$$U(x_A,t) = 0, U_x(x_A,t) = 0, \quad U_{xx}(x_B,t) = 0, \quad U_{xxx}(x_B,t) = 0, \tag{6.1.13}$$

where the last two conditions are equivalent to $M(x_B,t) = V(x_B,t) = 0$, because of (6.1.6) and (6.1.7).

The external forces and moments at the ends F_A, M_A, F_B and M_B can be known loads or unknown reactive forces. In both cases, the relationships between external forces and shear force and the bending moment (internal forces) are expressed as (Fig. 6.1.3)

$$F_A = V(x_A,t), \quad F_B = -V(x_B,t), \tag{6.1.14}$$

$$M_A = -M(x_A,t), \quad M_B = M(x_B,t). \tag{6.1.15}$$

These relationships are clear from Fig. 6.1.3. However, they can be rigorously derived from Eqs. (6.1.1) and (6.1.3). Indeed, consider Eq. (6.1.1) on $(x_A, x_A + \Delta x)$ (Fig. 6.1.4, left)

$$\int_{x_A}^{x_A+\Delta x} (F - \rho U_{tt}) \, dx + F_A - V(x_A + \Delta x, t) = 0.$$

For $\Delta x \to 0$, this equation gives $(6.1.14)_1$. Next, consider Eq. (6.1.3) on $(x_B - \Delta x, x_B)$ (Fig. 6.1.4, right)

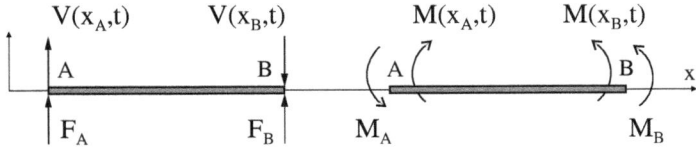

Figure 6.1.3.   Forces and moments at the ends.

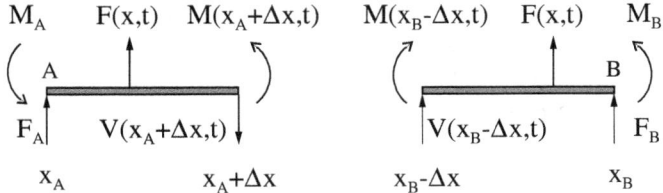

Figure 6.1.4.   Relationships between external and internal forces at the ends.

$$\int_{x_B - \Delta x}^{x_A} (x - x_B)(\rho U_{tt} - F)\, dx + M_B - M(x_B - \Delta x, t) - \Delta x V(x_B - \Delta x, t) = 0.$$

For $\Delta x \to 0$, we get $(6.1.15)_2$. Similar reasonings lead to Formulas $(6.1.14)_2$ and $(6.1.15)_1$.

6.1.2 *Shape Functions*

Consider the grid of punts $x_1, \ldots, x_n$, with step h, in Fig. 6.1.5 (right), and refer to the interval $e_1 = [x_1, x_2] = [0, h]$, the named element in this context. The shape functions are defined as (Fig. 6.1.5, left)

$$N_1(x) = 1 - \frac{3}{h^2}x^2 + \frac{2}{h^3}x^3, \Rightarrow \quad \begin{array}{l} N_1'(x) = -6x/h^2 + 6x^2/h^3, \\[6pt] N_1''(x) = -6/h^2 + 12x/h^3, \end{array} \tag{6.1.16}$$

$$N_2(x) = x - \frac{2}{h}x^2 + \frac{1}{h^2}x^3, \Rightarrow \quad \begin{array}{l} N_2'(x) = 1 - 4x/h + 3x^2/h^2, \\[6pt] N_2''(x) = -4/h + 6x/h^2, \end{array} \tag{6.1.17}$$

$$N_3(x) = \frac{3}{h^2}x^2 - \frac{2}{h^3}x^3, \Rightarrow \quad \begin{array}{l} N_3'(x) = 6x/h^2 - 6x^2/h^3, \\[6pt] N_3''(x) = 6/h^2 - 12x/h^3, \end{array} \tag{6.1.18}$$

$$N_4(x) = -\frac{1}{h}x^2 + \frac{1}{h^2}x^3, \Rightarrow \quad \begin{array}{l} N_4'(x) = -2x/h + 3x^2/h^2, \\[6pt] N_4''(x) = -2/h + 6x/h^2. \end{array} \tag{6.1.19}$$

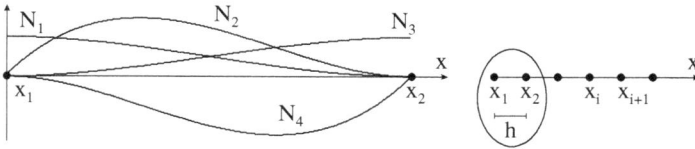

Figure 6.1.5.   Hermitian functions.

These cubic polynomials are known as *hermitian functions* or *hermitian polynomials*. The expression of the hermitian functions on the element $e_i = [x_i, x_{i+1}]$ is obtained by replacing x with $x - x_i$ in Formulas (6.1.16) to (6.1.19).

Remark 6.1.1   *The hermitian functions have an interesting meaning in Statics that is outlined below. Consider the Euler–Bernoulli equation*

$$EIU^{iv} = F(x). \tag{6.1.20}$$

Apply Eq. (6.1.20) to discuss the fixed end-fixed end beam in Fig. 6.1.6 (left). Assume that the beam is subjected to no external load and the constraint in A is subjected to a vertical displacement $U_A = 1$ (Fig. 6.1.6, right). The deformed configuration is obtained by solving Eq. (6.1.20), where $F = 0$, with the following boundary conditions

$$U(0) = 1, \quad U'(0) = 0, \quad U(h) = 0, \quad U'(h) = 0, \tag{6.1.21}$$

where h is the length of the beam. A simple calculation shows that $U = N_1$ and it explains the physical meaning of N_1 defined in (6.1.16). Now, suppose that the constraint in A is subjected to the vertical displacement U_A. After replacing the first condition in (6.1.21) with $U(0) = U_A$, an easy calculation leads to $U = U_A N_1$. Very often, it is $U_A < 0$. The shear force and bending moment are derived from Formulas (6.1.6) and (6.1.7), (6.1.16). Then, the reactive forces follow from Formulas (6.1.14) and (6.1.15). See Exercise 6.4.1 for the physical meaning of N_2, N_3 and N_4.

The importance of the hermitian polynomials is due to the fact that any cubic polynomial defined on the element can be expressed as a linear combination of those polynomials. Indeed, consider the generic cubic polynomial

$$u(x) = a_1 x^3 + a_2 x^2 + a_3 x + a_4, \quad x \in e_1, \tag{6.1.22}$$

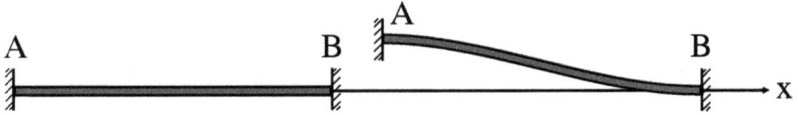

Figure 6.1.6. Meaning of N_1 in Statics.

and introduce the notations

$$u_1 = u(0), \quad u_2 = u'(0), \quad u_3 = u(h), \quad u_4 = u'(h). \tag{6.1.23}$$

Use (6.1.23) in (6.1.22)

$$\begin{cases} a_4 = u_1, \\ a_3 = u_2, \\ a_1 h^3 + a_2 h^2 + a_3 h + a_4 = u_3, \\ 3a_1 h^2 + 2a_2 h + a_3 = u_4. \end{cases} \tag{6.1.24}$$

Solving System (6.1.24) with respect to a_i and the result substituted into (6.1.22) yields

$$u(x) = u_1 N_1(x) + u_2 N_2(x) + u_3 N_3(x) + u_4 N_4(x), \tag{6.1.25}$$

that is the desired result. See Exercise 6.4.2. In addition, the functions $\{N_i, \ i = 1, \ldots, 4\}$ are a linearly independent system, a basis, for the cubic polynomials defined on the element. It must be proved that

$$0 = u_1 N_1(x) + u_2 N_2(x) + u_3 N_3(x) + u_4 N_4(x), \quad x \in [0, h], \tag{6.1.26}$$

if and only if

$$u_i = 0, \quad i = 1, \ldots, 4. \tag{6.1.27}$$

It is enough to show that (6.1.26) implies (6.1.27), since the opposite statement is apparent. For $x = 0$ and $x = h$, from Formulas (6.1.16) to (6.1.19), it follows that

$$0 = u_1 N_1(0) = u_1, \quad 0 = u_3 N_3(h) = u_3.$$

In addition, let us differentiate (6.1.26)

$$0 = u_1 N_1'(x) + u_2 N_2'(x) + u_3 N_3'(x) + u_4 N_4'(x), \quad x \in [0, h].$$

For $x = 0$ and $x = h$, from Formulas (6.1.16) to (6.1.19), it follows that

$$0 = u_2 N_2'(0) = u_2, \quad 0 = u_4 N_4'(h) = u_4,$$

that completes the proof.

Let $U(x,t)$ be a function defined on $e_1 \cup e_2$, $e_1 = [x_1, x_2]$, $e_2 = [x_2, x_3]$, differentiable with respect to x. Introduce the notations

$$u_1(t) = U(x_1, t), \quad u_3(t) = U(x_2, t), \quad u_5(t) = U(x_3, t),$$
$$u_2(t) = U_x(x_1, t), \quad u_4(t) = U_x(x_2, t), \quad u_6(t) = U_x(x_3, t),$$

and consider the function

$$u(x,t) = u_1(t)N_{1,1}(x) + u_2(t)N_{2,1}(x) + u_3(t)N_{3,1}(x)$$
$$+ u_4(t)N_{4,1}(x), \quad x \in e_1. \tag{6.1.28}$$

This function is named the *hermitian approximating* of U in e_1. Therefore, the function

$$u(x,t) = u_3(t)N_{1,2}(x) + u_4(t)N_{2,2}(x) + u_5(t)N_{3,2}(x)$$
$$+ u_6(t)N_{4,2}(x), \quad x \in e_2, \tag{6.1.29}$$

is the hermitian approximating of U in e_2. A second subscript, related to the element, was added to the hermitian functions, as their expressions depend on the element. As it is immediately deduced that

$$u(x_1, t) = u_1(t), \quad u(x_2, t) = u_3(t), \quad u(x_3, t) = u_5(t),$$
$$u_x(x_1, t) = u_2(t), \quad u_x(x_2, t) = u_4(t), \quad u_x(x_3, t) = u_6(t),$$

we realize that the values of U and U_x on the nodes are the same as hermitian approximating. Combine (6.1.28) and (6.1.29) in one formula

$$u(x,t) = \begin{cases} u_1N_{1,1} + u_2N_{2,1} + u_3N_{3,1} + u_4N_{4,1}, & x \in e_1, \\ u_3N_{1,2} + u_4N_{2,2} + u_5N_{3,2} + u_6N_{4,2}, & x \in e_2, \end{cases} \tag{6.1.30}$$

that is the piecewise hermitian approximating of U in $e_1 \cup e_2$. Lastly, if the following definitions

$$\Phi_1 = N_{1,1}, \quad \Phi_2 = N_{2,1}, \quad \Phi_3 = \begin{cases} N_{3,1}, & x \in e_1, \\ N_{1,2}, & x \in e_2, \end{cases}$$

$$\Phi_4 = \begin{cases} N_{4,1}, & x \in e_1, \\ N_{2,2}, & x \in e_2, \end{cases} \quad \Phi_5 = N_{3,2}, \quad \Phi_6 = N_{4,2}, \tag{6.1.31}$$

are used, Formula (6.1.30) is rewritten as (Fig. 6.1.7)

$$u = u_1\Phi_1 + u_2\Phi_2 + u_3\Phi_3 + u_4\Phi_4 + u_5\Phi_5 + u_6\Phi_6. \tag{6.1.32}$$

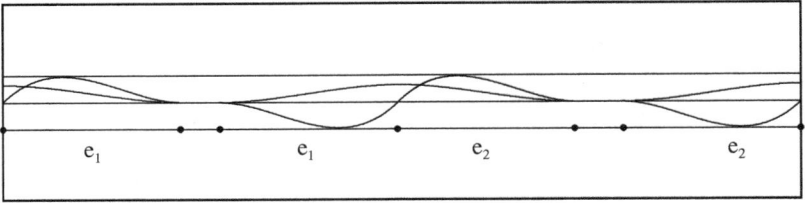

Figure 6.1.7.   Functions Φ_1, Φ_2 (left), Φ_3, Φ_4 (center), and Φ_5, Φ_6 (right).

This formula is related to two elements and it suggests the generalization to any number of elements.

6.1.3   Weak Form

Consider the Euler–Lagrange equation

$$\rho U_{tt} + (EIU_{xx})_{xx} = F(x,t), \quad 0 < x < L, \quad t > 0. \tag{6.1.33}$$

Multiply Eq. (6.1.33) by the smooth test function $v = v(x)$ and integrate over $[0, L]$

$$\int_0^L \rho U_{tt} v \, dx + \int_0^L (EIU_{xx})_{xx} v \, dx = \int_0^L Fv \, dx.$$

Integrate the second integral by parts twice to obtain

$$\int_0^L \rho U_{tt} v \, dx + \int_0^L EIU_{xx} v'' \, dx = \int_0^L Fv \, dx + [v'M - vV]_0^L, \tag{6.1.34}$$

where Formulas (6.1.6) and (6.1.7) related to moment and shear were considered in the finite term. Equation (6.1.34) is the *weak form of the beam equation* (6.1.33) that is named the strong form by contrast. Since integrating by parts lowered the order of the derivative with respect to x, the set of the solutions is wider. Solutions can exist to the weak form and do not have the necessary regularity to be solutions to the strong form.

The FEM considers the weak form and the piecewise hermitian approximating functions on $[0, L]$. Let us begin to discuss the first element. The hermitian approximating u of U on e_1 is given by

$$u(x,t) = u_1(t)N_1(x) + u_2(t)N_2(x)u_3(t)N_3(x) + u_4(t)N_4(x), \tag{6.1.35}$$

where $u_i(t)$, $i = 1, \ldots, 4$, are unknown functions. Substituting u into Eq. (6.1.34) and assuming $v = N_i \ i = 1, \ldots, 4$, we arrive at the following system of second-order ordinary differential equations

$$\sum_{j=1}^{4} \ddot{u}_j \int_0^h \rho N_i N_j dx + \sum_{j=1}^{4} u_j \int_0^h EIN_i'' N_j'' \, dx = \int_0^h FN_i dx$$

$$+ [N_i'M - N_iV]_0^h, \quad i = 1,\ldots,4. \tag{6.1.36}$$

Hence,

$$\sum_{j=1}^{4} M_{ij}\ddot{u}_j + \sum_{j=1}^{4} k_{ij}u_j = f_i, \quad i = 1,\ldots,4, \tag{6.1.37}$$

where

$$M_{ij} = \int_0^h \rho N_i N_j \, dx, \quad K_{ij} = \int_0^h EIN_i'' N_j'' \, dx, \quad i,j = 1,\ldots,4, \tag{6.1.38}$$

$$f_i = \int_0^h FN_i \, dx + [N_i'M - N_iV]_0^h, \quad i = 1,\ldots,4, \tag{6.1.39}$$

In compact matrix form, Formula (6.1.37) is written as

$$M\ddot{\mathbf{u}} + k\mathbf{u} = \mathbf{f}. \tag{6.1.40}$$

The matrix k is the *stiffness matrix*. The matrix M is named the *mass matrix*, or more precisely, *consistent mass matrix*, as there is another mass matrix that will be introduced later. The vector $\mathbf{f}$ is the *load vector*. For constant EI, k is given by

$$k = \frac{EI}{h^3} \begin{bmatrix} 12 & 6h & -12 & 6h \\ 6h & 4h^2 & -6h & 2h^2 \\ -12 & -6h & 12 & -6h \\ 6h & 2h^2 & -6h & 4h^2 \end{bmatrix}. \tag{6.1.41}$$

For constant ρ, M is expressed as

$$M = \frac{\rho h}{420} \begin{bmatrix} 156 & 22h & 54 & -13h \\ 22h & 4h^2 & 13h & -3h^2 \\ 54 & 13h & 156 & -22h \\ -13h & -3h^2 & -22h & 4h^2 \end{bmatrix}. \tag{6.1.42}$$

Example 6.1.1   A function is presented that returns the stiffness matrix. A way to call the function is illustrated in the comment lines of the listing.

```
function K = stiffness(h, EI)
% This is the function file stiffness.m.
% The function returns the stiffness matrix. The input variables are:
% length of the element h and the product E*I. For example, the function
% can be called with the commands: h = 1; EI = 1; K = stiffness(h,EI).
K = zeros(4,4);
for i=1:4
    for j=i:4
        K(i,j) = EI*integral(@(x)Nxx(i, x, h).*Nxx(j, x, h), 0, h);
        K(j,i) = K(i,j);
    end
end
end
% Local function
function f = Nxx(i, x, h)
switch i
    case 1
        f = -6/h^2 + 12*x/h^3;
    case 2
        f = -4/h + 6*x/h^2;
    case 3
        f = 6/h^2 - 12*x/h^3;
    case 4
        f = -2/h + 6*x/h^2;
end
end
```

Another application is suggested in Exercise 6.4.3.

Example 6.1.2 A function is presented that returns the mass matrix. A simple way to call the function is illustrated in the comment lines of the listing.

```
function M = mass(h, rho)
% This is the function file massa.m.
% The function returns the mass matrix. The input variables are:
% length of the element h and the density rho. For example, the function
% can be called with the commands: h = 1; rho = 1; M = mass(h, rho).
```

```
M = zeros(4,4);
for i=1:4
    for j=i:4
        M(i,j) = rho*integral(@(x)N(i, x, h).*N(j, x, h), 0, h);
        M(j,i) = M(i,j);
    end
end
end
% Local function
function f = N(i, x, h)
switch i
    case 1
        f = 1 - 3*x.^2/h^2 + 2*x.^3/h^3;
    case 2
        f = x - 2*x.^2/h + x.^3/h^2;
    case 3
        f = 3*x.^2/h^2 - 2*x.^3/h^3;
    case 4
        f = -x.^2/h + x.^3/h^2;
end
end
```

Another application is suggested in Exercise 6.4.4.

Lastly, let us provide another expression of the mass matrix that is much more used in the programs for beams in Dynamics,

$$M = \frac{\rho h}{2} \begin{bmatrix} 1 & 0 & 0 & 0 \\ 0 & 0 & 0 & 0 \\ 0 & 0 & 1 & 0 \\ 0 & 0 & 0 & 0 \end{bmatrix}. \tag{6.1.43}$$

This expression is derived by discretizing the distributed inertial forces $-\rho \ddot{U}(x,t)$.

6.2 Statics

In Statics, the inertial forces are zero and the Euler–Lagrange equation simplifies to the Euler–Bernoulli equation

$$(EIU'')'' = F(x), \quad 0 < x < L, \tag{6.2.1}$$

where the unknown displacement depends on x only: $U = U(x)$. The weak form is derived from (6.1.34) as a special case

$$\int_0^L EIU''v'' \, dx = \int_0^L Fv \, dx + [v'M - vV]_0^L . \tag{6.2.2}$$

The FEM solution on e_1 is expressed as

$$u(x) = u_1 N_1(x) + u_2 N_2(x) + u_3 N_3(x) + u_4 N_4(x), \tag{6.2.3}$$

where, now, u_i are unknown coefficients that must satisfy the algebraic system

$$k\mathbf{u} = \mathbf{f}, \tag{6.2.4}$$

that follows from (6.1.40) when the inertial forces are zero. Consider the load vector (6.1.39), composed by two terms

$$\mathbf{f} = \mathbf{f}_d + \mathbf{f}_b. \tag{6.2.5}$$

The first vector depends on the special distributed load $F(x,t)$ on the beam. The second vector refers to the boundary conditions on the element, and it has the following interesting expression

$$\mathbf{f}_b = \begin{bmatrix} [N_1'M - N_1 V]_0^h \\ [N_2'M - N_2 V]_0^h \\ [N_3'M - N_3 V]_0^h \\ [N_4'M - N_4 V]_0^h \end{bmatrix} = \begin{bmatrix} V(0) \\ -M(0) \\ -V(h) \\ M(h) \end{bmatrix}, \tag{6.2.6}$$

where the last equality follows from the properties of the shape functions.

Example 6.2.1   Consider a fixed end-free end beam subjected to the tip load p (Fig. 6.2.1). The following boundary conditions hold

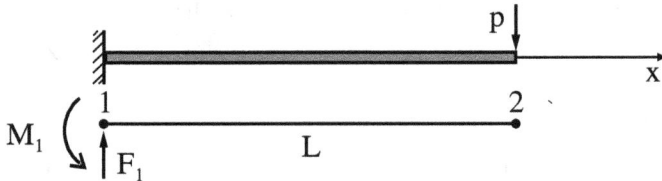

Figure 6.2.1.   Fixed end-free end beam subjected to a tip load.

$$\begin{cases} U(0) = 0, \quad U'(0) = 0, \\ M(L) = 0 \Leftrightarrow U''(L) = 0, \quad V(L) = p \Leftrightarrow EIU'''(L) = p. \end{cases} \tag{6.2.7}$$

Consider one element $e_1 = [0, L]$. The approximating Solution (6.2.3) simplifies to

$$u(x) = u_3 N_3(x) + u_4 N_4(x), \tag{6.2.8}$$

because of boundary Conditions (6.2.7). In addition, the vector $\mathbf{f}$ in System (6.2.4) simplifies to $\mathbf{f}_b$, as $F = 0$. The expression of $\mathbf{f}_b$ is derived from the general formula

$$\mathbf{f}_b = [V(0) \quad -M(0) \quad -V(L) \quad M(L)]^T = [F_1 \quad M_1 \quad F_2 \quad M_2]^T, \tag{6.2.9}$$

by considering boundary Conditions (6.2.7),

$$\mathbf{f}_b = [F_1 \quad M_1 \quad -p \quad 0]^T, \tag{6.2.10}$$

where F_1 and M_1 are the unknown reactive force and moment. Considering all that, System(6.2.4) gives the following four equations

$$k_{13}u_3 + k_{14}u_4 = F_1, \quad k_{23}u_3 + k_{24}u_4 = M_1, \tag{6.2.11}$$

$$k_{33}u_3 + k_{34}u_4 = -p, \quad k_{43}u_3 + k_{44}u_4 = 0. \tag{6.2.12}$$

The last two equations are a simple algebraic system with unknowns u_3 and u_4. Solving it yields the unknowns

$$u_3 = -pL^3/(3EI), \quad u_4 = -pL^2/(2EI). \tag{6.2.13}$$

Insert these values into (6.2.11) to find the reactive force and moment

$$F_1 = p, \quad M_1 = pL.$$

In addition, substitute (6.2.13) into (6.2.8) to obtain u. Using Definitions (6.1.16) to (6.1.19) of the shape functions, we find that u is the cubic polynomial

$$u(x) = p(x^3 - 3Lx^2)/(6EI). \tag{6.2.14}$$

An easy calculation shows that this solution is the same as the analytical solution U of the Euler–Bernoulli equation. This result is due to the fact that the hermitian approximating of a cubic polynomial is the same polynomial. See Exercise 6.4.5.

The situation of the previous example is very rare, although interesting. Generally, accurate results need many elements. Therefore, let us begin to discuss two elements: $[0, L] = e_1 \cup e_2 = [x_1, x_2] \cup [x_2, x_3]$. From Formula (6.1.32), we know that the approximate solution in this case is expressed as

$$u = u_1 \Phi_1 + u_2 \Phi_2 + u_3 \Phi_3 + u_4 \Phi_4 + u_5 \Phi_5 + u_6 \Phi_6, \qquad (6.2.15)$$

with Φ_i, $i = 1, ..., 6$, defined in (6.1.31). Substituting (6.2.15) into Eq. (6.2.2) and assuming $v = \Phi_i$, $i = 1, \ldots, 6$, we arrive at a system of six equations that can be obtained more easily by considering the two elements separately. For the first element, we get

$$
\begin{aligned}
k_{11}u_1 + k_{12}u_2 + k_{13}u_3 + k_{14}u_4 &= \int_{x_1}^{x_2} FN_{1,1}\, dx + V(x_1), \\
k_{21}u_1 + k_{22}u_2 + k_{23}u_3 + k_{24}u_4 &= \int_{x_1}^{x_2} FN_{2,1}\, dx - M(x_1), \\
k_{31}u_1 + k_{32}u_2 + k_{33}u_3 + k_{34}u_4 &= \int_{x_1}^{x_2} FN_{3,1}\, dx - V(x_2), \\
k_{41}u_1 + k_{42}u_2 + k_{43}u_3 + k_{44}u_4 &= \int_{x_1}^{x_2} FN_{4,1}\, dx + M(x_2),
\end{aligned}
\qquad (6.2.16)
$$

and for the second element we get

$$
\begin{aligned}
k_{11}u_3 + k_{12}u_4 + k_{13}u_5 + k_{14}u_6 &= \int_{x_2}^{x_3} FN_{1,2}\, dx + V(x_2), \\
k_{21}u_3 + k_{22}u_4 + k_{23}u_5 + k_{24}u_6 &= \int_{x_2}^{x_3} FN_{2,2}\, dx - M(x_2), \\
k_{31}u_3 + k_{32}u_4 + k_{33}u_5 + k_{34}u_6 &= \int_{x_2}^{x_3} FN_{3,2}\, dx - V(x_3), \\
k_{41}u_3 + k_{42}u_4 + k_{43}u_5 + k_{44}u_6 &= \int_{x_2}^{x_3} FN_{4,2}\, dx + M(x_3).
\end{aligned}
\qquad (6.2.17)
$$

Summing Eqs. $(6.2.16)_{3,4}$ to $(6.2.17)_{1,2}$, related to the same nodes, yields the desired system

$$
\begin{aligned}
k_{11}u_1 + k_{12}u_2 + k_{13}u_3 + k_{14}u_4 &= \int_{x_1}^{x_2} F\Phi_1\, dx + V(x_1), \\
k_{21}u_1 + k_{22}u_2 + k_{23}u_3 + k_{24}u_4 &= \int_{x_1}^{x_2} F\Phi_2\, dx - M(x_1), \\
k_{31}u_1 + k_{32}u_2 + (k_{33} + k_{11})u_3 + (k_{34} + k_{12})u_4 & \\
+\, k_{13}u_5 + k_{14}u_6 &= \int_{x_1}^{x_3} F\Phi_3\, dx, \\
k_{41}u_1 + k_{42}u_2 + (k_{43} + k_{21})u_3 + (k_{44} + k_{22})u_4 & \\
+\, k_{23}u_5 + k_{24}u_6 &= \int_{x_1}^{x_3} F\Phi_4\, dx, \\
k_{31}u_3 + k_{32}u_4 + k_{33}u_5 + k_{34}u_6 &= \int_{x_2}^{x_3} F\Phi_5\, dx - V(x_3), \\
k_{41}u_3 + k_{42}u_4 + k_{43}u_5 + k_{44}u_6 &= \int_{x_2}^{x_3} F\Phi_6\, dx + M(x_3).
\end{aligned}
\qquad (6.2.18)
$$

Now, we can be more rigorous. The first equation of System (6.2.18) is derived by assuming $v = N_{1,1} = \Phi_1$ in (6.2.2), noting that the support of Φ_1 is e_1 and using Definitions (6.1.16) to (6.1.19). The third equation is found by assuming $v = \Phi_3$ in (6.2.2), noting that the support of Φ_3 is $e_1 \cup e_2$ (Fig. 6.1.7) and using Definitions (6.1.16) to (6.1.19). The other equations are derived with similar reasonings. The matrix K of the unknowns of System (6.2.18) is given by (Fig. 6.2.2, left)

$$K = \begin{bmatrix} k_{11} & k_{12} & k_{13} & k_{14} & 0 & 0 \\ k_{21} & k_{22} & k_{23} & k_{24} & 0 & 0 \\ k_{31} & k_{32} & k_{33}+k_{11} & k_{34}+k_{12} & k_{13} & k_{14} \\ k_{41} & k_{42} & k_{43}+k_{21} & k_{44}+k_{22} & k_{23} & k_{24} \\ 0 & 0 & k_{31} & k_{32} & k_{33} & k_{34} \\ 0 & 0 & k_{41} & k_{42} & k_{43} & k_{44} \end{bmatrix}.$$

The generalization of matrix K to more elements is illustrated in Fig. 6.2.2 (right). Thus, we arrive at the algebraic system

$$K\mathbf{u} = \mathbf{f}, \tag{6.2.19}$$

that allows us to find beam displacements and rotations on the nodes. Subsequently, shear forces and bending moments are derived from the equations of each element. For example, for $i = 1, 2$, these equations follow from $(6.2.16)_{1,2}$ and $(6.2.17)_{1,2}$

$$V(x_i) = k_{11}u_{2i-1} + k_{12}u_{2i} + k_{13}u_{2i+1} + k_{14}u_{2i+2} - \int_{x_i}^{x_{i+1}} FN_{1,i}dx,$$
$$M(x_i) = \int_{x_i}^{x_{i+1}} FN_{2,i}dx - k_{11}u_{2i-1} - k_{12}u_{2i} - k_{13}u_{2i+1} - k_{14}u_{2i+2}. \tag{6.2.20}$$

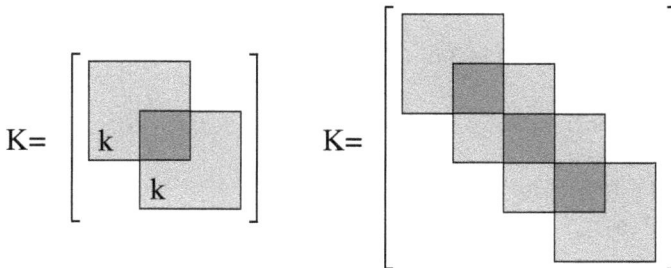

Figure 6.2.2.   Matrix K for 2 and 4 elements.

Example 6.2.2   The following listing presents a function that applies the FEM to solve a fixed end-free end beam subjected to a distributed load. The boundary conditions are:

$$U(0) = 0, \quad U'(0) = 0, \quad M(L) = 0, \quad V(L) = 0 \;\Leftrightarrow\; M'(L) = 0. \quad (6.2.21)$$

```
function [U, V, M, F1, M1] = beam_fx_fr(L, EI, F, n)
% This is the function file beam_fx_fr.m.
% The FEM is applied to solve the fixed end-free end beam. The input variables
% are: length of the beam L, product E*I, distributed load F and number of
% elements n. The function returns the displacement U, the shear force V,
% the bending moment M and the reactive force and moment F1, M1.
% Initialization
h = L/n;
N1 = @(xi, xn)  1 - 3*(xi - xn).^2/h^2 + 2*(xi - xn).^3/h^3;
N2 = @(xi, xn)  (xi - xn) - 2*(xi - xn).^2/h + (xi - xn).^3/h^2;
N3 = @(xi, xn)  3*(xi - xn).^2/h^2 - 2*(xi - xn).^3/h^3;
N4 = @(xi, xn)  -(xi - xn).^2/h + (xi - xn).^3/h^2;
x = linspace(0,L,n+1); V = zeros(n+1,1); M = zeros(n+1,1);
k = stiffness(h,EI);
fd = zeros(n*4,1);
for i=1:n
    fd((i-1)*4+1) = integral(@(xi)F(xi).*N1(xi,x(i)), x(i), x(i+1));
    fd((i-1)*4+2) = integral(@(xi)F(xi).*N2(xi,x(i)), x(i), x(i+1));
    fd((i-1)*4+3) = integral(@(xi)F(xi).*N3(xi,x(i)), x(i), x(i+1));
    fd((i-1)*4+4) = integral(@(xi)F(xi).*N4(xi,x(i)), x(i), x(i+1));
end
nn = n*2 + 2;
K = zeros(nn,nn);% Global matrix
for i=1:2:nn-3
    K(i:i+3,i:i+3) = K(i:i+3,i:i+3)+k;
end
K = K(3:nn,3:nn);
f = zeros(n*2+2,1);
f(1) = fd(1); f(2) = fd(2);
for i=2:2:2*(n-1)
    f(i+1) = fd(2*i-1) +fd(2*i +1);
    f(i+2) = fd(2*i) +fd(2*i+2);
end
```

```
f(2*n+1) = fd(n*4-1); f(2*n+2) = fd(n*4);
f = f(3:2*n+2,1);
% FEM
ur(3:nn,1) = K\f;% Displacements and rotations.
U = ur(1:2:nn,1); % Displacements.
for i=1:n
    V(i) = k(1,1)*ur(2*i-1) + k(1,2)*ur(2*i) + k(1,3)*ur(2*i+1)...
        + k(1,4)*ur(2*i+2) - fd((i-1)*4+1);
    M(i) = -(k(2,1)*ur(2*i-1) + k(2,2)*ur(2*i) + k(2,3)*ur(2*i + 1)...
        + k(2,4)*ur(2*i + 2)) + fd((i-1)*4+2);
end
F1 = V(1); M1 = -M(1); % Reactive force and moment.
end
```

Example 6.2.3 The following listing illustrates a way to call the beam_fx_fr function and solves the fixed end-free end beam in Fig. 6.2.3, subjected to the triangular load

$$F = -q_0(L - x)/L, \quad q_0 > 0, \quad 0 \le x \le L. \tag{6.2.22}$$

The graphs of the displacement, shear force and bending moment are shown in Fig. 6.2.4.

```
function beam_fx_fr_ex
% This is the function file beam_fx_fr_ex.m.
% Beam_fx_fr function is called to solve the fixed end-free end beam subjected
% to a triangular load. The lengths are in [mm] and the forces are in [N].
L = 3*10^3; E = 3*10^4; I = 9*10^8; EI = E*I;
q0 = 100; n = 10;
```

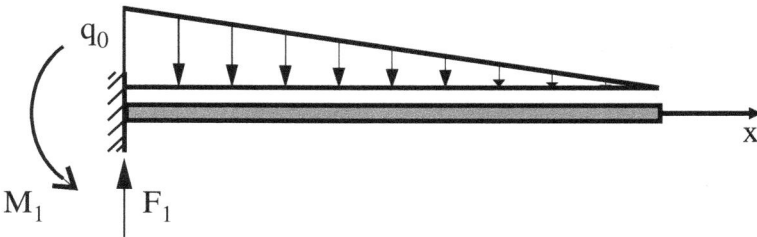

Figure 6.2.3. Fixed end-free end beam subjected to a triangular load.

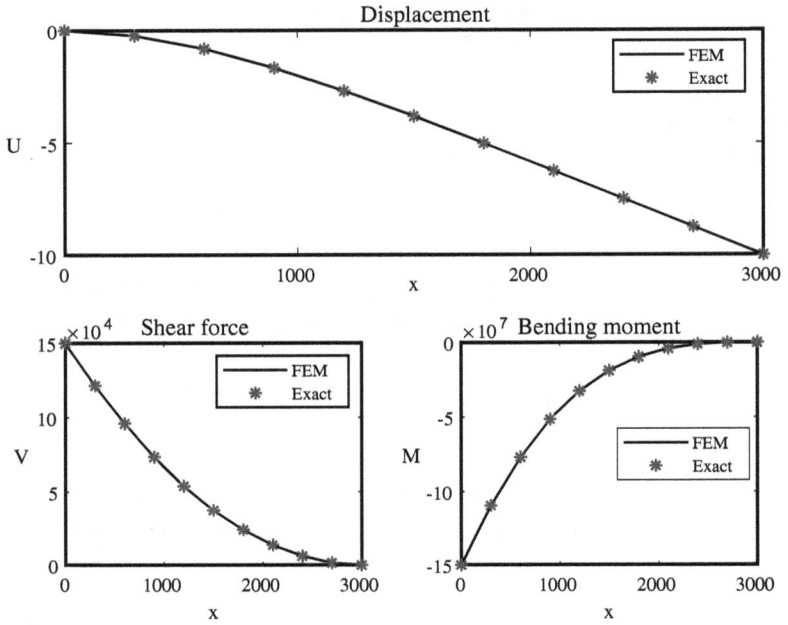

Figure 6.2.4.   Deformed beam, shear force and the bending moment.

```
F = @(xi)  -q0*(L - xi)/L; % Triangular load.
[u, v, m, F1, M1] = trave_m(L, EI, F, n);
x = linspace(0,L,n+1);
     % Exact displacement, bending moment and shear force.
U = q0/120/L*(x'.^5 - 5*L*x'.^4 + 10*L^2*x'.^3 - 10*L^3*x'.^2)/EI;
M = q0/6/L*(x' - L).^3;
V = q0/2/L*(x' - L).^2;
subplot (2,2,1:2)
plot(x,u,'k',x,U,'r*','LineWidth',1);
xlabel('x'); ylabel('U'); title('Displacement'); legend('FEM','Exact');
subplot (2,2,3)
plot(x,v,'k',x,V,'r*','LineWidth',1);
xlabel('x'); ylabel('V'); title('Shear force'); legend('FEM','Exact');
subplot (2,2,4)
plot(x,m,'k',x,M,'r*','LineWidth',1);
xlabel('x'); ylabel('M'); title('Bending moment');
legend('FEM','Exact','Location','Best');
```

```
fprintf('Reactive force F1 = %g\n', F1)
fprintf('Reactive moment M1 = %g\n', M1)
fprintf( 'Maximum error U = %g\n',max(abs(U-u)))
fprintf( 'Maximum error V = %g\n',max(abs(V-v)))
fprintf( 'Maximum error M = %g\n',max(abs(M-m)))
end
```

Another application is suggested in Exercise 6.4.6.

Example 6.2.4   A function is presented that applies the FEM to solve a pinned-pinned beam subjected to a distributed load. The boundary conditions are:

$$U(0) = 0, \quad M(0) = 0, \quad U(L) = 0, \quad M(L) = 0. \tag{6.2.23}$$

```
function [U, V, M, F1, F2] = beam_pn_pn(L, EI, F, n)
% This is the function file beam_pn_pn.m.
% The FEM is applied to solve the pinned-pinned beam. The input variables
% are: length of the beam L, product E*I, distributed load F, the number
% of elements n. The function returns the displacement U, the shear
% force V, the bending moment M and the reactive forces F1, F2.
% Initialization
h = L/n; x = linspace(0,L,n+1);
N1 = @(xi, xn)  1 - 3*(xi - xn).²/h² + 2*(xi - xn).³/h³;
N2 = @(xi, xn)  (xi - xn) - 2*(xi - xn).²/h + (xi - xn).³/h²;
N3 = @(xi, xn)  3*(xi - xn).²/h² - 2*(xi - xn).³/h³;
N4 = @(xi, xn)  -(xi - xn).²/h + (xi - xn).³/h²;
U = zeros(n+1,1); V = zeros(n+1,1); M = zeros(n+1,1);
k = stiffness(h,EI);
fd = zeros(n*4,1);
for i=1:n
    fd((i-1)*4+1) = integral(@(xi)F(xi).*N1(xi,x(i)), x(i), x(i+1));
    fd((i-1)*4+2) = integral(@(xi)F(xi).*N2(xi,x(i)), x(i), x(i+1));
    fd((i-1)*4+3) = integral(@(xi)F(xi).*N3(xi,x(i)), x(i), x(i+1));
    fd((i-1)*4+4) = integral(@(xi)F(xi).*N4(xi,x(i)), x(i), x(i+1));
end
nn = n*2 + 2;
K = zeros(nn,nn);
for i=1:2:nn-3
```

```
      K(i:i+3,i:i+3) = K(i:i+3,i:i+3)+k;
end
K=K([2:nn-2 nn],[2:nn-2 nn]);
f = zeros(n*2+2,1);
f(1) = fd(1); f(2) = fd(2);
for i=2:2:2*(n-1)
      f(i+1) = fd(2*i-1) +fd(2*i +1);
      f(i+2) = fd(2*i) +fd(2*i+2);
end
f(2*n+1) = fd(n*4-1); f(2*n+2) = fd(n*4);
f = f([2:nn-2 nn],1);
% FEM
ur = K\f;% Displacements and rotations.
for i=1:n-1
      U(i+1) = ur(2*i);% Displacements.
end
ur = [0; ur(1:n*2-1); 0;ur(n*2)];
V(1) = k(1,1)*ur(1) + k(1,2)*ur(2) + k(1,3)*ur(3) + k(1,4)*ur(4) - fd(1);
for i=2:n
      V(i) = k(1,1)*ur(2*i-1) + k(1,2)*ur(2*i) + k(1,3)*ur(2*i+1)...
            + k(1,4)*ur(2*i+2) - fd((i-1)*4+1);
      M(i) = -(k(2,1)*ur(2*i-1) + k(2,2)*ur(2*i) + k(2,3)*ur(2*i + 1)...
            + k(2,4)*ur(2*i + 2))+ fd((i-1)*4+2);
end
V(n+1) = -(k(3,1)*ur(2*n-1) + k(3,2)*ur(2*n) + k(3,3)*ur(2*n+1)...
      + k(3,4)*ur(2*n+2))+fd(4*n-1);
F1 = V(1); F2 = -V(n+1); % Reactive forces.
end
```

Example 6.2.5   The following listing illustrates a way to call the beam_pn_pn function and solves the pinned-pinned beam in Fig. 6.2.5, subjected to the trapezoidal load

$$F(x) = -(q_2 - q_1)x/L - q_1 = -q_3 x/L - q_1, \quad q_2 > q_1 \geq 0, \quad 0 \leq x \leq L.$$
(6.2.24)

For $q_1 = 0$, F simplifies to a triangular load and for $q_3 = 0$ to a uniform load. The graphs of the displacement, shear force and bending moment are shown in Fig. 6.2.6.

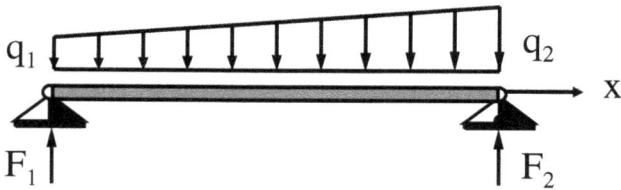

Figure 6.2.5.   Pinned-pinned beam subjected to a trapezoidal load.

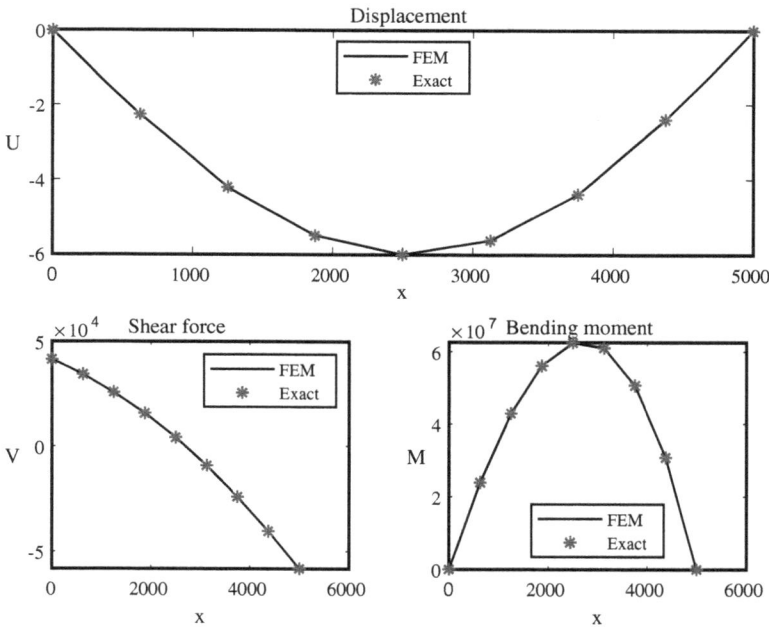

Figure 6.2.6.   Graphs of displacement, shear force and the bending moment.

```
function beam_pn_pn_ex
% This is the function file beam_pn_pn_ex.m.
% The beam_pn_pn function is called to solve the pinned-pinned beam sub-
% jected to a trapezoidal load. The lengths are in [mm] and the forces are in
% [N].
L = 5*10^3; E = 3*10^4; I = 9*10^8; EI = E*I;
n = 8; q2 = 30; q1 = 10; q3 = q2-q1;
F = @(xi) -q3*xi/L-q1; % Trapezoidal load.
```

```
[u, v, m, F1, F2] = beam_pn_pn(L, EI, F, n);
x = linspace(0,L,n+1);
       % Exact displacement, bending moment and shear force.
U =(-q3*x'.^5/120/L - q1*x'.^4/24 + (q3/36+q1/12)*L*x'.^3-...
       (7*q3/360 + q1/24)*L^3*x')/EI;
M = -q3*x'.^3/6/L - q1*x'.^2/2 + (q3/6 + q1/2)*L*x';
V = -q3*x'.^2/2/L - q1*x' + q3*L/6 + q1*L/2;
subplot (2,2,1:2)
plot(x,u,'k',x,U,'r*','LineWidth',1); xlabel('x'); ylabel('U');
title('Displacement'); legend('FEM','Exact','Location','North');
subplot (2,2,3)
plot(x,v,'k',x,V,'r*','LineWidth',1); xlabel('x'); ylabel('V');
title('Shear force'); legend('FEM','Exact');
subplot (2,2,4)
plot(x,m,'k',x,M,'r*','LineWidth',1); xlabel('x'); ylabel('M');
title('Bending moment'); legend('FEM','Exact','Location','South');
fprintf('Reactive force F1 = %g\n', F1)
fprintf('Reactive force F2 = %g\n', F2)
fprintf( 'Maximum error U = %g\n',max(abs(U-u)))
fprintf( 'Maximum error V = %g\n',max(abs(V-v)))
fprintf( 'Maximum error M = %g\n',max(abs(M-m)))
end
```

Another application is suggested in Exercise 6.4.7.

Example 6.2.6 A function is presented that applies the FEM to solve the fixed end-pinned end beam, subjected to a distributed load. The boundary conditions are:

$$U(0) = 0, \quad U'(0) = 0, \quad U(L) = 0, \quad M(L) = 0. \qquad (6.2.25)$$

```
function [U, V, M, F1, F2, M1] = beam_fx_pn(L, EI, F, n)
% This is the function file beam_fx_pn.m.
% The FEM is applied to solve the fixed end-pinned end beam. The input
% variables are: length of the beam L, product E*I, distributed load F,
% number of elements n. The function returns the displacement U, the
% shear force V, the bending moment M and the reactive forces and
% moment F1, F2, M1.
```

```
% Initialization
h = L/n;
N1 = @(xi, xn)  1 - 3*(xi - xn).^2/h^2 + 2*(xi - xn).^3/h^3;
N2 = @(xi, xn)  (xi - xn) - 2*(xi - xn).^2/h + (xi - xn).^3/h^2;
N3 = @(xi, xn)  3*(xi - xn).^2/h^2 - 2*(xi - xn).^3/h^3;
N4 = @(xi, xn)  -(xi - xn).^2/h + (xi - xn).^3/h^2;
x = linspace(0,L,n+1); V = zeros(n+1,1); M = zeros(n+1,1);
k = stiffness(h,EI);
fd = zeros(n*4,1);
for i=1:n
    fd((i-1)*4+1) = integral(@(xi)F(xi).*N1(xi,x(i)), x(i), x(i+1));
    fd((i-1)*4+2) = integral(@(xi)F(xi).*N2(xi,x(i)), x(i), x(i+1));
    fd((i-1)*4+3) = integral(@(xi)F(xi).*N3(xi,x(i)), x(i), x(i+1));
    fd((i-1)*4+4) = integral(@(xi)F(xi).*N4(xi,x(i)), x(i), x(i+1));
end
nn = n*2 + 2;
K = zeros(nn,nn);
for i=1:2:nn-3
    K(i:i+3,i:i+3) = K(i:i+3,i:i+3)+k;
end
K = K([3:nn-2 nn],[3:nn-2 nn]);
f = zeros(n*2+2,1);
f(1) = fd(1); f(2) = fd(2);
for i=2:2:2*(n-1)
    f(i+1) = fd(2*i-1) +fd(2*i +1);
    f(i+2) = fd(2*i) +fd(2*i+2);
end
f(2*n+1) = fd(n*4-1); f(2*n+2) = fd(n*4);
f = f([3:nn-2 nn],1);
% FEM
ur = K\f;
ur = [0; 0; ur(1:end-1,1); 0; ur(end,1)];
U = ur(1:2:nn,1);
for i=1:n
    V(i) = k(1,1)*ur(2*i-1) + k(1,2)*ur(2*i) + k(1,3)*ur(2*i+1)...
           + k(1,4)*ur(2*i+2) - fd((i-1)*4+1);
    M(i) = -(k(2,1)*ur(2*i-1) + k(2,2)*ur(2*i) + k(2,3)*ur(2*i + 1)...
           + k(2,4)*ur(2*i + 2))+ fd((i-1)*4+2);
```

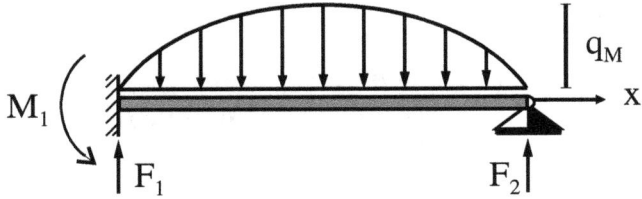

Figure 6.2.7. Fixed end-pinned end beam subjected to a parabolic load.

```
end
V(n+1) = -(k(3,1)*ur(2*n-1)+k(3,2)*ur(2*n)+k(3,3)*ur(2*n+1)+...
    k(3,4)*ur(2*n+2))+fd(4*n-1);
M(n+1) = k(4,1)*ur(2*n-1)+k(4,2)*ur(2*n)+k(4,3)*ur(2*n+1)+...
    k(4,4)*ur(2*n+2)-fd(4*n);
F1 = V(1); F2 = -V(n+1); M1 = -M(1);
end
```

Example 6.2.7 The following listing illustrates a way to call the beam_fx_pn function and solves the fixed end-pinned end beam in Fig. 6.2.7, subjected to the parabolic load

$$F = -4q_M(Lx - x^2)/L^2, \quad q_M > 0. \tag{6.2.26}$$

The graphs of the displacement, shear force and bending moment are shown in Fig. 6.2.8.

```
function beam_fx_pn_ex
% This is the function file beam_fx_pn_ex.m.
% The beam_fx_pn function is called to solve the fixed end-pinned end beam
% subjected to a parabolic load. The lengths are in [mm] and the forces are in
% [N].
L = 6*10^3; E = 3*10^4; I = 9*10^8; EI = E*I;
qM = 10; n = 10;
F = @(xi) -4*qM*(L*xi-xi.^2)/L^2;
[u, v, m, F1, F2, M1] = beam_fx_pn(L, EI, F, n);
x = linspace(0,L,n+1);
U = qM/180/L^2*(2*x'.^6 -6*L*x'.^5+13*L^3*x'.^3 -9*L^4*x'.^2)/EI;
M = qM/30/L^2*(10*x'.^4 - 20*L*x'.^3 + 13*L^3*x' - 3*L^4);
V = qM*(40*x'.^3-60*L*x'.^2+13*L^3)/30/L^2;
```

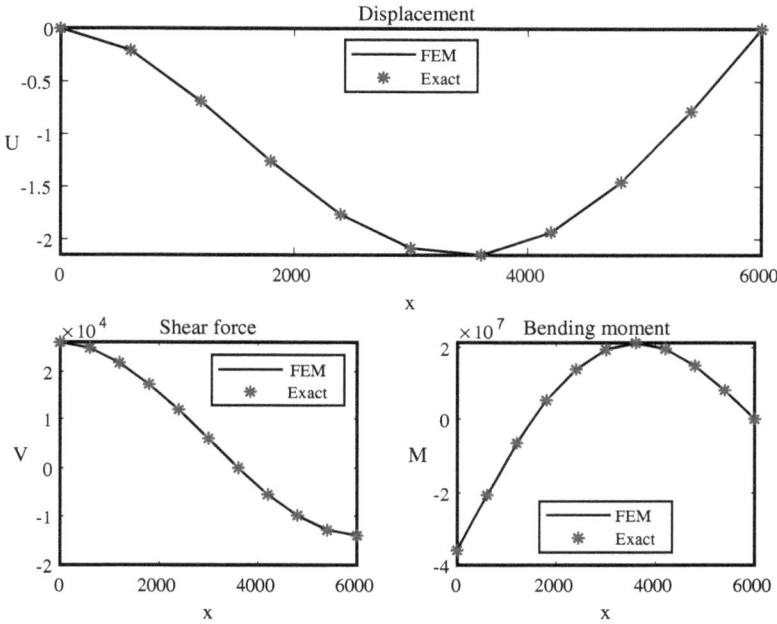

Figure 6.2.8. Graphs of displacement, shear force and the bending moment.

```
subplot (2,2,1:2)
plot(x,u,'k',x,U,'r*','LineWidth',1); xlabel('x'); ylabel('U');
title('Displacement'); legend('FEM','Exact','Location','North');
subplot (2,2,3)
plot(x,v,'k',x,V,'r*','LineWidth',1); xlabel('x'); ylabel('V');
title('Shear force'); legend('FEM','Exact');
subplot (2,2,4)
plot(x,m,'k',x,M,'r*','LineWidth',1); xlabel('x'); ylabel('M');
title('Bending moment'); legend('FEM','Exact','Location','South');
fprintf('Reactive force F1 = %g\n', F1)
fprintf('Reactive force F2 = %g\n', F2)
    fprintf('Reactive moment M1 = %g\n', M1)
fprintf( 'Maximum error U = %g\n',max(abs(U-u)))
fprintf( 'Maximum error V = %g\n',max(abs(V-v)))
fprintf( 'Maximum error M = %g\n',max(abs(M-m)))
end
```

Another application is suggested in Exercise 6.4.8.

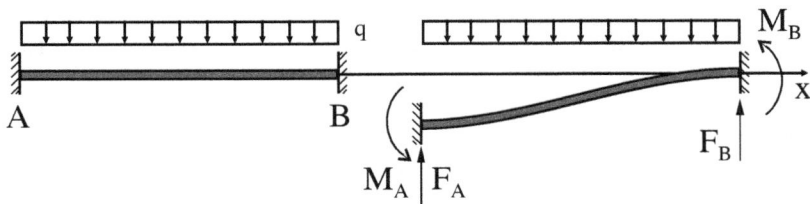

Figure 6.2.9.   Fixed end-fixed end beam with the constraint in A subjected to a vertical displacement.

Consider the fixed end-fixed end beam of length L in Fig. 6.2.9 (left), subjected to the uniform load $F = -q$. Suppose that the constraint in A is subjected to the vertical displacement U_A (Fig. 6.2.9, right). The statical problem is governed by the Euler–Bernoulli Eq. (6.2.1) with the following boundary conditions

$$U(0) = U_A, \quad U'(0) = 0, \quad U(L) = 0, \quad U'(L) = 0. \tag{6.2.27}$$

Straightforward calculations show that the displacement is given by

$$U(x) = \frac{F}{24EI}(x^4 - 2Lx^3 + L^2x^2) + \frac{U_A}{L^3}(2x^3 - 3x^2L + L^3). \tag{6.2.28}$$

Hence, the bending moment, shear force and reactive forces are

$$M(x) = \frac{F}{12}(6x^2 - 6Lx + L^2) + \frac{EIU_A}{L^3}(12x - 6L), \tag{6.2.29}$$

$$V(x) = F(2x - L)/2 + 12EIU_A/L^3, \tag{6.2.30}$$

$$F_A = -FL/2 + 12EIU_A/L^3, \quad M_A = -FL^2/12 + 6EIU_A/L^2,$$
$$F_B = -FL/2 - 12EIU_A/L^3, \quad M_B = FL^2/12 + 6EIU_A/L^2. \tag{6.2.31}$$

Consider the FEM. Since boundary Conditions (6.2.27) are inhomogeneous, the form of the algebraic system that solves the problem is different from (6.2.19) and similar to that considered in Sec. 4.2.4, where other inhomogeneous problems were discussed. Therefore, the algebraic system is

$$k\mathbf{u} = \mathbf{f} - \mathbf{g}, \tag{6.2.32}$$

where

$$g_i = k_{i1}U_A, \quad i = 1, 2, 3, 4, \quad g_i = 0, \quad i > 4. \tag{6.2.33}$$

Formulas (6.2.32) and (6.2.33) will be used in the FEM applications below.

Example 6.2.8 A function is presented that applies the FEM to solve the beam in Fig. 6.2.9.

```
function [U, V, M, F1, F2, M1 M2] = beam_fx_fx_uA(L, EI, F, n, uA)
% This is the function file beam_fx_fx_uA.m.
% The FEM is applied to solve the fixed end-fixed end beam subjected to a
% vertical displacement of the first constraint.
% Initialization
h = L/n;
N1 = @(xi, xn)  1 - 3*(xi - xn).^2/h^2 + 2*(xi - xn).^3/h^3;
N2 = @(xi, xn)  (xi - xn) - 2*(xi - xn).^2/h + (xi - xn).^3/h^2;
N3 = @(xi, xn)  3*(xi - xn).^2/h^2 - 2*(xi - xn).^3/h^3;
N4 = @(xi, xn)  -(xi - xn).^2/h + (xi - xn).^3/h^2;
x = linspace(0,L,n+1); V = zeros(n+1,1); M = zeros(n+1,1);
k = stiffness(h,EI);
fd = zeros(n*4,1);
for i=1:n
    fd((i-1)*4+1) = integral(@(xi)F(xi).*N1(xi,x(i)), x(i), x(i+1));
    fd((i-1)*4+2) = integral(@(xi)F(xi).*N2(xi,x(i)), x(i), x(i+1));
    fd((i-1)*4+3) = integral(@(xi)F(xi).*N3(xi,x(i)), x(i), x(i+1));
    fd((i-1)*4+4) = integral(@(xi)F(xi).*N4(xi,x(i)), x(i), x(i+1));
end
nn = n*2 + 2; K = zeros(nn,nn);
for i=1:2:nn-3
    K(i:i+3,i:i+3) = K(i:i+3,i:i+3)+k;
end
K = K(3:2*n,3:2*n);
f = zeros(n*2+2,1);
f(1) = fd(1); f(2) = fd(2);
for i=2:2:2*(n-1)
    f(i+1) = fd(2*i-1) +fd(2*i +1);
    f(i+2) = fd(2*i) +fd(2*i+2);
end
f(2*n+1) = fd(n*4-1); f(2*n+2) = fd(n*4);
f(1:4,1)= f(1:4,1)- k(1:4,1)*uA; % Inhomogeneous boundary conditions.
f = f(3:2*n,1);
% FEM
ur(3:n*2,1) = K\f;
ur = [uA; 0; ur(3:n*2,1); 0; 0]; U = ur(1:2:nn,1);
```

```
for i=1:n
    V(i) = k(1,1)*ur(2*i-1) + k(1,2)*ur(2*i) + k(1,3)*ur(2*i+1)...
        + k(1,4)*ur(2*i+2) - fd((i-1)*4+1);
    M(i) = -(k(2,1)*ur(2*i-1) + k(2,2)*ur(2*i) + k(2,3)*ur(2*i + 1)...
        + k(2,4)*ur(2*i + 2)) + fd((i-1)*4+2);
end
V(n+1) = -(k(3,1)*ur(2*n-1)+k(3,2)*ur(2*n)+k(3,3)*ur(2*n+1)+...
        k(3,4)*ur(2*n+2))+fd(4*n-1);
M(n+1) = k(4,1)*ur(2*n-1)+k(4,2)*ur(2*n)+k(4,3)*ur(2*n+1)+...
        k(4,4)*ur(2*n+2)-fd(4*n);
F1 = V(1); F2 = -V(n+1); M1 = -M(1); M2 = M(n+1);
end
```

Example 6.2.9   The following listing illustrates a way to call the beam_fx_fx_uA function. It solves the fixed end-fixed end beam in Fig. 6.2.9, subjected to the uniform load, with the first constrain subjected to the vertical displacement U_A. The graphs of the displacement, shear force and bending moment are shown in Fig. 6.2.10.

```
function beam_fx_fx_uA_ex
% This is the function file beam_fx_fx_uA_ex.m.
% The beam_fx_fx_uA function is called to solve the fixed end-fixed end beam
% subjected to a uniform load and to a displacement of the first
% constraint. The lengths are in [mm] and the forces are in [N].
L = 4*10^3; E = 3*10^4; I = 9*10^8; EI = E*I; uA = -.1;
n = 18; q = 10;
F = @(xi) -q;
[u, v, m, F1, F2, M1, M2] = beam_fx_fx_uA(L, EI, F, n, uA);
x = linspace(0,L,n+1);
    % Exact displacement, bending moment and shear force.
U = -q/24*(x'.^4 - 2*L*x'.^3 + L^2*x'.^2)/EI...
    + uA*(2*x'.^3/L^3 - 3*x'.^2/L^2 + 1);
M = -q/12*(6*x'.^2 - 6*L*x' + L^2)+EI*uA*(x'*12/L^3 - 6/L^2);
V = -q/2*(2*x' - L)+12*uA*EI/L^3;
subplot (2,2,1:2);
plot(x,u,'k',x,U,'r*'); xlabel('x'); ylabel('U'); title('Displacement');
legend('FEM','Exact','Location','North');
subplot (2,2,3);
```

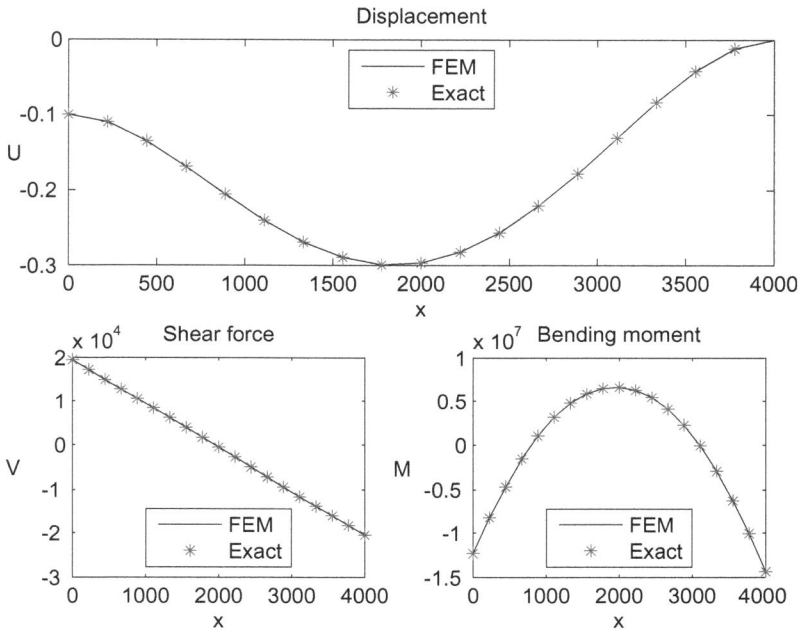

Figure 6.2.10.   Graphs of displacement, shear force and the bending moment.

```
plot(x,v,'k',x,V,'r*'); xlabel('x'); ylabel('V');
title('Shear force'); legend('FEM','Exact','Location','South');
subplot (2,2,4);
plot(x,m,'k',x,M,'r*');
xlabel('x'); ylabel('M'); title('Bending moment');
legend('FEM','Exact','Location','South');
fprintf('Reactive force F1 = %g\n', F1)
fprintf('Reactive force F2 = %g\n', F2)
fprintf('Reactive moment M1 = %g\n', M1)
fprintf('Reactive moment M2 = %g\n', M2)
fprintf( 'Maximum error U = %g\n',max(abs(U-u)))
fprintf( 'Maximum error V = %g\n',max(abs(V-v)))
fprintf( 'Maximum error M = %g\n',max(abs(M-m)))
end
```

Another application is suggested in Exercise 6.4.9.

6.3   Beam Subjected to Concentrated Forces

Consider a beam subjected to a distributed load and concentrated forces

$$\{(x_h, F_h), \ h = 1, \ldots, N\}.$$

Equation (6.1.1) is easily modified to include the new forces

$$\int_{x_1}^{x_2} (F - \rho U_{tt}) \ dx + V(x_1, t) - V(x_2, t) + \sum_{i=n_1}^{n_2} F_i = 0, \qquad (6.3.1)$$

where F_i, $i = n_1, \ldots, n_2$ are the concentrated forces within the interval $[x_1, x_2]$. Now, the discussion is delicate. Indeed, consider a single concentrated force, say (x_h, F_h), and consider the interval $[x_1, x_2]$ with $x_1 = x_h - \Delta x$ and $x_2 = x_h + \Delta x$. If Δx is sufficiently small, only F_h is in $[x_1, x_2]$ and Eq. (6.3.1) is written as

$$\int_{x_h - \Delta x}^{x_h + \Delta x} (F - \rho U_{tt}) \ dx + V(x_h - \Delta x, t) - V(x_h + \Delta x, t) + F_h = 0.$$

When $\Delta x \to 0$, the previous equation gives

$$V(x_h^+, t) - V(x_h^-, t) = F_h, \qquad (6.3.2)$$

proving that V is discontinuous on $x = x_h$. Consequently, V cannot be differentiated for $x = x_h$ and U cannot be a (classical) solution to the Euler–Lagrange Eq. (6.1.8). Let us show that U is a weak solution with a suitable forcing term that depends on F_h. For simplicity, suppose that the beam is subjected to a single concentrated force (x_h, F_h) and consider the weak form (6.1.34) on the intervals, $0 < x < x_h$ and $x_h < x < L$, respectively,

$$\int_0^{x_h} \rho U_{tt} v \ dx + \int_0^{x_h} EIU_{xx} v'' \ dx = \int_0^{x_h} Fv \ dx + [v'M - vV]_0^{x_h} , \qquad (6.3.3)$$

$$\int_{x_h}^{L} \rho U_{tt} v \ dx + \int_{x_h}^{L} EIU_{xx} v'' \ dx = \int_{x_h}^{L} Fv \ dx + [v'M - vV]_{x_h}^{L} . \qquad (6.3.4)$$

Note that

$$[v'M - vV]_{x_h}^{L} + [v'M - vV]_0^{x_h} = v(x_h)(V(x_h^+, t) - V(x_h^-, t)) + [v'M - vV]_0^{L} .$$

Therefore, summing (6.3.3) and (6.3.4) yields

$$\int_0^L \rho U_{tt} v \ dx + \int_0^L EIU_{xx} v'' \ dx = \int_0^L Fv \ dx + v(x_h)F_h + [v'M - vV]_0^L ,$$

(6.3.5)

where the step Relationship (6.3.2) was used. Equation (6.3.5) shows that U is a weak solution of the Euler–Lagrange, with a new forcing term depending on F_h and the Dirac δ function

$$\rho U_{tt} + (EIU_{xx})_{xx} = F + F_h \delta(x - x_h).$$

(6.3.6)

In the general case, where the beam is subjected to many concentrated forces, the last term in Eq. (6.3.6) is replaced by a sum. In Statics, Eq. (6.3.6) simplifies to the Euler–Bernoulli equation

$$EIU^{iv} = F + F_h \delta(x - x_h).$$

(6.3.7)

Let us discuss the previous situation with the FEM. Since this method considers the weak equation, it is able to provide an approximating solution to the problem. It remains to understand the role of the new forcing term in FEM equations. Let us refer to $h = 2$, as the equations related to e_1 and e_2 are explicitly written in Formulas (6.2.16) and (6.2.17). When the third equation in (6.2.16) is added to the first in (6.2.17), the following new forcing term appears

$$V(x_2^+) - V(x_2^-) = F_2,$$

(6.3.8)

due to the discontinuity of V. All the other equations remain unchanged. In the general case of a concentrated force (x_h, F_h), only the forcing term f_{2h-1} (related to the distributed load) has to be modified and the new term added

$$f_{2h-1} + F_h.$$

(6.3.9)

See Exercise 6.4.10.

Example 6.3.1 The following listing presents a function that applies the FEM to solve the pinned end-pinned end beam subjected to a distributed load and a concentrated force.

```
function [U, V, M, F1, F2] = beam_c_pn_pn(L, EI, F, n, ih, Fh)
% This is the function file beam_c_pn_pn.m.
% The FEM is applied to solve the pinned end-pinned end beam subjected to
```

```
% a distributed load and a concentrated force. The input variables are: length
% of the beam L, product E*I, distributed load F, number of elements n, node
% ih, and concentrated force Fh. The function returns the displacement U,
% the shear force V, the bending moment M, and the reactive forces F1, F2.

% Initialization
h = L/n;
N1 = @(xi, xn)  1 - 3*(xi - xn).^2/h^2 + 2*(xi - xn).^3/h^3;
N2 = @(xi, xn)  (xi - xn) - 2*(xi - xn).^2/h + (xi - xn).^3/h^2;
N3 = @(xi, xn)  3*(xi - xn).^2/h^2 - 2*(xi - xn).^3/h^3;
N4 = @(xi, xn)  -(xi - xn).^2/h + (xi - xn).^3/h^2;
x = linspace(0,L,n+1);
U = zeros(n+1,1); V = zeros(n+1,1); M = zeros(n+1,1);
k = stiffness(h,EI);
fd = zeros(n*4,1);
for i=1:n
    fd((i-1)*4+1) = integral(@(xi)F(xi).*N1(xi,x(i)), x(i), x(i+1));
    fd((i-1)*4+2) = integral(@(xi)F(xi).*N2(xi,x(i)), x(i), x(i+1));
    fd((i-1)*4+3) = integral(@(xi)F(xi).*N3(xi,x(i)), x(i), x(i+1));
    fd((i-1)*4+4) = integral(@(xi)F(xi).*N4(xi,x(i)), x(i), x(i+1));
end
nn = n*2 + 2;
K = zeros(nn,nn);
for i=1:2:nn-3
    K(i:i+3,i:i+3) = K(i:i+3,i:i+3)+k;
end
K = K([2:nn-2 nn],[2:nn-2 nn]);
f = zeros(n*2+2,1);
f(1) = fd(1); f(2) = fd(2);
for i=2:2:2*(n-1)
    f(i+1) = fd(2*i-1) +fd(2*i +1);
    f(i+2) = fd(2*i) +fd(2*i+2);
end
f(2*n+1) = fd(n*4-1); f(2*n+2) = fd(n*4);
f(ih*2-1) = f(ih*2-1) + Fh;% Concentrated force.
f = f([2:nn-2 nn],1);
% FEM
ur = K\f;
for i=1:n-1
```

```
    U(i+1) = ur(2*i);
end
ur = [0; ur(1:n*2-1); 0; ur(n*2)];
V(1) = k(1,1)*ur(1) + k(1,2)*ur(2) + k(1,3)*ur(3) + k(1,4)*ur(4) - fd(1);
for i=2:n
    V(i) = k(1,1)*ur(2*i-1) + k(1,2)*ur(2*i) + k(1,3)*ur(2*i+1)...
        + k(1,4)*ur(2*i+2) - fd((i-1)*4+1);
    M(i) = -(k(2,1)*ur(2*i-1) + k(2,2)*ur(2*i) + k(2,3)*ur(2*i + 1)...
        + k(2,4)*ur(2*i + 2))+ fd((i-1)*4+2);
end
V(n+1) = -(k(3,1)*ur(2*n-1) + k(3,2)*ur(2*n) + k(3,3)*ur(2*n+1)...
    + k(3,4)*ur(2*n+2))+fd(4*n-1);
F1 = V(1); F2 = -V(n+1); % Reactive forces.
end
```

Example 6.3.2   The following listing illustrates a way to call the beam_c_pn_pn function to solve the pinned end-pinned end beam in Fig. 6.3.1, subjected to a concentrated force and the uniform load

$$F(x) = -q, \quad 0 \le x \le L.$$

The graphs of the displacement, shear force and bending moment are shown in Fig. 6.3.2.

```
function beam_c_pn_pn_ex
% This is the function file beam_c_pn_pn_ex.m.
% Beam_c_pn_pn function is called to solve the pinned end-pinned end beam
% subjected to a uniform load and a concentrated force. The lengths are in
% [mm] and the forces are in [N].
L = 8*10^3; E = 3*10^4; I = 9*10^8; EI = E*I;
n = 40; q = 1;
```

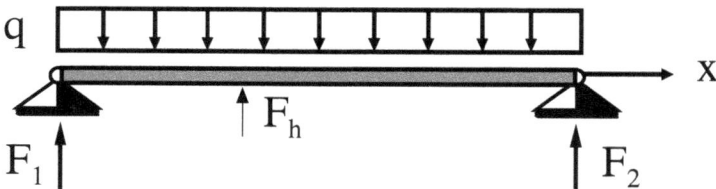

Figure 6.3.1.   Pinned end-pinned end beam subjected to a concentrated force.

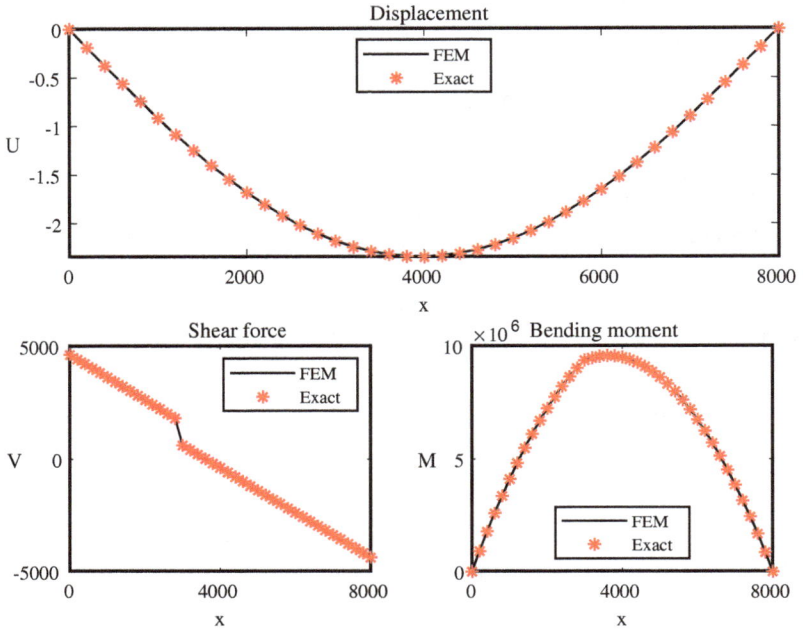

Figure 6.3.2.   Graphs of displacement, shear force and bending moment.

```
F = @(xi)  0*xi-q; % Uniform load.
ih = 16; % 2 < ih < n.
Fh = -1000; % Concentrated Force.
[u, v, m, F1, F2] = trave_c_cc(L, EI, F, n, ih, Fh);
x = linspace(0,L,n+1); xh = x(ih);
     % Displacement, bending moment and shear force related to uniform load.
U = -q*(x'.^4 - 2*L*x'.^3 + L^3*x')/24/EI;
M = -q*(x'.^2 - L*x')/2; V = -q*(2*x' - L)/2;
C1 = Fh*(xh - L)/6/L/EI; C2 = Fh*xh*(xh - L)*(xh - 2*L)/6/L/EI;
C3 = Fh*xh/6/L/EI; C4 = Fh*xh*(xh^2 - L^2)/6/L/EI;
     % Displacement, moment and shear force related to concentrated force.
Uh(1:ih-1,1) = C1*x(1:ih-1)'.^3 + C2*x(1:ih-1)';
Uh(ih:n+1,1) = C3*(x(ih:n+1)'- L).^3 + C4*(x(ih:n+1)'- L);
Mh(1:ih-1,1) = EI*6*C1*x(1:ih-1)'; Mh(ih:n+1,1) = EI*6*C3*(x(ih:n+1)'-L);
Vh(1:ih-1,1) = EI*6*C1; Vh(ih:n+1,1) = EI*6*C3;
U = U + Uh; M = Mh + M; V = V + Vh;
subplot (2,2,1:2)
```

```
plot(x,u,'k',x,U,'r*','LineWidth',1); xlabel('x'); ylabel('U');
title('Displacement'); legend('FEM','Exact','Location','North');
subplot (2,2,3)
plot(x,v,'k',x,V,'r*','LineWidth',1); xlabel('x'); ylabel('V');
title('Shear force'); legend('FEM','Exact');
subplot (2,2,4)
plot(x,m,'k',x,M,'r*','LineWidth',1); xlabel('x'); ylabel('M');
title('Bending moment'); legend('FEM','Exact','Location','South');
fprintf('Reactive force F1 = %g\n', F1)
fprintf('Reactive force F2 = %g\n', F2)
fprintf( 'Maximum error U = %g\n',max(abs(U-u)))
fprintf( 'Maximum error V = %g\n',max(abs(V-v)))
fprintf( 'Maximum error M = %g\n',max(abs(M-m)))
end
```

Another application is suggested in Exercise 6.4.11.

Example 6.3.3 Consider the overhanging beam of length L in Fig. 6.3.3, subjected to the tip load p. The boundary conditions are:

$$U(0) = 0, \quad U'(0) = 0, \quad M(L) = 0, \quad V(L) = p. \tag{6.3.10}$$

The roller support is located at $x_h < L$, and it is cinematically modeled by the equation

$$U(x_h) = 0. \tag{6.3.11}$$

In this problem, the concentrated force is the unknown reactive force F_2. Of course, the step Relationship (6.3.2) holds

$$V(x_h^+) - V(x_h^-) = F_h = F_2, \tag{6.3.12}$$

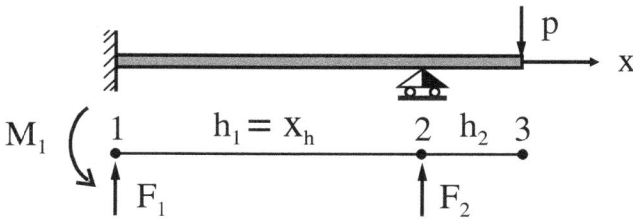

Figure 6.3.3.   Overhanging beam.

but both the left and right sides are unknown. The solution U to Problem (6.3.10) and (6.3.12) is found by solving the homogenous Euler–Bernoulli equation on the two intervals $x < x_h$ and $x > x_h$. An easy calculation gives

$$U = \begin{cases} -C_1(x^3 - x_h x^2), & x \in e_1, \\ C_2(x - x_h)^3 - C_3(x - x_h)^2 - C_4(x - x_h), & x \in e_2, \end{cases} \tag{6.3.13}$$

where

$$C_1 = \frac{p(L - x_h)}{4x_h EI}, \quad C_2 = \frac{p}{6EI}, \quad C_3 = \frac{p(L - x_h)}{2EI}, \quad C_4 = \frac{px_h(L - x_h)}{4EI}.$$

Shear force and the bending moment are deduced by deriving (6.3.13) on the related intervals. Hence, the reactive forces and moments are

$$F_1 = -3p(L - x_h)/2x_h, \quad M_1 = -p(L - x_h)/2, \quad F_2 = p(3L - x_h)/2x_h.$$

Let us apply the FEM. Consider two elements of lengths $h_1 = x_h$ and $h_2 = L - x_h$, respectively. The approximating solution is

$$u = \begin{cases} u_4 N_{4,1}, & x \in e_1, \\ u_4 N_{2,2} + u_5 N_{3,2} + u_6 N_{4,2}, & x \in e_2, \end{cases} \tag{6.3.14}$$

since

$$u_1 = U(0) = 0, \quad u_2 = U'(0) = 0, \quad u_3 = U(x_2) = 0. \tag{6.3.15}$$

Refer to Formulas (6.2.16) and (6.2.17), where the equations for two elements are explicitly written. Adding the last two equations in (6.2.16) to the first two in (6.2.17) and considering (6.3.12) and (6.3.15), we arrive at the following system

$$\begin{cases} k_{14}^1 u_4 = V(x_1), \\ k_{24}^1 u_4 = -M(x_1), \\ (k_{34}^1 + k_{12}^2)u_4 + k_{13}^2 u_5 + k_{14}^2 u_6 = F_2, \\ (k_{44}^1 + k_{22}^2)u_4 + k_{23}^2 u_5 + k_{24}^2 u_6 = 0, \\ k_{32}^2 u_4 + k_{33}^2 u_5 + k_{34}^2 u_6 = -p, \\ k_{42}^2 u_4 + k_{43}^2 u_5^2 + k_{44}^2 u_6 = 0, \end{cases} \tag{6.3.16}$$

where subscript 1, or 2, indicates that k must be evaluated for $h = h_1$, or $h = h_2$. Solve the system formed by the last three equations in (6.3.16). We get the unknown coefficients

$$\begin{cases} u_4 = -px_h(L - x_h)/4EI, \\ u_5 = -p(L - x_h)^3/3EI - px_h(L - x_h)^2/4EI, \\ u_6 = -p(L - x_h)^2/2EI - px_h(L - x_h)/4EI, \end{cases} \quad (6.3.17)$$

that, inserted in (6.3.14), give u. A rapid calculation shows that $u = U$, as expected. Substituting u_4, u_5 and u_6 into the first three equations of System (6.3.16) yields the shear force, bending moment and reactive forces.

The following listing presents a function that applies the FEM to solve the overhanging beam.

```
function [U, V, M, F1, M1, F2] = beam_ov(L, EI, F, n, ih, Fp)
% This is the function file beam_ov.m.
% The FEM is applied to solve an overhanging beam with a roller support
% located at xh (< L). The input variables are: length of the beam L,
% product E*I, distributed load F, number of elements n, node ih where
% the roller support is located, and tip load Fp. The function returns the
% displacement U, shear force V, bending moment M, and reactive forces
% and moment F1, F2, M1.

% Initialization
h = L/n;
N1 = @(xi, xn)  1 - 3*(xi - xn).²/h² + 2*(xi - xn).³/h³;
N2 = @(xi, xn)  (xi - xn) - 2*(xi - xn).²/h + (xi - xn).³/h²;
N3 = @(xi, xn)  3*(xi - xn).²/h² - 2*(xi - xn).³/h³;
N4 = @(xi, xn)  -(xi - xn).²/h + (xi - xn).³/h²;
x = linspace(0,L,n+1); V = zeros(n+1,1); M = zeros(n+1,1);
k = stiffness(h,EI);
fd = zeros(n*4,1);
for i=1:n
    fd((i-1)*4+1) = integral(@(xi)F(xi).*N1(xi,x(i)), x(i), x(i+1));
    fd((i-1)*4+2) = integral(@(xi)F(xi).*N2(xi,x(i)), x(i), x(i+1));
    fd((i-1)*4+3) = integral(@(xi)F(xi).*N3(xi,x(i)), x(i), x(i+1));
    fd((i-1)*4+4) = integral(@(xi)F(xi).*N4(xi,x(i)), x(i), x(i+1));
end
nn = n*2 + 2;
K = zeros(nn,nn);
for i=1:2:nn-3
    K(i:i+3,i:i+3) = K(i:i+3,i:i+3)+k;
```

```
end
K = K([3:2*ih-2 2*ih:nn],[3:2*ih-2 2*ih:nn]);
f = zeros(n*2+2,1);
f(1) = fd(1);
f(2) = fd(2);
for i=2:2:2*(n-1)
    f(i+1) = fd(2*i-1) +fd(2*i +1);
    f(i+2) = fd(2*i) +fd(2*i+2);
end
f(2*n+1) = fd(n*4-1)+ Fp;
f(2*n+2) = fd(n*4);
f = f([3:2*ih-2 2*ih:nn]);

% FEM
ur = K\f;
ur = [0; 0; ur];
ur = [ur(1:2*ih-2,1); 0; ur(2*ih-1:nn-1,1)];
U = ur(1:2:nn,1);
for i=1:n
    V(i) = k(1,1)*ur(2*i-1) + k(1,2)*ur(2*i) + k(1,3)*ur(2*i+1)...
        + k(1,4)*ur(2*i+2) - fd((i-1)*4+1);
    M(i) = -(k(2,1)*ur(2*i-1) + k(2,2)*ur(2*i) + k(2,3)*ur(2*i + 1)...
        + k(2,4)*ur(2*i + 2)) + fd((i-1)*4+2);
end
    V(n+1) = -(k(3,1)*ur(2*n-1)+k(3,2)*ur(2*n)+k(3,3)*ur(2*n+1)+...
        k(3,4)*ur(2*n+2))+fd(4*n-1);
    M(n+1) = k(4,1)*ur(2*n-1)+k(4,2)*ur(2*n)+k(4,3)*ur(2*n+1)+...
        k(4,4)*ur(2*n+2)-fd(4*n);
F1 = V(1); % Reactive force.
M1 = -M(1); % Reactive moment.
F2 = V(ih) - V(ih-1);% Reactive force.
end
```

See Exercise 6.4.12. If the following data

$$L = 3*10^3,\ E = 3*10^4,\ I = 9*10^8,$$
$$F = 0,\ p = 10,\ n = 15,\ ih = 11,\ Fp = -p,$$

$$(6.3.18)$$

are used, Fig. 6.3.4 should be produced.

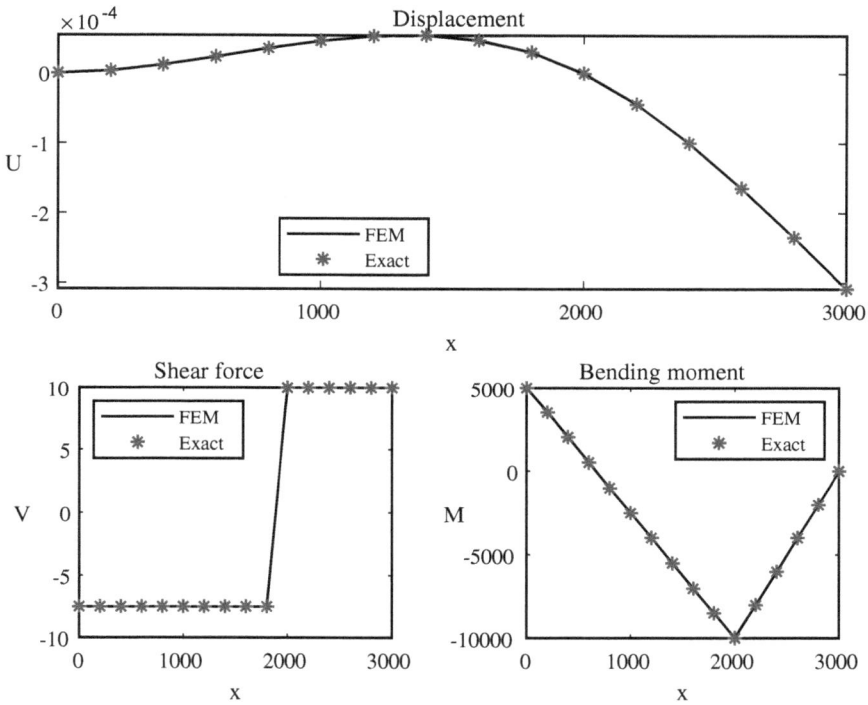

Figure 6.3.4. Overhanging beam: displacement, shear force and bending moment.

6.4 Exercises

Exercise 6.4.1   Explain the physical meaning of N_2, N_3 and N_4.

Hint. If the constraint in A is subjected to the rotary displacement $U'(0) = 1$, it is

$$EIU^{iv} = 0, \quad U(0) = 0, \quad U'(0) = 1, \quad U(h) = 0, \quad U'(h) = 0, \quad \Rightarrow \quad U = N_2.$$

Exercise 6.4.2   Solve the algebraic System (6.1.24) and find Formula (6.1.25).

Exercise 6.4.3   Write a function that returns the symbolic stiffness matrix.

Answer.

```
function K = stiffness_s
% This is the function file stiffness_s.m.
% Symbolic stiffness matrix is returned.
syms x h EI;
K(4,4) = h;
for i=1:4
    for j=i:4
        K(i,j) = EI*int(Nxx(i, x, h).*Nxx(j, x, h), x, 0, h);
        K(j,i) = K(i,j);
    end
end
end
% Local function
function f = Nxx(i, x, h)
switch i
    case 1
        f = -6/h^2 + 12*x/h^3;
    case 2
        f = -4/h + 6*x/h^2;
    case 3
        f = 6/h^2 - 12*x/h^3;
    case 4
        f = -2/h + 6*x/h^2;
end
end
```

Exercise 6.4.4   Write a function, say mass_s, that returns the symbolic mass matrix.

Hint. See the stiffness_s function.

Exercise 6.4.5   Consider the pinned-pinned beam subjected to the external moments M_1 and M_2 at the ends, shown in Fig. 6.4.1 (left). The following boundary conditions hold

$$U(0) = 0, \quad U''(0) = -M_1/EI, \quad U(L) = 0, \quad U''(L) = M_2/EI. \quad (6.4.1)$$

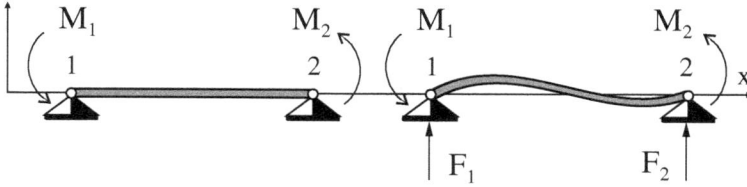

Figure 6.4.1.   Pinned-pinned beam subjected to external moments at the ends.

Find the approximating Solution by considering one element $e_1 = [0, L]$.

Answer. The approximating Solution (6.2.3) simplifies to

$$u(x) = u_2 N_2(x) + u_4 N_4(x), \tag{6.4.2}$$

because of boundary Conditions (6.4.1). Moreover, the vector $\mathbf{f}$ in System (6.2.4) simplifies to $\mathbf{f}_b$, as $F = 0$. The expression of $\mathbf{f}_b$ is derived from general Formula (6.2.6) by considering boundary Conditions (6.4.1),

$$\mathbf{f}_b = [F_1 \ \ M_1 \ \ F_2 \ \ M_2]^T, \tag{6.4.3}$$

where F_1 and F_2 are the unknown reactive forces. Considering all that, System(6.2.4) gives the following four equations

$$k_{12}u_2 + k_{14}u_4 = F_1, \quad k_{32}u_2 + k_{34}u_4 = F_2, \tag{6.4.4}$$

$$k_{22}u_2 + k_{24}u_4 = M_1, \quad k_{42}u_2 + k_{44}u_4 = M_2. \tag{6.4.5}$$

The last two equations are a simple algebraic system with unknowns u_2 and u_4. Solving it yields the unknowns

$$u_2 = \frac{M_1 L}{3EI} - \frac{M_2 L}{6EI}, \quad u_4 = \frac{M_2 L}{3EI} - \frac{M_1 L}{6EI}. \tag{6.4.6}$$

Insert these values into (6.4.4) to find the reactive forces

$$F_1 = (M_1 + M_2)/L, \quad F_2 = -(M_1 + M_2)/L.$$

In addition, substitute (6.4.6) into (6.4.2) to obtain u. Using Definitions (6.1.16) to (6.1.19) of the shape functions, we find that u is the cubic polynomial (Fig. 6.4.1, right)

$$u(x) = \frac{M_1 + M_2}{6LEI}x^3 - \frac{M_1}{2EI}x^2 + \frac{2M_1 - M_2}{6EI}Lx. \tag{6.4.7}$$

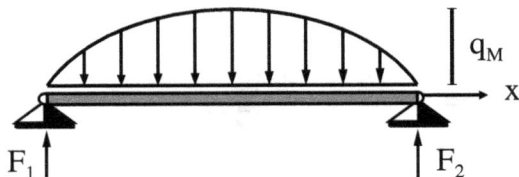

Figure 6.4.2.   Pinned-pinned beam subjected to a parabolic load.

Figure 6.4.3.   Fixed end-fixed end beam.

An easy calculation shows that this solution is the same as the analytical solution U of the Euler–Bernoulli equation, as expected.

Exercise 6.4.6   Write a listing that calls the beam_fx_fr function to solve the fixed end-free end beam subjected to the uniform load $F(x) = -q$.

Exercise 6.4.7   Write a listing that calls the beam_pn_pn function to solve the pinned-pinned beam in Fig. 6.4.2, subjected to a parabolic load

$$F = -4q_M(Lx - x^2)/L^2, \quad q_M > 0.$$

Exercise 6.4.8   Write a function, say beam_fx_fx, to solve the fixed end-fixed end beam in Fig. 6.4.3.

Exercise 6.4.9   Consider the fixed end-fixed end beam of length L in Fig. 6.4.4 (left), subjected to a uniform load. Suppose that the constraint in A is subjected to a rotational displacement (Fig. 6.4.4, right). The statical problem is governed by the Euler–Bernoulli Eq. (6.2.1) with the following boundary conditions

$$U(0) = 0, \quad U'(0) = r_A, \quad U(L) = 0, \quad U'(L) = 0.$$

Write a function, say beam_fx_fx_rA, that applies the FEM to solve the problem above.

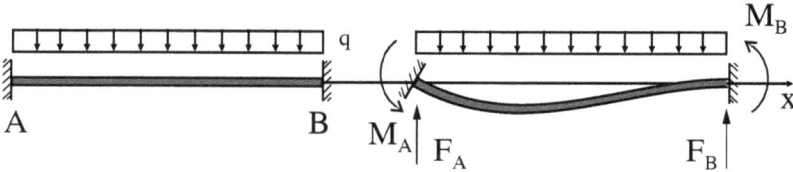

Figure 6.4.4. Fixed end-fixed end beam with the constraint in A subjected to a rotational displacement.

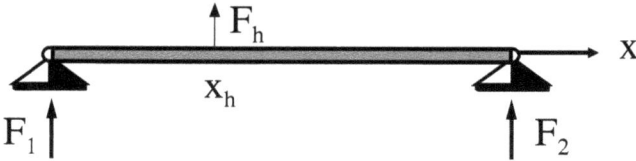

Figure 6.4.5. Pinned end-pinned end beam subjected to concentrated force.

Exercise 6.4.10 Consider the pinned end-pinned end beam in Fig. 6.4.5, subjected to the concentrated force (x_h, F_h). The boundary conditions are:

$$U(0) = 0, \quad M(0) = 0, \quad U(L) = 0, \quad M(L) = 0. \tag{6.4.8}$$

Find the displacement, shear force and bending moment.

Answer. The function U can be found by solving the homogeneous Euler–Bernoulli equation on the two intervals $x < x_h$ and $x > x_h$ with boundary Conditions (6.4.8) and two more conditions

$$U(x_h^+) = U(x_h^-), \quad V(x_h^+) - V(x_h^-) = F_h. \tag{6.4.9}$$

These conditions follow from the beam continuity on x_h and step Relationship (6.3.2). An easy calculation gives

$$U = \begin{cases} C_1 x^3 + C_2 x, & x < x_h, \\ C_3 (x - L)^3 + C_4 (x - L), & x > x_h, \end{cases} \tag{6.4.10}$$

where

$$C_1 = F_h(x_h - L)/6LEI, \, C_2 = F_h x_h(x_h - 2L)(x_h - L)/6LEI,$$
$$C_3 = F_h x_h/6LEI, \quad\quad C_4 = F_h x_h(x_h^2 - L^2)/6LEI.$$

The analytical expressions of the shear force and bending moment are obtained by differentiating (6.4.10) on the related intervals.

Exercise 6.4.11   Write a function that applies the FEM to solve a fixed end-pinned end beam subjected to a distributed load and a concentrated force.

Exercise 6.4.12   Write a listing to call the beam_ov. Use Data (6.3.18) and obtain Fig. 6.3.4. Moreover, introduce the distributed load, which was assumed to be zero in (6.3.18).

Bibliography

Cannon, J. R. (1984). *The One-Dimensional Heat Equation* (Addison-Wesley, London, UK).

Carslaw, H. S. and Jager, J. C. (1959). *Conduction of Heat in Solids* (Clarendon Press, London, England).

Clough, R. W. (1960). The finite element method in plane stress analysis, *Second ASCE Conference on Electronic Computation*, Pittsburg, USA, September 8–9, pp. 345–378.

Collatz, L. (1966). *The Numerical Treatment of Differential Equations* (Springer-Verlag, New York, NY, USA).

Cooper, J. M. (1998). *Introduction to Partial Differential Equations with Matlab* (Birkhauser, Boston, Ma, USA).

Courant, R., Friedrichs, K. and Lewy, H. (1928). *Uber die partiellen Differenzen-gleichunghen der Mathematischen Physik*, *Math. Ann.*, Vol. 100, pp. 32–74.

Crank, J. (1979). *The Mathematics of Diffusion* (Clarendon Press, Oxford, UK).

Crank, J. (1984). *Free and Moving Boundary Problems* (Oxford Science Publications, Clarendon Press, Oxford, UK).

D'Acunto, B. (2004). *Computational Methods for PDE in Mechanics* (World Scientific, Singapore).

D'Acunto, B. and Massarotti, P. (2016). *Meccanica Razionale per Ingegneria* (Maggioli Editore, Santarcangelo di Romagna, Italy).

de Vahl Davis, G. (1986). *Numerical Methods in Engineering & Science* (Chapman & Hall, London, UK).

Fenner, R. T. (2005). *Finite Element Method for Engineers* (Imperial College Press, London, UK).

Forsythe, G. E. and Wasov, W. R. (1960). *Finite Difference Methods for Partial Differential Equations* (Wiley, New York, NY, USA).

Hutton, D. V. (2004). *Fundamentals of Finite Element Analysis* (McGraw-Hill, New York, NY, USA).

Kharab, A. and Guenther, R. B. (2002). *Introduction to Numerical Methods. A Matlab Approach* (Chapman & Hall/CRC, Boca Raton, Fl, USA).

Knabner, P. and Angermann, L. (2003). *Numerical Methods for Elliptic and Parabolic Partial Differential Equations* (Springer, New York, NY, USA).

Kwon, Y. W. and Bang, H. (2000). *The Finite Element Method Using Matlab* (CRC Press, Boca Raton, Fl, USA).

Lapidus, L. and Pinter, G. F. (1982). *Numerical Solutions of Partial Differential Equations in Science and Engineering* (J. Wiley & Sons, New York, NY, USA).

Mitchell, A. R. and Griffiths, D. F. (1995). *The Finite Difference Method in Partial Differential Equations* (J. Wiley & Sons, New York, NY, USA).

Moler, C. (2011). *Experiments with MATLAB* (E-book: The MathWorks, Inc., Natick, MA, USA).

Necati Ozisik, M. (1994). *Finite Difference Methods in Heat Transfer* (CRC Press, London, UK).

Rao, S. S. (2005). *The Finite Element Method in Engineering* (Elsevier, Oxford, UK).

Richtmyer, R. D. and Morton, K. W. (1967). *Difference Methods for Initial-value Problems* (J. Wiley & Sons, New York, NY, USA).

Rubinstein, L. I. (1971). *The Stefan Problem* (Translation of Mathematical Monographs, American Mathematical Society, Providence, RI, USA).

Schwartz, L. (1950). *Theorie des Distributions* (Hermann, Paris, France).

Index

< (less), Matalab operator, 20

<= (less or equal), Matalab operator, 20

>= (greater or equal), Matalab operator, 20

>= (greater), Matalab operator, 20

== (equal), Matalab operator, 20

~= (not equal), Matalab operator, 20

abs, Matlab command, 42

all, logical function, Matlab, 30

anonymous function, Matlab, 26

any, logical function, Matlab, 31

axis, Matlab command, 13

backward, function file, 93

beam_c_pn_pn, function file, 297

beam_fx_fr, function file, 282

beam_fx_fx_uA, function file, 293

beam_fx_pn, function file, 288

beam_ov, function file, 303

beam_pn_pn, function file, 285

Bessel, historical note, 158

besselj, Matlab command, 158

break, Matlab command, 24

burgers, function file, 128

Burgers, historical note, 127

central, function file, 43

central_implicit, function file, 108

clc, Matlab command, 12

consolidation1, function file, 121

consolidation2, function file, 125

crank, function file, 86

Crank, historical note, 85

dam, function file, 235

dam_5p, function file, 254

Darcy, historical note, 49

diag, Matlab command, 5

diffusion_d, function file, 192

diffusion_n, function file, 195

Dirac δ function, 184

Dirac, historical note, 184

Dirichlet boundary condition, 52

Dirichlet, historical note, 52

disp, Matlab command, 13

error, Matlab command, 20

errordlg, Matlab command, 20

Euler, historical note, 39

Euler–Bernoulli equation, 268

Euler–Lagrange equation, 268

euler_e, function file, 60

euler_edn, function file, 72

euler_el, function file, 92

euler_em, function file, 62

euler_en, function file, 70

euler_enm, function file, 71

euler_i, function file, 81

eye, Matlab command, 4

fem_dd, function file, 174

fem_dn, function file, 177

feval, Matlab command, 26
Fick's law, 55
Fick, historical note, 49
find, logical function, Matlab, 31
five_point, function file, 243
five_point_f1, function file, 246
for, Matlab structure, 10
forward, function file, 41
Fourier's law, 49
Fourier, historical note, 49
fplot, Matlab command, 27
fprintf, Matlab command, 42
ftbs, function file, 113
ftfs, function file, 115
full, Matlab command, 7
function file, Matlab, 15

Gauss, historical note, 51
gradient, Matlab command, 17
Green's identities, 214
Green, historical note, 214

hermitian functions, 271
hold off, Matlab command, 122
hold on, Matlab command, 121
homogeneous boundary condition, 53
Hooke, historical note, 165

if-elseif-else, Matlab structure, 19
int, Matlab command, 155, 160
integral, Matlab command, 155
integral2, Matlab command, 160
integral_1, function file, 155
integral_2, function file, 156
integral_2D_1, function file, 160
integral_2D_2, function file, 205
integral_2D_3, function file, 163
integral_2D_4, function file, 163
integral_3, function file, 157
integral_4, function file, 158
ismember, logical function, Matlab, 32
isspace, Matlab command, 25

Kronecker δ function, 169
Kronecker, historical note, 169

Lagrange, historical note, 169
laplace, function file, 229
Laplace, historical note, 52
layers, function file, 77
left division, Matlab, 10
legend, Matlab command, 13, 21
lines_c_d_1, function file, 137
lines_c_d_2, function file, 138
lines_heat1, function file, 123
linspace, Matlab command, 3
load, Matlab command, 142
local function, Matlab, 25
logical operators, Matlab, 27

mass, function file, 276
max, Matlab command, 42

Neumann boundary condition, 52
Neumann, historical note, 52
Newton, historical note, 53
Nicolson, historical note, 85
nonlinear, function file, 131
norm, Matlab command, 96
num2str, Matlab command, 13

ode15s, Matlab command, 132
ode45, Matlab command, 121
ones, Matlab command, 4

Péclet, historical note, 100
pause(s), Matlab command, 13
pause, Matlab command, 15
plot, Matlab command, 13
Poisson, historical note, 214
pure Neumann problem, 179
pyramid, function file, 223
pyramid_phi, function file, 225

quiver, Matlab command, 18

Rectangle Rule, 153
rectangle, Matlab command, 18
relational operators, Matlab, 20
repmat, Matlab command, 7
reshape, Matlab command, 5
right division, Matlab, 9

Robin boundary condition, 53
Robin, historical note, 53

save, Matlab command, 141
sheet_pile_wall, function file, 262
Simpson Rule, 154
Simpson, historical note, 154
simpson, function file, 156
simpson2d, function file, 161
size, Matlab command, 29
spdiags, Matlab command, 6
spy, Matlab command, 7
sqr, function file, 15
Stefan condition, 57
Stefan problem, 58
Stefan, historical note, 57
stiffness, function file, 276
stiffness_s, function file, 306
strcat, Matlab command, 23
stress_1, function file, 181
stress_2, function file, 210
surf, Matlab command, 15
switch, Matlab structure, 22
sym, Matlab command, 156
symbol O, 40

Taylor, nota storica, 40
test function, 167

tic, Matlab command, 123
title, Matlab command, 13
toc, Matlab command, 124
transpose matrix, Matlab, 1
Trapezium Rule, 154
triangulation_plot1, function file, 217
triangulation_plot2, function file, 218
type, Matlab command, 142

upper, Matlab command, 23
upwind, function file, 101

variable1, function file, 134
Von Neumann criterium, 89
Von Neumann, historical note, 89

wave_d, function file, 200
wave_n, function file, 203
while, Matlab structure, 24
whos, Matlab command, 7

xlabel, Matlab command, 13

ylabel, Matlab command, 13
Young, historical note, 165

zeros, Matlab command, 4
zlabel, Matlab command, 15